International Terrorism and World Politics

Fundamentalism's Challenge to 'New World Order'

International Terrorism and World Politics

Fundamentalism's Challenge to 'New World Order'

Sharda Jain

REGAL PUBLICATIONS
New Delhi - 110 027

INTERNATIONAL TERRORISM AND WORLD POLITICS
Fundamentalism's Challenge to 'New World Order'

ISBN 978-81-8484-193-0

Typeset by
THE LASER PRINTERS
8/15, 3rd Floor, Subhash Nagar, New Delhi-110027

Printed in India at
MAYUR ENTERPRISES,
WZ Plot No. 3, Gujjar Market, Tihar Village, New Delhi-110018

Published by
REGAL PUBLICATIONS
F-159, Rajouri Garden, New Delhi-110027 • Phone : 45546396
E-mail : regalbookspub@yahoo.com

Dedicated to
My Gurus (The Teachers)
who came in various forms in all stages of my life

One can...never create [freedom] by an invading force.

—Maximilien Robespierre

What difference does it make to the dead, the orphans, and the homeless, whether the mad destruction is wrought under the name of totalitarianism or the holy name of liberty and democracy?

—Mahatma Gandhi

Where is the justice of political power if it ... marches upon neighbouring lands, killing thousands and pillaging the very hills?

—Kahlil Gibran

Great is the guilt of an unnecessary war.

—John Adams

Ours is a world of nuclear giants and ethical infants.

—General Omar N. Bradley

We have guided missiles and misguided men.

—Martin Luther King, Jr.

Contents

Preface

The international relations and world politics in last decade of 20th century and at the beginning of the 21st century owes its roots to the real world events in late 1960s and into 1970s and constitute the main stuff of the International Relations (IR) theory and practice and also of Foreign Policy studies. By mid-1960s, most of the world's colonies had become independent, sovereign and members of various international organizations and bodies, such as United Nations (UN), and brought new kinds of problems of nation-building as well as of security to the forefront. So, the IR scholars needed new framework to build new paradigms and therefore, developed more diverse and conceptual tools to study this reality. The second-generation IR studies reflected upon the complex issues of their times, such as, the impact of the latest wave of democratization, the implications of the relative decline of United States' economic power from early to mid-1980s, collapse of Soviet empire and Soviet-style Communism, the unprecedented international co-operation in the war in Persian-Gulf in 1991 and the December-1991 disintegration of Soviet Union.

The years between mid-1990s and 2001 are marked by some monumental political events on the international stage, such as: the resurgence of US economy and the world economic globalization, use of United Nations to promote human rights and to tackle various humanitarian crises with moderate success, resurgence of North Atlantic Treaty Organization (NATO) in conflict situations, sharp increase in illegal drugs and small-arms trade and trafficking, two arch-rival in the Indian sub-continent, India and Pakistan, pushing their way into the elite nuclear weapon club in 1998, rise of religious fundamentalism, particularly the Islamic fundamentalism and the attacks on US diplomatic and military personnel in late 1990s culminating in the catastrophic attacks on the American soil on 11 September, 2001. These and many other realities of 21st century include not only the dangers and frustrations of the new millennium some of which may be underlined as, (1) the unparallel US hegemony in economic, political and military strength; (2) globalization of economies and culture, creating strong undercurrents and backlashes among the peoples attempting to assert their differences and their rights to these differences; (3) beginning of the process of democratization which released some unrealistic and dangerous demands on governments for which these governments were not prepared; and (4) intrastate warfare having inter-state connections that emerged in the form of international terrorism and the resultant 'war on terror'. Hence, a critical analysis of the post-cold war international politics is required to address these new challenges and realities.

The end of Soviet Union in 1991 marked the end of communism, of socialism (though may be for the time being till it might reemerge and bounce back) and also of cold-war that remained the dominant paradigm for all international relations and politics and also for the foreign policies of most of the nations for nearly forty-five years after World War II. Yet, at the same time, it also marked the beginning of a new, and perhaps an uncomfortable or even dangerous world dominated by a single superpower unchecked by any rival or ideology and with a decisive reach in every corner of the globe so much so that some scholars go as far as saying that even Roman empire in its hay days was no match to the America of today. In other words, if the defining challenges of the twentieth century ended with the fall of the Berlin Wall, full recognition of the first great challenge of the twenty-first century came with the attacks of September 11, 2001, even though Islamist terrorists had begun their assault on a world order decades before which was largely defined and driven by the American-style neo-liberal-capitalist ideology. Since then, US abandoned a decade-long strategy of defensive reaction in favor of a vigorous offense and set in motion changes to the international system that promised a safer and better world for generations to come, though such a strategy seems to be counterproductive at the moment and the promise for a safer world seems a distant dream. The challenges before the world today are to set a course for victory on the terrorists' war on global order, to strengthen the international system that the terrorists seek to destroy and to extend the benefits of the international system in an ever-widening arc of security and stability across the globe. Islamic fundamentalism is simply one variety of a new global phenomenon in world politics. It is not the cause of current crisis in the world, but an expression of it and response of it. Fundamentalism, be it of any variety, is not a solution; rather it deepens the ongoing cultural fragmentation of the world and leads to disorder.

It may also be noticed that the breakdown of communism and the end of bipolarity has also proved costly to the West which seems to have lost the political factor that secured its unity for a period of four decades from 1945 onwards. The primary trajectory of events in this post-cold era is marked by confrontations between the Muslim world and the West—United States in particular—and the subsequent rise of international terrorism on the one hand and the 'war on terror' on the other. These developments have become a crucial and significant stuff of all academic and intellectual discourses on international relations and politics. Any serious observer of international politics would find that in post-1991 period, multiple and variegated Islamic movements have emerged and have been trying to enhance their grip over the society and politics of most of the nations in the Middle-East. Their main point of reference being Islam, these movements are largely termed as 'religious' though their thrust extends in political as well as social arena too. Interchangeably termed as Islamist movements or 'political Islam', these movements are projected as representing the mainstream Muslim nationalism [see: *Harvard International Review*: Jan. 15, 2007]. Within a broad spectrum of Islamist movements, there exist a number of variants ranging from moderate to radical to violent—all of these nevertheless having one thing in common: they all search their roots in history, culture and religion, which gives them all a distinctive Islamic characteristic.

In the root of these radical and violent Islamist movements are the historical grievances against the West that include: centuries of Western imperialism and domination of the Middle-East; a currently US-driven effort for global economic, political and cultural hegemony to weaken the political power of Islam; post-9/11 US-led 'War on Terrorism'—perceived by Muslims as a global war against Islam; military action in Afghanistan and preemptive war on Iraq without any substantiated proof of weapons of mass destruction (WMD) alleged to be in possession of Saddam Hussein; arm-twisting of Iran; presence of US troops in many Muslim countries including Saudi Arabia—the home to most sacred holy mosques revered by the

Muslims all over the world; US-Israel campaign to deny Palestinians their rightful place in West Asia and a vocal and globally permeating anti-Islamic Western media—all help exacerbate Muslim anxieties, anger and frustrations, lending sufficient credence to these anti-US Muslim fears as well as to the widely discredited 'clash of civilizations' thesis.

An ultra-Right Muslim opinion against West/US is fed with the similar expressions from a well-developed cottage industry in the US which selectively picks passages from Koran, totally out of context, and then make pronouncements on 'eternal character' of Islam and Muslims—obviously offending their religious and cultural sensitivities. In fact, the neo-conservatives in the Muslim world, in the West and in Israel—all feed on each other's fears and thus create a spiral of suspicion and hatred against each other.

There is no doubt that today, the contemporary Islam is in a volatile state, going through an internal and external struggle over its values, its identity, and its place in the world. Rival versions are contesting for spiritual and political supremacy. Though with the death of Osama bin-Laden in 2011, al Qaeda is weaker and with revolutions in several Arab countries due to frustrations with unpopular autocratic rulers, recruiting grounds for terrorist groups seem to have been mitigated, tensions associated with this kind of development has serious costs not only for the Islam as a religion and as a political ideology but also for the rest of the world for its social, economic, political and security implications. West, in its efforts to come to terms with these developments related to Islam, releases in the process such corroding elements that the clash between the two seems imminent. The US and the West and the rest of the industrialized world try to mould Islam so as to make it compatible with democratic and pluralistic values, with international peace and modernity. This effort, however, is viewed in the Muslim world as an effort to westernize it and therefore the resentment and also the violence.

It is these kinds of issues and confrontations that have inspired this project. It tries to answer the following questions: Does terrorism occur in a vacuum or are there some pressing historical precipitants at the root of its origin?; Is there something inherently wrong with religion in general and with Islam in particular that exacerbates, endorses or condones violence and terrorism in the name of God?; Is it the thrust and urge for hegemony by the sole superpower in a unipolar world that invites terrorism?; Is the world really entering into a phase of 'Clash of Civilizations' or a 'Clash of Fundamentalisms': Political and Religious?; Can preemptive militarism be a solution to the problem of terrorism?; and What long-term strategies are required for a more stable and secured world order?. The pertinent question here is why these questions are important and why a search for answers to these questions is important and urgent? It is because the current confrontation between the religious fundamentalism and American aggressiveness in pursuing a unilateralist foreign policy, particularly during George W. Bush Presidency, has not only overshadowed every other aspect of international relations and politics, it has also brought the world on the disastrous path of a politics of hatred wherein the proponents of each side help to feed and sustain the confrontation through their seemingly convincing arguments and attacks over the other. Such a tendency and trend needs to be contained and things to be put in proper perspective so as to explode various myths that surround this confrontation.

The phenomenal increase in global terrorism in recent years, and a heightened threat perception by the governments and the public has, of necessity, focused keen interest in this area. In addition with an overall increase in militancy, what is even more worrisome is the fact that the complex and sinister ties are emerging between globally and widely dispersed militant organizations, characterized by flows of funds, personnel and ideological and military support. Terrorist and criminal organizations use network forms of organizations that tend to be intrinsically more stable and less liable to be disrupted by simple measures. Being covert organizations, with largely invisible membership, it can be extremely difficult to obtain accurate information of their ties and linkages. These phenomena and the related issues of interplay between terrorism and politics need to be probed.

To search answers to these questions, a senior research project (post-doctoral) titled: *International Terrorism and World Politics: A Study of the Post-Cold War Dynamics* was undertaken. In the course of research, various new paradigms and insights were encountered. A dispassionate analysis of these dimensions motivated me to reframe the topic which is also the theme and topic of this book. It is: ***International Terrorism and World Politics: Fundamentalism's Challenge to 'New World Order'***. The rationale of this theme is, as has also been mentioned in the preceding paragraphs, the post-cold war world is facing a challenge of international terrorism which in turn is the outburst of various forms of fundamentalisms, both religious and political. It is a clash between the violent expression of political and radicalized Islam against the rigidity, universalism, unilateralism and fundamentalism of United States' foreign policy that after the 'victory' in cold-war was too obsessed with reordering the rest of the world in its own image, the so-called—'New World Order' which was accepted by some but rejected by some others, particularly the Islamist fundamentalists who saw in this New World Order a threat to their values, religion and culture. The result was terrorism and more terrorism in the name of countering it. To explain issues related with this formulation, the following Chapter Plan has been devised:

1. Introduction: Transformations in World Politics: Post-Cold-War to War on Terror
2. Religious Fundamentalism and its impact on International Relations: Special Reference to Rise of Militant Islam
3. Political Fundamentalism: Mapping Fundamentalist Trends in US Foreign Policy: 1991-2001 and onwards
4. Expanding Terror Networks: Asian Context
5. Conclusion: Addressing Clash of Fundamentalisms

This book is the product of this above mentioned post-doctoral research project undertaken independently with assistance of University of Delhi which granted me a two-year study leave from July 2006-08. In the course of this research, I have made tremendous use of both primary and secondary sources that would find mention in the bibliographical notes and references. I am indebted to all those authors, intellectuals and researchers whose works have been used in course of my research and preparation of this book. With all humility, I express a deep sense of gratitude to all of them as it is not possible to acknowledge their contribution individually. Additionally, let me admit that in this age of internet, a flood of very useful and relevant information comes handy with the help of some popular search engines, such as, Google. My heartfelt thanks to the wonders presented by revolution in Information and Communication Technology (ICT) and to all those internet sources without whose rich and variegated pool of data, this research could not claim to have tapped wide range of sources.

I owe my thanks to Professor Bidyut Chakraborty, the then Head of the Department of Political Science at the University of Delhi and Dr. Malti, the ex-Principal of Kalindi college for granting me study leave (July 2006-08) to take up this project. Many colleagues and friends have also contributed by way of their useful inputs during informal discussions in the staff room and elsewhere. Dr. Rakhi Chauhan, a colleague in the Department of Political Science deserves my special thankfulness for going through the script and for giving useful suggestions some of which I readily incorporated in the text. Of course, to a teacher, the young and dynamic students' fresh and unbiased thinking expressed without fear or fervor and without any preconceived notions serves always as a very original source in any research process. My thanks to all those students who in my more-than-four-decades' teaching career emerged as my best teachers at times.

Both (Late) Professor Madan M. Sankhdher who taught me first lessons in pursuing a research and Mrs. Mandakini Sankhdher who had tremendous energy and managerial skills and had always been a guiding force in motivating me to do something of research constantly remained a source of inspiration during the entire process of this work. Their absence was felt each moment. I owe deeply felt gratitude to these wonderful souls. My thanks are also due to my teacher, Prof. Randhir B. Singh who has always inspired me for hard work by his own example. Despite limitations of health, he is never lacking in support.

In the family, special thanks are due to my son-in-law, Nitin Bansal, for his very sharp and clear understanding of dilemmas facing Islamic world and for engaging in long and useful discussions pertaining to the subject under study. I owe a special thankfulness to all members of my family: S.K. Jain, Aparna, Vikesh, Prerna, and grand children for assisting in ways more than one. All of them and my mother-in-law, Mrs. Saroj Jain, took a very kind care of me when a cancer intercepted my research work in its initial stage. I wish I had words to express my deep sense of gratitude for those who so lovingly and readily stood by me in testing times.

My publisher, Regal Publications did a wonderful job in accepting my challenge to bring out the book in the least possible time. I would be failing in my duty if I do not express my heartfelt thanks to them.

Despite my sincere efforts to acknowledge the resources and to present the facts to the best of my understanding, some errors in interpretation might have crept in inadvertently. I owe all errors of omission and commission.

New Delhi SHARDA JAIN

Glossary

Burqa: The voluminous, all-covering outer garment worn by Afghan women.

fatwa: A formal pronouncement on a doctrinal or legal matter by an Islamic scholar or scholarly body.

Hadith: A narrated story relating to the actions or sayings of the prophet Muhammad and his closest followers, presumed to reflect the correct way of doing things and to supplement the guidance given in the Quran. An exacting science has been created around the need to substantiate and verify hadith, but the very hugeness of the body of hadith makes it subject to accidental or intentional misuse.

Hanafi: One of the schools of Islamic law; more liberal on most matters.

Hanbali: One of the schools of Islamic law; more conservative on most matters.

hijab: Literally, the Islamic "dress code" for women; the term can be used to refer to the simple headscarf or to more elaborate coverings.

hudud: Specific Islamic criminal punishments.

ijma: Community consensus as a tool of modifying and interpreting Islamic law.

ijtihad: The practice of informed interpretation, another tool for establishing and modifying correct Islamic practice.

Khilafa: Another spelling for *Caliphate.*

Kufr: Non-Islamic disbelief.

Madrassah: Generic term for an Islamic religious school, whether of the traditional non-political variety or as a politicized source of radical fundamentalist indoctrination.

mullah: An Islamic preacher, regardless of the level of training and education.

Quran : The Islamic holy book.

Sharia: Also commonly spelled *shariah* or *shariat*; the entire body of Islamic law and guidance, based on the Quran, hadith, and scholarly judgments and open to selective use and interpretation.

Shi'a Islam: Literally, *faction* or *party*; a dissident version of Islam that began with a dispute over the leadership succession shortly after the death of Muhammad and then developed further doctrinal and political differences *vis-à-vis* orthodox, Sunni Islam.

Sufisim: Islamic mysticism, either in its variant as a populist folk religion.

Sunni Islam: The orthodox version of Islam adhered to by the overwhelming majority, although Shi'a Islam is dominant in some countries and regions.

sunnah: The body of tradition complementing the Quran.

sura: A section or verse of the Quran and the organizing principle structuring the revelations.

Ulama: Body of scholars, scholarly community.

ummah: The community of believers.

Wahhabi: An extremist, puritanical, and aggressive form of Islamic fundamentalism founded in the 18th century and adopted by the house of Saud; disrespecting other versions of Islam, including Sufi Islam, Shi'a Islam, and moderate Islam in general as incorrect aberrations of the true religion. Its expansionist ambitions are heavily funded by the Saudi government.

Chapter 1

Introduction Transformations in World Politics: Post-Cold War to War on Terror

The history of international politics until World War II has been largely Eurocentric due mainly to the fact that the international political system was born in Europe in the modern age with the rise of modern nation-states which gradually was extended over the world through colonization and imperialism which afterwards was replaced by the fact of 'bipolarism' and the post-1945 process of decolonization that culminated into the loss of centrality of Europe in international politics. Throughout the latter-half of the twentieth century, except for the last decade, the list of world's great powers was small comprising of only a few countries, such as, United States, Soviet Union, Japan and some countries of north-western Europe, such as, Great Britain, France and Germany or the countries of G-7 at the maximum. In 21st century, the list is smaller, consisting by and large of only one superpower—United States of America—though some would disagree with this kind of opinion and would like to include a number of great powers previously existing and also the newly emerging ones, such as, China and India and plead for the incorporation of these countries into the great power framework in the emerging 'multipolar era' in world politics. But the recent unipolar impulses vividly displayed by American behaviour in Iraq and setbacks in its war on terrorism raises questions about the validity of the principle of 'unipolarity' and invites criticism for US foreign policy in the post-cold war world, particularly after 9/11, to the extent of overshadowing some of its genuine efforts at multilateralism and at creation of a 'new world order' which are interpreted by strategic analysts as merely the efforts to further US goals and its universalistic impulses.

Since the demise of the ideologically based bipolar division of the world, the nature of international relations and its study as a discipline has drastically changed. The traditional IR theory is confronted with ever-emerging new challenges, such as, the phenomenon of religion and politicization of religion and its impact on actual relations amongst nations. Of course, religion was studied as an important component of society in sociology and anthropology. But the use of religion in the background of the contemporary developments in the Islamic world, such as, Islamic revolution in Iran and Soviet invasion of Afghanistan and the subsequent turn of events in which religion has been openly used for political ends, even to the extent of using terrorist methods, is a new dimension in IR theory. After the 9/11 terrorist attacks, the focus on political Islam has become the significant point of reference in IR studies, particularly in studies on terrorism. Let it be stressed here that Islam, as opposed to fundamentalist or neo-fundamentalist Islam, posits a worldview that can deal with and selectively integrate modernity. In contrast, fundamentalist Islam calls for a return to an ontological form of Islam that rejects modernity; groups such as Al-Qaeda are representative of fundamentalist Islam. But any study of the rise of fundamentalist Islam in post-cold war era cannot be understood without situating it in the currents and dynamics of US foreign policy, with special focus on its policies in the oil rich Middle-East, a region where not only Islam was born but also a land where the large populations of followers of Islam live. In addition to Islam, oil is what makes the Middle-East geo-politically relevant. Two-thirds of global oil (some 690 billion barrels of proved reserves) is located in the unstable, autocrats-ruled Persian Gulf region: Saudi Arabia sits atop the world's biggest oil reserves, followed by Iran, Iraq and Kuwait, respectively. As a result, the Middle-East has been one of the main priorities of American foreign policy for the past half century.

The United States is the largest and most technologically advanced economy in the world, yet fossil fuels remain the lifeblood of the American system. The United States is, in fact, the world's largest consumer of oil, but it depends on foreign producers to supply most of its demand. Because energy security is inextricably intertwined with economic prosperity and national security, oil is one of the primary factors shaping US foreign and military policies. Indeed, from Iraq to Iran and Nigeria to Venezuela, oil is omnipresent in the calculations of American foreign policy. Even the centre-piece of America's foreign policy, encouraging democracy in the Middle-East, is being jeopardized in the current environment of rising wave of anti-American terrorist violence directed against US and its interests elsewhere, much before and also after the 9/11 catastrophe. More the US gets involved in the region, so does its exposure to trouble in the world's volatile oil-producing regions.

The Post-Cold War Features of World Politics

The phrase *'New World Order'* gained currency and came to be widely used to define the nature of the world that emerged or was emerging after the demise of Soviet Union and its basic ideology of communism. Francis Fukuyama declared the 'End of History' and concluded it as the triumph of the western liberal, capitalist and democratic values. But the new euphoria was short-lived as the internal crises in the collapsing Soviet Union and the second Gulf War intercepted the trouble free march towards a new world order, perhaps in the image of unipolarist leader of the world: United States. However, some cooperative gestures were also seen as George H.W. Bush linked the success of Gulf War to international community, including the cold war rival, the former Soviet Union. Initial agreement by the Soviets to allow action against Saddam highlighted this linkage in the western and American media. But no one could perhaps imagine at that time that a collective response against Saddam Hussein was putting this emergence of a New World Order at stake, which became increasingly clear in the years to come.

The idea that the Gulf War would usher in the new world order began to take shape in which the United States would be obligated to lead the world community to an unprecedented degree and would pursue its national interest uninterruptedly within a framework of concert with its friends and the international community as a whole. President Bush Sr. made an American commitment to lead the world towards rule of law rather than of force, and though military force matters, it should be made less important in the 'New World Order'. His idealism included the efforts like making the world safe for democracy, reducing and containing the instability in the Third World through mutual cooperation and efforts of the First and

Second world, helping United Nations to resurge and play a vital role in maintaining peace and security in the world, fighting religiously motivated terrorism, drug trafficking, poverty, disease and other social and economic scourges that impair the balanced growth of the world. The critics however were not very sure about this naïve idealism as also of what the 'new world order' actually entailed. The *New York Times* observed that the American 'left' was calling the new world order a "rationalization for imperial ambitions" in the Middle-East, while the 'right' rejected new security arrangements altogether and fulminated about any possibility of UN revival ["George Bush Meet Woodrow Wilson", *New York Times,* November 20, 1990]. Patrick Buchanan even predicted that the Gulf War would in fact be the demise of the new world order, the concept of UN peacekeeping, and the US's role as global policeman. Critics went as far as calling the rhetoric of this dream of 'New World Order' as a US road map for emerging as the single greatest power in the unipolar world as Moscow was crippled by internal problems, and thus unable to project power abroad. The United States, while hampered by economic malaise, was militarily unconstrained for the first time since the end of WW-II. *The Economist* even wrote that the drive toward the Gulf War was the run-up for Iraq war in 2003 ["New World Order Inc", *The Economist,* November 10, 1990].

Key characteristics of 'New World Order': John Lewis Gaddis, a cold war historian, described the key characteristics of the potential new order as unchallenged American primacy, increasing integration, resurgent nationalism and religiosity, a diffusion of security threats, and collective security. Changes in communications, the international economic system, the nature of security threats, and the rapid spread of new ideas would prevent nations from retreating into isolation. In light of this, Gaddis saw a chance for the democratic peace predicted by liberal international relations theorists to come closer to reality. However, he illustrated that a revitalized Islam could play both integrating and fragmenting roles—emphasizing common identity, but also contributing to new conflicts that could resemble the Lebanese Civil War. The integration coming from the new order could also aggravate ecological, demographic, and epidemic threats. National self-determination, leading to the breakup and reunification of states (such as Yugoslavia on one hand, and Germany on the other) could signal abrupt shifts in the balance of power, with a destabilizing effect. Integrated markets, especially energy markets, would be a security liability for the world economic system, as events affecting energy security in one part of the globe could threaten countries far removed from potential conflicts [Gaddis: 1991]. The "new world order" was also described as an ideological tool of legitimating of the global exercise of power by the US in a unipolar environment.

Of course, the world became different in 1990s, but not necessarily peaceful. The illusion of harmony was soon dissipated by the rise of multiple ethnic conflicts and ethnic cleansing, such as, in Yugoslavia where after the Hitler's holocaust during inter-war period, the ethnic Muslims were subjected to genocide and ultimate breakdown of Yugoslavia in early 1990s. The US-led NATO intervention[1], termed as 'humanitarian intervention', was a direct assault on the sovereignty and territorial integrity of another nation and it marked the beginning of the end of post-cold war euphoria of a harmonious world. The most dangerous development was the increasing role of unipolarist United States as the unchallengeable arbiter of world affairs.

One of the most controversial of all the post-cold war theories was ***The Clash of Civilizations theory*** proposed by political scientist Samuel P. Huntington[2], in response to Francis Fukuyama's thesis of The End of History, arguing that people's cultural and religious identities will be the primary source of conflict in the post-cold war world. Though the 'Clash' thesis was rejected by the mainstream academia and media and even by strategic analysts, the rise of Islamic fundamentalism and international terrorism, largely due to US foreign policy orientations in post-cold war world politics holds some basis and needs probe dispassionately. In the post-cold war world in which Islamic fundamentalism is internationally more pronounced than any other variety of religious revivalism and fundamentalism, giving some credence to the Clash thesis in some academic and strategic debates. Although it is not a new phenomenon, in recent times it has attained a vicious and virulent character. Modern fundamentalism in reality is a reactionary culmination of the trends of Islamic revivalism in an epoch of modern world economy and politics.

Globalization was another feature of post-cold war world politics in which western democracy and Anglo-American model of capitalism made an unstoppable sort of sway which later was resented by some religiously driven ideologies like Islamic fundamentalism. Globalization unleashed a process that is associated with the growth of international linkages and reduction in the capacity of nation-states to act independently which also meant weakening effect on exclusive and absolute national sovereignty. The primacy of the state-centric international structure does not sit comfortably with the fact that globalization helps facilitate a multi-centric world of transnational actors, ranging from multinational corporations to terrorist groups. The proliferation of these non-state actors would seem directly or indirectly to affect the capacity of the sovereign state to serve as a single and autonomous political authority within a territory. Somewhat appreciable part of the process of globalization includes the extension of an international rule of law beyond the domain of the state and thus encroaching upon or limiting the ability of the sovereign states to act, both internally and externally, in an autonomous and authoritative manner.

Now, the question may arise as to ***how does economic globalization contribute to terrorism?*** In fact, globalization has great potential to provide economic benefits that can be realized by disadvantaged groups and provides openings for the incorporation of these groups. Yet, the process of globalization has also vastly increased incentives and opportunities for terrorism and makes it easier to organize, finance, and sustain terrorist campaigns. But, the weak globalisers who integrate late into global economy are less competitive; their populations have falling or stagnant incomes, resulting in growing unemployment, political tension, and religious fundamentalism. A number of African and Muslim countries may be put in the category of 'weak globalisers'. This has resulted into an increase in inequalities and social polarization among states. Such growing inequality may lead to terrorist acts justified by the perpetrators in the name of a more equitable distribution of wealth. The permeability of cultural boundaries and the global spread of market culture is interpreted by some militants as the 'infiltration of alien and corrupt culture', which is then used as a justification for nationalist and radical religious movements that aim at cleansing their societies and cultures of foreign influence.

As globalization facilitates the movement of workers and refugees across borders, it fosters new minority groups in 'settled' societies, many of whom are linked politically to kindred elsewhere. Moreover, the cross-border movement of activists, information, and money from supporters (governments, diasporas, political sympathizers) to terrorist groups is also facilitated by globalization. Simultaneously, the sinews of globalization—from pipelines to communication networks—become 'soft targets' for transnational terrorists. This factor helps in setting up linkages between political and criminal networks. Organized crime and terrorist groups use similar—sometimes the same—means for moving materials, people and funds across boundaries. Underground banking networks developed by criminal groups are also used by terrorist groups. Some proceeds from illegal businesses end up funding terrorist groups. These developments blur the distinctions between political and criminal, increase the capacity of the linked groups to resist international action, and increase the terrorists' incentives for continuing their campaigns. When militant groups turn to crime to finance their political activities, politics may gradually become a cover for crime for profit. At times, the leaders or factions within the militant movement oppose political solutions to the conflict because it would undermine their vested 'business interests'.

Global Resurgence of Religion as a factor in terrorism: Though *religion* is seldom the only cause of terrorism, it has great potential to contribute to terrorism. Islam does not cause terrorism, nor does any other religion with which terrorist acts have been associated in history. As John Esposito explained, usually 'political and economic grievances are primary causes or catalysts, and religion becomes a means to legitimate and mobilize'. Religion brings to a situation of conflict images of grand struggle and an abiding absolutism. It is often 'centered around themes that can be inherently polarizing — concepts of truth, notions of good, of absolutes and ultimate realities'. For this reason, religion can contribute to a culture of violence where violence becomes 'a defining issue' in the identity of activist groups. However, though at times, the religious causes may be intertwined with economic and political grievances/causes too, in some cases, such as of Al-Qaeda in recent times, the preeminence of religious factor cannot be denied. Religion was foremost in the minds of the Jewish extremists who attempted to blow up the Dome of the Rock and wanted to rebuild the ancient temple on Temple Mount. Esposito highlights the fact that the Islamic attacks of 9/11 were largely against the secular state and not against religious targets. It has been noticed that monotheistic religions (Christianity, Judaism and Islam) have a greater tendency to authoritarianism and intolerance due to 'their centrist emphasis on a single deity'. The exclusivist theologies/ worldviews of the 'three Abrahamic traditions' have proved to be more prone to imperialist expansion, violence, and terror [see: Madrid Agenda: 2005]. The distinction between 'monotheism' and 'polytheism' is much debated by scholars of comparative religion who point out that a supposedly polytheistic religion such as Hinduism has a strong sense of unity behind its diversity, and that notions such as the Christian Trinity and concepts of sainthood introduce a complexity into the idea of God in the supposed monotheisms of Western religions.

But merely pinpointing religion as the causatory factor for terrorism is not enough unless one probes "what combinations of factors brought a religious response and that too at a particular time, say for example, why Islamic fundamentalism became an important factor for causing international terrorism in the post-cold war world. In fact, a variety of factors intersected in post-1991 period due to which the forces of Islamic fundamentalism got instigated and facilitated to execute the terrorist acts and that too against the rising wave of unipolarist tendencies in the US foreign policy behavior, globalization and the prospects of ushering in of a 'New World Order' that would have imposed everything western on the traditional societies such as those in the oil-rich Middle-East.

One important fact about the acts of terrorism motivated by religion is that these acts are largely the rebellions against the secularization of state and society, against the the excesses of modernity and globalization. It may be viewed in terms of the failures of modern states and societies to accommodate religious identity and values. The global resurgence of religion in post-cold war world reveals how radical forms religion has taken and becomes a primary vehicle for both government and anti-government legitimization (India, Iran, Israel, Egypt, Sudan, Indonesia, Thailand, Uzbekistan, etc.). It must also be remembered that even though religious groups may espouse an anti-modernist ideology they are themselves quite modern in their organization and outlook of the world.

Political factors as a cause of Terrorism: Martha Crenshaw and her team at Club DE Madrid discussion [2005] points out that terrorism is a form of political action. It cannot be taken out of specific historical contexts or treated as a generic phenomenon. It is a strategy rooted in political discontent, used in the service of many different beliefs and doctrines that help legitimize and sustain violence. Ideologies associated with nationalism, revolution, religion, and defense of the *status quo* have all inspired terrorism. Even saying 'yes' to democracy does not necessarily mean saying 'no' to terrorism. We have to look at the opportunities, resources, intentions, and perceptions of actors for whom terrorism is useful to intimidate opponents, communicate goals, advertise the cause, recruit followers, and mobilize popular support.

It is also possible that the goals of some modern terrorism (*Al-Qaeda* and its offshoots) may include overturning the international order, perceived as a manifestation of Western domination of the Muslim world. Another source of concern at the level of the international system is state weakness, whether collapse or involvement in extensive civil conflict (the former often a result of the latter). Some failed or failing states—those without central governments or with governments that cannot maintain control over their territory or populations—become hosts for radical conspiracies that both impede stabilization and export terrorism to other targets and audiences. Prolonged civil conflict and instability produce waves of refugees and immigrants who form alienated Diasporas in which terrorist groups may find shelter. Economic weakness and political repression may also contribute to immigration. Dissatisfaction with local conditions is displaced onto the international system. These conditions are thus a serious problem for the international community. It may be noted that any type of regime, democratic or authoritarian, may be involved in an asymmetric conflict outside its borders. Stable and well-developed democracies may not face a serious threat from internally-generated terrorism, but external intervention or political, economic, and cultural presence may provoke terrorism from the outside. **Thus, a state's susceptibility to terrorism is determined not just by how it treats its citizens at home but by its actions abroad.** When such actions lack international legitimacy and local populations perceive them as unjust, radical groups come to see terrorism as an appropriate response. Not all interventions are the same, of course, and some are perceived positively by the populations in question, particularly if interventions are genuinely multilateral.

Several of these factors mentioned above may be tested as the causing factors for rise of militant Islam in the post-cold war period with a special analysis of the foreign policy orientations of the unipolarist US which became the direct and main target of terrorists' attacks since 1991 onwards ultimately culminating into the catastrophic attacks of 9/11. But before coming to the post-cold war evolution of Islamic fundamentalism and terrorism, let us briefly trace the early origins of this violent phenomenon as a means of achieving political goals.

History of Terrorism

Terror in Antiquity: 1st-14th Century AD: The earliest known organization that exhibited aspects of a modern terrorist organization was the ***Zealots*** of Judea. Known to the Romans as sicarii, or dagger-men, they carried on an underground campaign of assassination of Roman occupation forces, as well as any Jews they felt had collaborated with the Romans. Their motive was an uncompromising belief that they could not remain faithful to the dictates of Judaism while living as Roman subjects. Eventually, the Zealot revolt became open, and they were finally besieged and committed mass suicide at the fortification of Masada. The ***Assassins*** were the next group to show recognizable characteristics of terrorism, as we know it today. A breakaway faction of Shia Islam called the Nizari Ismalis adopted the tactic of assassination of enemy leaders because the cult's limited manpower prevented open combat. Their tactic of sending a lone assassin to successfully kill a key enemy leader at the certain sacrifice of his own life (the killers waited next to their victims to be killed or captured) inspired fearful awe in their enemies. Even though both the Zealots and the Assassins operated in antiquity, they are relevant today, first as forerunners of modern terrorists in aspects of motivation, organization, targeting, and goals and second, as sustainers of a psychological impact which is severely felt till date. Although both were ultimate failures, the fact that they are remembered hundreds of years later, demonstrates the deep psychological impact they caused.

Early Origins of Terrorism: 14th-18th Century: From the time of the Assassins (late 13th century) to the 1700s, terror and barbarism were widely used in warfare and conflict, but key ingredients for terrorism were lacking. Until the rise of the modern nation-state after the Treaty of Westphalia in 1648, the sort of central authority and cohesive society that terrorism attempts to influence barely existed. Communications were inadequate and controlled, and the causes that might inspire terrorism (religious schism, insurrection, ethnic strife) typically led to open warfare. By the time kingdoms and principalities became nations, they had sufficient means to enforce their authority and suppress activities such as terrorism.

The ***French Revolution*** provided the first usages of the words "Terrorist" and "Terrorism" in the context of the Reign of Terror initiated by the Revolutionary government. The agents of the Committee of Public Safety and the National Convention that enforced the policies of "The Terror" were referred to as 'Terrorists". The French Revolution provided an example to future states in oppressing their populations. It also inspired a reaction by royalists and other opponents of the Revolution who employed terrorist tactics such as assassination and intimidation in resistance to the Revolutionary agents. The Parisian mobs played a critical role at key points before, during, and after the Revolution. Such extra-legal activities as killing prominent officials and aristocrats in gruesome spectacles started long before the guillotine was first used.

During the late 19th century, radical political theories and improvements in weapons technology spurred the formation of small groups of revolutionaries who effectively attacked nation-states. ***Anarchists*** espousing belief in the "propaganda of the deed" produced some striking successes, assassinating heads of state from Russia, France, Spain, Italy, and the United States. However, their lack of organization and refusal to cooperate with other social movements in political efforts rendered anarchists ineffective as a political movement. In contrast, Communism's role as an ideological basis for political terrorism was just beginning, and would become much more significant in the 20th century.

Another trend in the late 19th century was the increasing tide of nationalism throughout the world, in which the nation (the identity of a people) and the political state were combined. As states began to emphasize national identities, peoples that had been conquered or colonized could, like the Jews at the times of the Zealots, opt for assimilation or struggle. Nationalism, like communism, became a much greater ideological force in the 20th century.

The terrorist group from this period that served as a model for future terrorist campaigns was the Russian Narodnya Volya (Peoples Will). It differed in some ways from modern terrorists, especially in that it would sometimes call-off attacks that might endanger individuals other than its intended target. Many of the traits of terrorism were witnessed for the first time; clandestine, cellular organization; impatience and inability for the task of organizing the constituents they claim to represent; and a tendency to increase the level of violence as pressures on the group mount.

20th Century Evolution of Terrorism: In the early years of the 20th Century nationalism and revolutionary political ideologies were the principal developmental forces acting upon terrorism. When the Treaty of Versailles redrew the map of Europe after World War I by breaking up the Austro-Hungarian Empire and creating new nations, it acknowledged the principle of self-determination for nationalities and ethnic groups. This encouraged minorities and ethnicities not receiving recognition to campaign for independence or autonomy. However, in many cases self-determination was limited to European nations and ethnic groups and denied to others, especially in the case of the colonial possessions of the major European powers, creating bitterness and setting the stage for the long conflicts of the anti-colonial period. In particular, Arab nationalists felt that they had been betrayed. Believing they were promised post-war independence, they were doubly disappointed; first, when the French and British were given authority over their lands; and then especially when the British allowed Zionist immigration into Palestine in keeping with a promise contained in the Balfour Declaration.

Since the end of World War II, terrorism has accelerated its development into a major component of contemporary conflict. Primarily in use immediately after the war as a subordinate element of anti-colonial insurgencies, it expanded beyond that role. In the service of various ideologies and aspirations, terrorism sometimes supplanted other forms of conflict completely. It also became a far-reaching weapon capable of effecting global course of events. It has also proven to be a significant tool of diplomacy and international power for states inclined to use it. The seemingly quick results and shocking immediacy of terrorism made some consider it as a short-cut to victory. Small revolutionary groups not willing to invest the time and resources to organize political activity would rely on the "propaganda of the deed" to energize mass action. This suggested that a tiny core of activists could topple any government through the use of terror alone. The result of this belief by revolutionaries in developed countries was the isolation of the terrorists from the population they claimed to represent, and the adoption of the Leninist concept of the "vanguard of revolution" by tiny groups of disaffected revolutionaries. In less developed countries small groups of foreign revolutionaries such as Che Guevara arrived from outside the country, expecting to immediately energize revolutionary action by their presence.

Evolution of Islamic fundamentalism and the dawn of modern international terrorism: One of the most threatening trends in the long history of terrorism is the emergence of fundamentalist Islam that became the main vehicle for using terrorism in the pre and post-cold war period. However, let it be made clear that the ***Islam, as opposed to fundamentalist or neo-fundamentalist Islam,*** posits a worldview that can deal with and selectively integrate modernity. In contrast, fundamentalist Islam calls for a return to an ontological form of Islam that rejects modernity; groups such as Al-Qaeda and the Egyptian Islamic Jihad are representative of fundamentalist Islam. In the wake of the Sept. 11, 2001 terrorist attacks on the US, the threat of militant Islamic terrorism—rooted in the Middle-East and South Asia—has taken center stage. While these extremely violent religious extremists represent a minority view, their threat is real. As pointed out by RAND's Bruce Hoffman, in 1980 two out of 64 groups were categorized as largely religious in motivation; in 1995 almost half of the identified groups, 26 out of 56, were classified as religiously motivated; the majority of these espoused Islam as their guiding force.

During the decades of 1960s and 1970s, the legacy of colonial era, failed post-colonial attempts at state formation, and the creation of Israel engendered a series of Marxist and anti-Western transformations and movements throughout the Arab and Islamic world. The growth of these nationalist and revolutionary movements, along with their view that terrorism could be effective in reaching political goals, generated the first phase of modern international terrorism.

In the late 1960s, Palestinian secular movements such as Al Fatah and the Popular Front for the Liberation of Palestine (PFLP) began to target civilians outside the immediate arena of conflict. Following Israel's 1967 defeat of Arab forces, Palestinian leaders realized that the Arab world was unable to militarily confront Israel. At the same time, lessons drawn from revolutionary movements in Latin America, North Africa, and Southeast Asia as well as during the Jewish struggle against Britain in Palestine saw the Palestinians move away from classic guerrilla, typically rural-based, warfare toward urban terrorism. Radical Palestinians took advantage of modern communication and transportation systems to internationalize their struggle. They launched a series of hijackings, kidnappings, bombings, and shootings, culminating in the kidnapping and subsequent deaths of Israeli athletes during the 1972 Munich Olympic Games.

These Palestinian groups became a model for numerous secular militants, and offered lessons for subsequent ethnic and religious movements. Palestinians created an extensive transnational extremist network—tied into which were various state sponsors such as the Soviet Union, certain Arab states, as well as traditional criminal organizations. By the end of the 1970s, the Palestinian secular network was a major channel for the spread of terrorist techniques worldwide.

While these secular Palestinians dominated the scene during the 1970s, religious movements also grew. The failure of Arab nationalism in the 1967 war resulted in the strengthening of both progressive and extremist Islamic movements. In the Middle-East, Islamic movements increasingly came into opposition with secular nationalism, providing an alternative source of social welfare and education in the vacuum left by the lack of government-led development—a key example is The Muslim Brotherhood. Islamic groups were supported by anti-nationalist conservative regimes, such as Saudi Arabia, to counter the expansion of nationalist ideology. Yet political Islam, more open to progressive change, was seen as a threat to conservative Arab regimes and thus support for more fundamentalist—and extremist—groups occurred to combat both nationalist and political Islamist movements.

1979-91: The Afghan Jihad and State-sponsors of Terrorism

The year 1979 was a turning point in the history and evolution of international terrorism. Throughout the Arab world and the West, the Iranian Islamic revolution sparked fears of a wave of revolutionary Shia Islam. Meanwhile, the Soviet invasion of Afghanistan and the subsequent anti-Soviet *mujahedeen* war, lasting from 1979 to 1989, stimulated the rise and expansion of terrorist groups. Indeed, the growth of a post-jihad pool of well-trained, battle-hardened militants is a key trend in contemporary international terrorism and insurgency-related violence. Volunteers from various parts of the Islamic world fought in Afghanistan, supported by conservative countries such as Saudi Arabia. In Yemen, for instance, the Riyadh-backed Islamic Front was established to provide financial, logistical, and training support for Yemeni volunteers. The so-called "Arab-Afghans" have—and are—using their experience to support local insurgencies in North Africa, Kashmir, Chechnya, China, Bosnia, and the Philippines.

In the West, attention was focused on state-sponsorship,

specifically the Iranian-backed and Syrian-supported Hezbollah; state sponsors' use of secular Palestinian groups was also of concern. Hezbollah pioneered the use of suicide bombers in the Middle-East, and was linked to the 1983 bombing and subsequent deaths of 241 US marines in Beirut, Lebanon, as well as multiple kidnappings of US and Western civilians and government officials. Hezbollah is understood to be the key trainer of secular, Shia and Sunni movements. As revealed during the investigation into the 1988 bombing of Pan Am Flight 103, Libyan intelligence officers were allegedly involved with the Palestinian Front for the Liberation of Palestine—General Command (PFLP-GC). Iraq and Syria were heavily involved in supporting various terrorist groups, with Baghdad using the Abu Nidal Organization on several occasions. These state sponsors used terrorist groups to attack Israeli as well as Western interests, in addition to domestic and regional opponents. However, it should also be noted that the American policy of listing state sponsors was heavily politicized, and did not include several countries—both allies and opponents of Washington—that, under US government definitions, were guilty of supporting or using terrorism.

1991-2001: The Globalization of Terror

The disintegration of almost the entire communist bloc of countries after the collapse of Soviet Union and the Cold War legacy of a world awash in advanced conventional weapons and know-how, has assisted the proliferation of terrorism worldwide. Vacuums of stability created by conflict and absence of governance in areas such as the Balkans, Afghanistan, Colombia, and certain African countries offer ready-made areas for terrorist training and recruitment activity, while smuggling and drug trafficking routes are often exploited by terrorists to support operations worldwide. With the increasing ease of transnational transportation and communication, the continued willingness of some states to provide training and support to suitable terrorist groups and dehumanizing ideologies that enable mass casualty attacks, the lethal potential of terrorist violence has reached new heights.

The region of Afghanistan has, particularly since the 1989 Soviet withdrawal, emerged as a fertile terrorist ground. Pakistan, struggling to balance its needs for political-economic reform with a domestic religious agenda, provides assistance to terrorist groups both in

Afghanistan and Kashmir while acting as a further transit area between the Middle-East and South Asia.

Since their emergence in 1994, the Pakistani-supported Taliban militia in Afghanistan has assumed several characteristics traditionally associated with state-sponsors of terrorism, providing logistical support, travel documentation, and training facilities. Although radical groups such as the Egyptian Islamic Jihad, Osama bin Laden's al-Qaeda, and Kashmiri militants were in Afghanistan prior to the Taliban, the spread of Taliban control has seen Afghan-based terrorism evolve into a relatively coordinated, widespread activity focused on sustaining and developing terrorist capabilities. Since the mid-1990s, Pakistani-backed terrorist groups fighting in Kashmir have increasingly used training camps inside Taliban-controlled areas. At the same time, members of these groups, as well as thousands of youths from Pakistan's Northwest Frontier Province (NWFP), have fought with the Taliban against opposition forces. This activity has seen the rise of extremism in parts of Pakistan neighboring Afghanistan, further complicating the ability of Islamabad to exert control over militants. Moreover, the intermixing of Pakistani movements with the Taliban and their Arab-Afghan allies has seen ties between these groups strengthen.

Since 1989 the increasing willingness of religious extremists to strike targets outside immediate country or regional areas underscores the global nature of contemporary terrorism. The 1993 bombing of the World Trade Center, and the Sept. 11, 2001, attacks on the World Trade Center and Pentagon, are representative of this trend. The superpower conflict in Afghanistan during 1980s became a formative period in the proliferation of weapons and emergence of militant, fundamentalist Islam. During 1990s, the trend toward directly targeting civilians continued, and gained even greater currency as ethno-nationalist, religious, and religio-nationalist actors filled the void left by the demise or decrease in leftist organizations. The end of the Cold War and the creation of new states, the leaving of certain states in unstable or anarchic conditions, gave impetus to the rise of a new set of extremists whose ideology or motivations allowed, or even called for, indiscriminate targeting [see Moore: Frontline].

One important aspect of the anti-Soviet *Jihad* was that along with native Afghan *mujahideen*, Muslim volunteers from other locations, particularly from Arab countries, were also participating to draw Soviet troops out of Afghanistan. One such Afghan-Arab was Osama bin Laden, a wealthy Saudi who provided his own money and helped raise millions from other wealthy Gulf Arabs in support of this movement. It was his organization: al-Qaeda that in coming years became the most dreaded terrorist organization with international terrorist networks and had a great impact on international politics in post-cold war years.

US as a factor in building up terror networks: The anti-Soviet *jihad* that was being spearheaded by Osama bin-Laden, who left Saudi Arabia and reached Afghanistan within a month of Soviet invasion in 1979, and by some other Afghan leaders, had the financial, military and logistical support of a multinational coalition organized by the US Central Intelligence Agency (CIA) and comprised of United States, Britain, Saudi Arabia, Pakistan, China and several other countries. This coalition was fully utilizing the Pakistani co-operation as Afghan and Arab fighters were being trained in its training camps and its Inter Service Intelligence (ISI) was serving as CIA's conduit to transfer sophisticated weapons to *Mujahideen*. In these early years of Afghan *jihad*, US was greatly in league with Osama who as part of his field work to raise funds and to lecture Muslims was touring west and US quite often [Gunaratana: 2002: p. 19]. Like Pakistan and US, Saudi Arabia was also a front runner in providing all sort of financial help in this *jihadi* project under the cover of philanthropy and relief work. In fact, al-Qaeda had a good start as it inherited a full-fledged training and operational infrastructure that had been funded by the US, European and Saudi Arabian and other governments throughout the 1980s. For recruitment, it drew on the vast *mujahideen* database originally created by Osama for tracking martyred or missing *mujahideens* in the later stages of the anti-Soviet *jihad*. American policy-makers in Afghanistan failed to realize the threat that this growing monster was about to pose soon, not even after attacks on US interests at different places. America was complacent in paying attention to constant complaints from the governments that had been victims of terror, such as, India where terrorism in Kashmir escalated exactly at the time of departure of Soviet troops as the availability of trained terrorists and operative infrastructure came handy to the hostile governments such as that of Pakistan. Failing to see a co-relation proved costly in 2001.

Moreover, as mentions Gunaratana, no weapons collection or buy-back program was started by the Western governments after the Soviet withdrawal. The weapons supplied by ISI to Taliban were later passed-on to al-Qaeda which already had a worldwide network for procurement of weapons. Also, the financial network of al-Qaeda flourished on finances provided not only from the Islamic philanthropists and foundations working in developing countries, particularly of Middle-East and Gulf but also from the overseas Muslims in US and West which later surprised even the CIA analysts. In terms of resources, al-Qaeda was a 'state within state' so much so that instead a host state controlling a terrorist organization, the later was controlling its host state. American government did very little to monitor the extremists within US who provided regular funding personnel for *jihadi* campaigns. US, instead, has remained one of the very important centres of procurement of sophisticated weaponry and other technological devices such as aircrafts and satellite phones which later were used by terrorists against US targets at home and abroad. US intelligence failed dismally to correctly and timely assess the dangerous potential of such activities. Its misplaced sense of security that by insulating itself from the rest of the world, American security will be ensured proved disastrous.

Reflecting on the state of ideological and physical penetration of Islamist groups in US, one of the leaders of Islamic Council of America said in 1999, "We can say that Islamists took over eighty percent of the mosques in US. There are more than 3000.... This means that the ideology of extremism has been spread to 80 percent of the Muslim population, mostly the youth and the new generation." [Kabbani: 1999] However, a little before the 9/11, FBI started infiltrating many Muslim organizations but al-Qaeda was aware of this monitoring and relocated its team for 9/11 attacks away from Muslim strongholds. Use of American aircraft and crashing them into the symbols of American supremacy splits open the shortfalls of American intelligence and excellence of terrorist planning in complete secrecy. A strong intelligence could have avoided not only the success of the novel tactic of using commercial passenger aircraft as improvised guided missiles by terrorists but it could also have tracked down much earlier their training in hijacking and aircraft flying in the American institutes itself.

Before opening hostilities in Afghanistan where the perpetrators of 9/11 catastrophe were hiding in safe havens, America persuaded many countries to break off diplomatic and commercial relations with Taliban regime in Kabul. UAE, the main channel to access financial and economic supplies to Taliban, was one such country. Pakistan, the fifty-year old recipient of American financial and military largesse throughout cold war years and still largely surviving on US support, came handy to provide its territory as the main launching pad for attacking terrorists' bases in nearby Afghanistan though there was a widespread public opposition to a pro-American gesture of President Musharraf who was caught between a dilemma of *realpolitik* where antagonizing a close and friendly Muslim neighbor, Afghanistan, could prove costly given the presence of many not-so-friendly surrounding neighbors in the region, such as, India, a Shiite Iran, Russia and former Soviet Republics of Central Asia—all viewed with mistrust by Islamabad. [Kepel: 2004, p. 3]

As the third largest recipient of US aid, including counter-terrorism subsidies that alone totaled $ 4.8 billion 2002 and 2006, Pakistan became critically dependent on such assistance and Washington played to that vulnerability. However, in initial years of war on terror, US had no intent to dump Musharraf as it needed Islamabad to employ Pakistan as a gateway to military operations in Afghanistan, a base for clandestine missions in Iran and a vehicle for other geopolitical objectives, including *vis-a-vis* India. Despite making democracy and freedom a rallying cry, Bush's democracy plank was merely to target regimes that defy US. In the case of friendly but dictatorial regimes, like in Pakistan and most Arab nations, US foreign policy recognizes that such governments can further American interests only if they stay insulated from the popular pressures of a democracy. Despite Bush calling Pakistan "wilder than the wild west", he kept giving primacy to the narrow, short-term geopolitical interests and relied on the very institution that was and still is the part of the problem—Pakistan's military. The United States could do so because the US is distant and the fallout of Pakistan's maneuvering and of US policies is still largely confined to the region. In fact, Pakistani military rulers have always been politically expedient to advance American interests and hence America's neglect of India's concerns suffering world's highest rate of terrorism. [Chellaney, Brahma, '***Out of Sync, Out of Mind***', *Hindustan Times*, Editorial, 6.03.07]

The catastrophic event of 9/11 shaped the entirely new concept of US national security—intercepting threats before they could damage American citizens, and hence the logic of the neo-conservatives' approval of preemption after 9/11, not only in the context of Taliban, al-Qaeda and Osama bin Laden, but the crusade against evildoers anywhere and everywhere [Kaplan and Kristol:

2003]. 'Operation Iraqi Freedom' in Iraq in 2003[3], without any substantial basis of possession of weapons of mass destruction by Iraqi dictator, freed the Bush administration from 'immorality' of using force and from the 'related obsession with avoiding casualties' reminiscent of Vietnam War and provided a compelling rationale for a sustained and proactive use of American power on a global scale justified as a necessary protective measure [Bacevich: 2003: pp. 229-31]. A number of foreign policy scholars are of the view that much before 9/11, Bush administration was working overtime to build a case against Saddam Hussein, and how to oust him and change Iraq into a new country and 9/11 provided a basis to go ahead. Even after America's Iraqi project, the "irrefutable proof" that Iraqi regime possessed WMD could not been found and evidence of Iraqi collaboration with Al-Qaeda has also not been substantiated. [Suskind: 2004; Herbert: 2004; Risen and Miller: 2003]. However, to provide a justification for going to war in Iraq, a series of rationalizations were offered ranging from alleged charge of possession of deadly weapons to claims of programs to acquire such weapons, an intention of Saddam to use them against West and America, to overriding objective of Bush Doctrine to export democracy in Middle-East initiating it with Iraq irrespective of weapons. Although the 'collateral damage' done in Iraq was tried to be erased from public memory by the 'embedded journalists' and mainstream media that hewed closely to the Bush administration's language and position, but the historical records of the US-Iraqi relations provided evidence of pro-Iraqi American policies pursued by Reagan and first Bush administrations in the aftermath of Islamic revolution in Iran in 1979 and their role, thenceforth, in providing Iraq the required 'material' to produce chemical, biological and nuclear weapons (CBN). President George W. Bush might have justified US intervention in Iraq on the grounds of stopping rogue states and their terrorist clients from threatening US and its allies, "in practice it sent a clear signal to its allies that in exchange for their support they would be free to act as they chose" [Gendzier: "Who Rules the Middle-East Agenda?" in Crotty, William: 2005: p. 61]. In other words, it was an approval for friendly states like Israel and its policies in West Bank and Gaza to the utter disregard for the concerns either of the Palestinians or for their pressing issues for getting a state of their own and for the military dictators like President Musharraf of Pakistan to continue to sponsor cross-border terrorism in Kashmir as long as he could be relevant in containing Taliban and al-Qaeda terrorists in Afghanistan.

One notable change that was visible after 9/11 was that a strategic alliance or realignment took place on two levels. At one level, and surprisingly so, most of the countries of the Muslim world irrespective of the region, including Pakistan—a state-sponsor of terrorism itself, came forward to join President Bush's War on Terrorism or quietly extended their cooperation in counter-terrorism efforts. But on the other level, a **polarization across regions was noticed amongst the majority of Muslims** who, openly or clandestinely, turned against the West and US and harbored a deep-rooted hatred against US perceiving it as anti-Muslim. A number of countries gained in multiple ways from this alliance with US. The monetary and strategic gains for Pakistan cannot be overemphasized. The *de facto* alliance of the newly independent Central Asian Republics with US not only brought much required money, opportunities, stature and international recognition of an isolated and neglected region, it also minimized the Taliban threat for the countries of the region.

Military Politics of US: The US political and military behavior in post-9/11 years, particularly under George W. Bush's presidency displayed its political and military prowess that is incomparable to the military or political might of any other nation independently or collectively. Gone are the days when imperialist agenda used to be measured by number of colonies an imperialist power had under its control. In 21st century, it is a state's military reach that determines a nation's influence, control and dominance. By this yardstick, Charmer Johnson in his latest work [Johnson: 2006] reveals that with 737 military bases of United States in other countries, as on in 2005, and with America's increasing deployments in Iraq, as was requested by the President Bush to the Congress in January 2007, the ever all-encompassing imperial footprints and militarism become too clear, particularly in the context of pursuit of President Bush's strategy of preemptive wars in a unipolar world, which were being profusely supported by his neocon (neoconservative) colleagues in administration and in mainstream media. The details of US militarism would be taken up in some details in Chapter 3 of this book, the fact remains that an ambitious militarism is too problematic domestically and internationally.

In addition with the immediate precipitants that facilitated rise of Islamic fundamentalism and subsequent events of international terrorism, particularly anti-American and anti-Western in nature, let us analyze some theoretical factors that converged to make international terrorism a reality, though a short-term reality, in the post-cold war era.

Islam factor: The American unipolarism, unilateralism and hegemonic behavior notwithstanding, these were not the isolated factors in precipitating terrorism. It has to be seen in the context of crisis of Islam. Islam's current crisis is related to failure to connect with the mainstream global society due to centuries of insulation from the outside world and refusal to think in ways other than religion. In frustration, many things have been tried—Pan-Arab nationalism, Islamic revolution, violence and terrorism in the name of God—but to no avail unless there is a balancing between traditional and modern that reforms Islam to make it compatible with the requirements of the age. The problem in Muslim society is that different groups hold different positions as to what to do to handle the current crisis related to the separation of religion and politics, individual and political freedom, human rights and the rights of women, question of how to modernize without losing Islamic characteristics.

There is a broad agreement in a large section of Muslim community as well as in the non-Muslim world globally that the more moderate, secular, women-friendly and humane elements in Islam should be encouraged to counter the influence of the rising volume of radical Islamists' fundamentalist interpretations that are detrimental not only to the progress and growth of Muslims as informed citizens of world family but also to world peace and harmony. Since Islam is an important world religion having vast following across continents, having enormous societal and political influence, a right direction in its world view is essential to dissipate the divisive and violent tendencies that have unfortunately got associated with it due to the violent activities of a small section of its followers.

The political revival of Islam has been the most salient feature of public life in countries belonging to Islamic civilization today. Traditional Islamic political thought is based on the religious universalism of Islamic revelation, and a political interpretation of this universalism seems to run counter to the modern institution of the nation-state. But although the current revival of political Islam is an expression of an Islamic revolt against the prevailing international order of nation-states and its local configurations, it is *not* a revival of traditional Islamic political thought. Islamic fundamentalists do not speak about the restoration of the traditional Islamic order of the caliphate, but rather of the *nizam Islami*/Islamic order, with clearly modern implications. The Western interpretations on the phenomenon of the repoliticization of Islam also suffer from failure to use primary Islamic sources, and thus fail to encounter fundamentalism on the ground. They suffer from the tendency to generalize from their one-country experience while lacking an appropriate conceptual framework. Only a very few of the authors of the large and continually growing body of writings on political Islam published in the West over the last two decades seem to make use of primary sources.

Discussing the challenge of Islamic fundamentalism, Bassam Tibi [Tibi: 2002] says that resorting to religion for the articulation of dissent results in translating religious beliefs into political convictions. The politicization of religion generally translates into religious fundamentalism. The new ideology poses a challenge to the world order of secular states. For rigidly religious societies and civilizations, it becomes a vehicle for the claim of both de-secularization and de-Westernization. Religious fundamentalism is a global political phenomenon observable in all major world religions, but the fact that only Islam and the West have universal claims explains this study's focus on Islam. Political Islam is not a religious belief. It is a political ideology which is anti-secular, and it charged with establishing a world order alternative to the prevailing one designed by Western norms and values. Instead of this envisaged world order, what is currently encountered is a disorder. Tibi argues that present age is characterized by simultaneity of structural globalization and cultural fragmentation and that the overall structural networking coincides with the revival of norms and values not compatible with one another giving rise to religious fundamentalism. Though the Gulf War of 1991 is nothing more than a fading memory for West, it has become a vital event in enhancing the anti-western attitudes in the Middle-Eastern region and in the world of Islam as a whole, resulting in politicization of Islam. The rise of political Islam leads to an increasing cultural fragmentation, a decline in consensus, and the diffusion of power in world politics. In the post-cold war period, the religious fundamentalism seems to be the greatest destabilizing challenge to the existing order of nation-states, and the ideology of the "Islamic state" is the framework for de-legitimization of existing secular states in the region of Middle East in particular. No doubt, Islam is compatible with democracy and human rights, but its abuse in the development of "al-Islam al-siyasi." or political Islam and the western universalism are the crux of the present day problems.

Terrorist versus Freedom Fighter Debate?

B. Raman, formerly with the Research and Analysis Wing (RAW), India feels that the pre-9/11 debate on the theme freedom-fighters *versus* terrorists has largely lost its relevance due to the traumatic experience of the international community after the horrendous acts of 9/11 and thereafter in different parts of the world. There is now growing realization in the world that the past attempts to rationalize the acts of terrorism of the organizations with national liberation as their objective by describing them as freedom-struggle have only played into the hands of terrorists. The cliché that one nation's terrorist is another nation's freedom-fighter, which was in vogue till the early 1980s, has been questioned seriously since the 1983 car bomb explosion in Beirut in which a large number of US Marines were killed. The senior George Bush, when he was the Vice-President, was appointed by President Ronald Reagan as the Chairman of a Presidential Task Force on Terrorism, which formulated the first clear distinction between freedom-fighters and terrorists in 1988. He said that any group or organization, which deliberately targeted innocent civilians, could not be categorized as freedom-fighters and had to be treated as terrorists. This distinction found increasing acceptance thereafter, particularly after the terrorists fighting for their so-called national liberation relied more and more on explosive devices instead of hand-held weapons in order to cause the maximum number of casualties amongst the civilians and shake the confidence of the people in the ability of the State to protect them. After the use of explosives by terrorists against the US Army barracks in Saudi Arabia in 1996 and against the US naval ship USS Cole in Aden in October 2000, this distinction has been further amplified by the US to precise that attacks on armed military personnel and installations would also constitute acts of terrorism if they take place in an area outside a zone of conflict. The US State Department's annual report on the Patterns of Global Terrorism during 2000 said: "We also consider as acts of terrorism attacks on military installations or on armed military personnel when a state of military hostilities does not exist at the site." After 9/11, there is unanimous agreement in the international community that terrorism should not be tolerated whatever be the cause.

The 21st century Terrorism: The *Jihadists* and environmental crises have replaced armies and missiles as the greatest threats, and globalization has eroded the significance of national borders. Many problems that were once national are now global, and dangers that once came only from states now come also from societies—not from hostile governments, but from hostile individuals. Despite this sea change of new challenges, there have been only ripples of new thinking about how to address them. While the problems have become largely global and societal, the solutions have not changed accordingly. The states must craft their foreign policy adapted to a world of complex global challenges which require thoughtful and global solutions. The political and economic stagnation in the Middle-East, the war in Iraq, the Arab-Israeli conflict, and other conflicts from Kashmir to Chechnya continue to produce the frustration and humiliation that cause terrorism, and with the right conditions, it only takes a small number of extremists to pose a serious threat.

The nature of crime and conflict has changed and continues to evolve. The current and future war is and will be influenced by irregular combatants—non-state soldiers—that utilize technology and networked doctrine to spread their influence across traditional geographic boundaries. This era shift in political and social organization, fueled by rapid developments of technology and exploitation of network organizational forms, blurs the distinctions between crime, terrorism and warfare. The resulting transitional period—a Dark Renaissance—benefits a range of non-state actors: drugs cartels, street gangs, terrorists, and warlords. Criminal organizations appear among the first to adapt to the new operational context resulting from this shift. These entities provide direct and indirect challenges to the solvency of nation-state institutions, potentially emerging as new war-making entities.

When the revolutionaries of France's Reign of Terror coined the term "terrorism" in 1793, they saw it as a positive tool, a "systematized attempt to weed out traitors" in the revolutionary ranks and one of the best methods to defend liberty. However, as the French Revolution unraveled, the term terrorism connoted strategic and systematic state violence with the use of the guillotine. Today, terrorism still evokes images of violence, but of more sophisticated and high-tech methodologies to carry out threats or actual bombing, hijacking, assassination or kidnapping. In an attempt to define what terrorism is, the US State Department describes it as "premeditated, politically motivated violence perpetrated against non-combatant targets by sub-national groups or clandestine agents, usually intended to influence an audience". In Asia, this deliberate use of violence against unarmed civilians for political or religious ends continues to hound many governments.

Defining Terrorism: Terrorism is not new, and even though it has been used since the beginning of recorded history it can be relatively hard to define. Terrorism has been described variously as both a tactic and strategy; a crime and a holy duty; a justified reaction to oppression and an inexcusable abomination. Obviously, a lot depends on whose point of view is being represented. Terrorism has often been an effective tactic for the weaker side in a conflict. As an asymmetric form of conflict, it confers coercive power with many of the advantages of military force at a fraction of the cost. Due to the secretive nature and small size of terrorist organizations, they often offer opponents no clear organization to defend against or to deter.

In some cases, terrorism has been a means to carry on a conflict without the adversary realizing the nature of the threat, mistaking terrorism for criminal activity. Because of these characteristics, terrorism has become increasingly common among those pursuing extreme goals throughout the world. But despite its popularity, terrorism can be a nebulous concept. The phrase "one man's terrorist is another man's freedom fighter" is a view terrorists themselves would accept. Terrorists do not see themselves as evil. They believe they are legitimate combatants, fighting for what they believe in, by whatever means possible. A victim of a terrorist act sees the terrorist as a criminal with no regard for human life. The general public's view is the most unstable. The terrorists take great pains to foster a "Robin Hood" image in hope of swaying the general public's point of view toward their cause. This sympathetic view of terrorism has become an integral part of their psychological warfare and needs to be countered vigorously.

The concept of moral equivalency is frequently used as an argument to broaden and blur the definition of terrorism. This concept argues that the outcome of an action is what matters, not the intent. Collateral or unintended damage to civilians from an attack by uniformed military forces on a legitimate military target is the same as a terrorist bomb directed deliberately at the civilian target with the intent of creating that damage. Simply put, any military action is simply terrorism by a different name. This is the reasoning behind the famous phrase "One man's terrorist is another man's freedom fighter". It is also a legacy of legitimizing the use of terror by successful revolutionary movements after the fact.

The very flexibility and adaptability of terror throughout the years has contributed to the confusion. Those seeking to disrupt, reorder or destroy the status quo have continuously sought new and creative ways to achieve their goals. Changes in the tactics and techniques of terrorists have been significant, but even more significant are the growth in the number of causes and social contexts where terrorism is used. Over the past 20 years, terrorists have committed extremely violent acts for alleged political or religious reasons. Political ideology ranges from the far left to the far right. For example, the far left can consist of groups such as Marxists and Leninists who propose a revolution of workers led by revolutionary elite. On the far right, we find dictatorships that typically believe in a merging of state and business leadership.

Many a times, nationalism is also found to be a powerful driver of violent movements, including terrorism. Typically, nationalists share a common ethnic background and wish to establish or regain a homeland. Religious extremists often reject the authority of secular governments and view legal systems that are not based on their religious beliefs as illegitimate. They often view modernization efforts as corrupting influences on traditional culture.

Almost eleven years after 9/11, the positive development is that though still there is no consensus on a definition of terrorism, a new global regime against terrorism has emerged, particularly in the field of disrupting terrorists' finances. Whereas, there were no coordinated global efforts before, today there exist 13 UN Conventions and various Security Council Resolutions on terrorism,[4] the number of countries joining fight against terror-finance rose from four to 158 after 9/11 though terrorism still continues to be funded by petty-crime, the money raised locally and a preferred system of illegal transfers like *hawala* markets. However, much needs to be done, such as, violations of financial regimes need to be checked, UN should also do more to build legal basis for fighting terrorism finance and to monitor its implementation. Terrorist attacks should be made more costly and therefore less likely. Fighting terror certainly is costly, not fighting it is costlier. Terrorism is expensive both in money terms as well as in terms of being a big risk to civil liberties. Another unfortunate feature is the growth of 'homegrown' terrorists, such as is strikingly evident after a number of terrorist acts in UK where according to the estimates of Security Service, there are perhaps 300 active cells or groups in UK and around 2000 terrorist-minded individuals—said Paul Beaver, a top security and intelligence analyst based in London [see: *Hindustan Times, Sept 8, 2007*]. In Britain the homegrown terrorists are becoming bolder and sophisticated, using freedom of Information and Racial Equality Legislation to their advantage which goes against the rest of the population; they have links with Pakistani training camps and with Afghan *mujahideens*, develop new explosives out of every-day-use chemical substances easily available from supermarkets. Such activities of terrorists have forced governments around the world to beef-up activity at airports investing huge amounts of money in airport and aviation security technology to sniff for explosives and their raw materials, resulting in more hectic security check-ups for the air-travellers—having costs in terms of curbs on civil liberties.

The war on terrorism has focused attention on the chaotic states that provide profit and sanctuary to nihilist outlaws, from Sudan and Afghanistan to Sierra Leone and Somalia. When such power vacuums threatened great powers in the past, they had a ready solution: imperialism. But the post-world war II anti-imperialist restraint is becoming harder to sustain, as the disorder in poor countries grows more threatening. Civil wars have grown nastier and longer. In a study of 52 conflicts since 1960, a recent World Bank study found that wars started after 1980 lasted three times longer than those beginning in the preceding two decades. Because wars last longer, the number of countries embroiled in them is growing. And the trend toward violent disorder may prove self-sustaining, for war breeds the conditions that make fresh conflict likely. Once a nation descends into violence, its people focus on immediate survival rather than on the longer-term solutions. Saving, investment, and wealth creation taper-off; government officials seek spoils for their cronies rather than designing policies that might build long-term prosperity. A cycle of poverty, instability, and violence emerges.

There is another reason why state failures may multiply. Violence and social disorder are linked to rapid population growth, and this demographic pressure shows no sign of abating. In the next 20 years, the world's population is projected to grow from around six billion to eight billion, with nearly all of the increase concentrated in poor countries. Some of the sharpest demographic stresses will be concentrated in Afghanistan, Pakistan, Saudi Arabia, Yemen, and the Palestinian territories—all Islamic societies with powerful currents of anti-Western extremism. Only sub-Saharan Africa faces a demographic challenge even sharper than that of the Muslim world. There, an excruciating combination of high birth rates and widespread AIDS infection threatens social disintegration and governmental collapse—which in turn offer opportunities for terrorists to find sanctuary. Terrorism is only one of the threats that dysfunctional states pose. Much of the world's illegal drug supply comes from such countries, whether opium from Afghanistan or cocaine from Colombia. Other kinds of criminal business flourish under the cover of conflict as well. Sierra Leone's black-market diamonds have benefited a rogues' gallery of thugs. Failed states also challenge orderly ones by boosting immigration pressures. And those pressures create a lucrative traffic in illegal workers, filling the war chests of criminals.

Centrality of Asia as a region facing and perpetrating terrorism: These facts bring into focus the centrality of Asia for any analysis of the causes and locale of terrorism in the current phase of history of international terrorism. The post-cold war events in Asian theatre show that Asia is not new for terrorism; what is new is its alarming proportions and connectivity to transnational terror networks in an age of globalization in which religion can be so easily politicized and misused for violence and terrorism. United States in its 'war on terror' rhetoric continuously mentions Asia as a key battleground for this anti-terror struggle. In fact, the post-1979 events reveal a strategic shift from Middle-East to South and South-East Asia. As Asian states confront ever newer economic and social challenges, combined with latent religious revivalism, they witness increased activities such as organized crime, drug and arms trade, outside interference and terrorism. Some major terrorism hubs, both with religious and secular agenda, are Afghanistan, Pakistan, Iran, Iraq, Lebanon, and Palestine, Sri Lanka: all part of larger Asia. Weak governance and widespread corruption have played important role in transforming Asia not only into terror network-hubs but also conducive ground for international crime.

A Rand study on ungoverned territories [Rabasa: 2007] brilliantly brings out the challenges that these areas pose to international security and serve as breeding grounds for terrorism and criminal activities and launching pads for attacks against the states. Ungoverned territories are unable to control routes that drug smugglers use to transport illegal drugs. The areas/states identified in this study include the Pakistani-Afghan border region; the Arabian Peninsula; the Sulawesi-Mindanao arc in Southeast Asia; the East African corridor from Sudan and the Horn of Africa to Mozambique and Zimbabwe; West Africa from Nigeria westward; the North Caucasus; the Colombian-Venezuelan border; and the Guatemala-Chiapas (Mexico) border.

The two dimensions: ***ungovernability,*** which means that the state is unable or unwilling to perform its primary functions of governance and maintaining law and order and ***conduciveness*** to terrorist or insurgent presence means that absence of adequate governance suits terrorists as they can conveniently use available infrastructure such as transportation and communication facilities, favorable demography, presence of extremist groups and the preexisting state of violence may be exploited and become breeding grounds for terrorists. Adequacy of infrastructure also means that a terrorist group must have a basic communications and transportation network and the means of transferring funds in order to operate and access its targets. When funding from external sources is not available, terrorists need to raise money locally in order to fund their operations.

In the Asian continent, the presence of organized armed groups outside of state's direct control is the characteristic of Pakistan-Afghan border (in South Asia), Yemen (in Arabian Peninsula), Somalia and Sudan (in East Africa). These territories are also among the areas of highest concern with regard to their potential for becoming sanctuaries for international terrorist groups. Afghanistan also makes a case of ***abdicated governance.*** The central government here abdicated its responsibility of producing public goods (such as safety, law and order and social services) in areas where the war lords control the difficult terrains. One of the very negative results of such an abdication of responsibility is that security forces collude with illegal drugs and arms trade which is so basic to sustain the terrorist campaigns.

Interference by external states has also been an important characteristic of international terrorism and many Asian locations which will be discussed in this study suffered from this syndrome. Soviet intervention in Afghanistan in 1979-89 and the American interference as a reaction to this in the cold war framework are case in point. Unwatched borders of ungoverned states are potential transit routes for their movement from one base to another and also to transport their logistics. To improve a state's ability to improve its control over its borders, vulnerable states need to be trained in detecting illegal entries, fake passports, and border crossing incidents and so on. Intelligence sharing and expedited communication between intelligence agencies may prove very effective in countering terrorism.

To eradicate this scenario of ungovernability, the international agencies like United Nations and its various social-economic bodies and programs and various states, such as US, must use development assistance as a tool to encourage recipient governments to invest in infrastructure and institutions and should also ensure accountability of these governments to use these assistance funds for the purpose they are given so as to check the incidents or possibility of siphoning-off the money for other purposes. The regional and local cooperation might be an effective tool to arrest the tendencies of bureaucratic lethargy and corruption and also to build the capacities of local military and counter terrorism forces. Improved governance with the help of international bodies and organizations will certainly boost the functional capacity of security forces to cut the terrorist networks from their lifelines: funding sources, illegal money transactions, safe havens and arms movement.

In next sections, the issue of rise of Islamic fundamentalism and international terrorism will be studied which needs to be situated in the foreign policy orientations of American hegemony in the post-cold war period as both have mutually spurred each other. In other words, the Islamist fundamentalism's clash with the concept and plan of a 'New World Order' caused much of international terrorism in the world. Since the post-cold war phase saw Asia as the central theatre for international terrorism, the issue will be studied in the Asian context. By way of conclusion, the focus would be on addressing the challenges, issues and problems that posed a very serious challenge to international peace and security.

Notes and References

1. The **NATO intervention in Bosnia and Herzegovina** comprised a series of actions undertaken by NATO to establish, and then preserve, peace during and after the Bosnian War. NATO's intervention began as largely political and symbolic, but gradually expanded to include large-scale air operations and the deployment of approximately 60,000 soldiers under Operation Joint Endeavor.
2. "It is my hypothesis that the fundamental source of conflict in this new world will not be primarily ideological or primarily economic. The great divisions

among humankind and the dominating source of conflict will be cultural. Nation states will remain the most powerful actors in world affairs, but the principal conflicts of global politics will occur between nations and groups of different civilizations. The clash of civilizations will dominate global politics. The fault lines between civilizations will be the battle lines of the future." [Huntington, Samuel P., "The Clash of Civilizations", *Foreign Affairs*, Summer 1993]

3. Operation Iraqi Freedom, in which a combined force of troops from the United States, the United Kingdom, Australia and Poland invaded Iraq and toppled the regime of Saddam Hussein in 21 days of major combat operations. The invasion phase consisted of a conventionally fought war which concluded with the capture of the Iraq capital Baghdad by United States forces. The United States supplied the majority of the invading forces, but also received support from Kurdish irregulars in Iraqi Kurdistan. The invasion of Iraq was strongly opposed by some traditional US allies, including the governments of France, Germany, New Zealand, and Canada as their argument was that there was no evidence of weapons of mass destruction in Iraq and that invading the country was not justified in the context of UNMOVIC's February 12, 2003 report.
4. **UN Conventions Open to All States**
   - 1961 *Vienna Convention on Diplomatic Relations*
   - 1963 *Vienna Convention on Consular Relations*
   - 1963 *Convention on Offences and Certain Other Acts Committed On Board Aircraft* (Tokyo Convention, agreed 9/63—safety of aviation)
   - 1970 *Convention for the Suppression of Unlawful Seizure of Aircraft* (Hague Convention)
   - 1971 *Convention for the Suppression of Unlawful Acts Against the Safety of Civil Aviation* (Montreal Convention)
   - 1979 *Convention on the Physical Protection of Nuclear Material* (Nuclear Materials Convention)
   - 1988 *Protocol for the Suppression of Unlawful Acts of Violence at Airports Serving International Civil Aviation*
   - 1988 *Convention for the Suppression of Unlawful Acts Against the Safety of Maritime Navigation*
   - 1988 *Protocol for the Suppression of Unlawful Acts Against the Safety of Fixed Platforms Located on the Continental Shelf*
   - 1991 *Convention on the Marking of Plastic Explosives for the Purpose of Identification*
   - 1997 *International Convention for the Suppression of Terrorist Bombings*
   - 1999 *International Convention for the Suppression of the Financing of Terrorism*
   - 2005 *International Convention for the Suppression of Acts of Nuclear Terrorism*

   A 14th international convention, a proposed *Comprehensive Convention on International Terrorism*, is currently under negotiations.

   Security Council Resolutions
   - *UN Security Council Resolution 731* (January 21, 1992)

- *UN Security Council Resolution 748* (March 31, 1992)
- *UN Security Council Resolution 883* (November 11, 1993)
- September 28, 2001 *United Nations Security Council Resolution 1373* adopted under *Chapter VII of the United Nations Charter*, which makes it legally binding to member-states. Among other provisions, it favored the exchange of intelligence between member-states and legislative reforms. It established the *United Nations Security Council Counter-Terrorism Committee* (CTC) to monitor state compliance with its provisions. Later resolutions concerning the same matter were UNSC resolutions *1390, 1456, 1535* (which restructured the CTC), 1566, and 1624.

REFERENCES

Addressing the Causes of Terrorism: The Club De Madrid Series on Democracy and Terrorism, Vol. March 2005.

Basevich, Andrew J., *American Empire: The Realities and Consequences of US Diplomacy*, Cambridge, Harvard University Press, 2003.

Crotty, William (ed.), *Democratic Development and Political Terrorism: The Global Perspective*, Boston: NE University, 2005.

Gaddis, John Lewis, "Toward the Post-Cold War World." *Foreign Affairs*, 1991, Spring.

Gaddis, John Lewis, *The United States and the Origins of the Cold War, 1941-1947* (1972; 2nd. ed., 2000).

Gaddis, John Lewis, *The Cold War: A New History*, Penguin Press, 2005.

Gunaratana, Rohan, *Inside Al-Qaeda: Global Network of Terror*, New Delhi: Roli Books, 2002.

Herbert, Bob, "This is Bush's Vietnam", *New York Times*, Sept 17, 2004.

Johnson, Charmer, *Nemesis: The Last Days of American Republic*, (reprint): Metropolitan, 2006, 2007.

Kabbani, Muhammad Hisham, "Islamic Extremism: A Viable Threat to US National Security", at Open Forum at the US Department of State, January 7, 1999.

Kepel, Gilles, *Jihad: The Trail of Political Islam*, London: I.B.Tauris, 2004.

Kepel, Gilles, *The War for Muslim Minds: Islam and the West*, Cambridge, London, 2004.

Kaplan, Lawrance F., William Kristol, *The War Over Iraq*, San Francisco: Encounter Books, 2003.

Moore, John, "The Evolution of Islamic Terrorism: An Overview", *Frontline*.

Rabasa, Angel *et. al*, *Ungoverned Territories: Understanding and Reducing Terrorism Risks*, RAND Project Air Force, 2007.

Suskind, Ron, *The Price of Loyalty*, NY: Simon & Schuster, 2004.

Tibi, Bassam, *The Challenge of Fundamentalism: Political Islam and the New World Disorder*, Berkeley, University of California Press, 1998, updated edition, 2002.

Chapter

2

Religious Fundamentalism and its Impact on International Relations: Special Reference to Rise of Militant Islam

The devil can cite Scripture for his purpose.
—William Shakespeare (1564-1616), The Merchant of Venice, I.iii

Science without religion is lame; religion without science is blind.

—Albert Einstein

Defining Religious Fundamentalism*:* Throughout human history, religions have played a major role in the lives of individuals as well in the integration of societies. This is still true in the post-Darwinian scientific West where religious truth and scientific truth have often clashed. Despite obvious contradictions between the modern science and literal interpretations of fundamentals of the religious texts, a large number of people across the world express their religious faith in some way or the other. One instant explanation to the question as to why in an age of scientific reasoning people still stick to some kind of religiosity is that human psychological characteristics shape religious systems in all their diversity. The recent advances in research in psychology, biology and social sciences help understand this almost inevitable phenomenon of religiosity in human behaviour. Discussing the positive as well as negative role of religion in human societies, Hinde highlights how religions have inculcated the values of love and tolerance for others that helped in the smooth running of societies [Hinde: 1999]. However, religions have also been widely used to perpetuate hatred, intolerance, inequities and torture leading to horrific religious wars. Positive role of religion notwithstanding, religious intolerance has resulted in violence against those who think differently.

The term 'fundamentalist' was coined in 1920 to describe conservative Evangelical Protestants who supported the principles expounded in *The Fnudamentals: Testimony to the Truth* (1910-15), a series of twelve pamphlets that attacked modernist theories of biblical criticism and reasserted the authority of the Bible. The central theme of *The Fundamentals* was that the Bible is the inerrant word of God. Scholars on 'Fundamentalism' insist that fundamentalism is nothing but a continuation of Christian orthodoxy. According to this theory, fundamentalism flourished for three centuries after Christ, went underground for twelve hundred years, surfaced again with the Reformation, took its knocks from various sources, and was alternately prominent or diminished in its influence and visibility. In short, according to its partisans, fundamentalism has always been the Christian remnant; the faithful who remain after the rest of Christianity has fallen into apostasy. Those people who today would be called fundamentalists formerly were Baptists, Presbyterians, or members of some other specific sect. But in the last decade of the nineteenth-century, issues came to the fore that made them start to withdraw from mainline Protestantism.

The term fundamentalism has rapidly entered the vocabulary of social science in the past two decades as a general designation for revivalist conservative religious orthodoxy. Though originally applied only to Christianity, the extension of the term to other religious traditions dates from the time of the Iranian Revolution in 1978-79. Today, it is used to describe Evangelical Christians, Iranian revolutionaries, ultra-orthodox Jews, militant Sikhs, and Buddhist resistance fighters and Hindu Right among others. Its categorical use is so widespread and so easily applied, that the misperception persists that it has always been with us. Religious fundamentalism is a 20th-21st century phenomenon, with roots reaching back to 19th century religious reactions against modern science and its adherents, whether Christian, Muslim, Jewish, Hindu or otherwise, share essentially the same primary characteristics as expressed in the concept of "enclave"—separation from the sinful world, demands for personal and social purity, adherence to a strict literal interpretation of a holy text, and loyalty to authoritarian religious group and its leaders.

Fundamentalism today is primarily expressed in a militant and sometimes bloody opposition to "liberalism" found both in secular culture and religious circles. It saps the vitality of the faith groups within which they reside by seeking total control over the lives of its adherents, threaten women's rights, either exercise or aspire for political power, have clear political agendas and are a hindrance to humanity's quest for more knowledge and understanding of world and self.

Martin and Appleby[1] identify certain characteristics that all religious fundamentalisms share. These are: religious idealism as basis for personal and communal identity; truth is understood to be revealed and unified which the 'outsiders' cannot understand; fundamentalists envision themselves as part of a cosmic struggle; they seize on historical moments and reinterpret them in light of this cosmic struggle; they demonize their opposition and are reactionary; fundamentalists are selective in that parts of their tradition and heritage that they stress; they are led by males and they all envy modernist cultural hegemony and try to overturn the distribution of power. ***Ideologically***, fundamentalists of all varieties feel highly 'concerned' with the erosion of religion and its proper role in society; they are selective in their choice of tradition which needs to be protected as well as modernity against which to react; they embrace some form of Manicheanism (dualism); stress absolutism and believe in impeccability in their sources of revelation and opt for some form of Millennialism or Messianism. ***Organizationally***, all fundamentalists have an elected or chosen membership, sharp group boundaries, charismatic authoritarian leaders and mandated behavioral requirements.

In short, fundamentalism is a growing, a worldwide and a modern religious phenomenon, which is found within all major world religions. It appears as the militant rejection of secular modernity and is not just traditional religiosity but an inherently political phenomenon, though this dimension may sometimes be dormant. Martin and Appleby also contended that fundamentalism is inherently totalitarian, insofar as it seeks to remake all aspects of society and government on religious principles. However, there are many objections to Martin and Appleby's theses particularly pertaining to the negative connotations of the term fundamentalism—usually including bigotry, zealotry, militancy, extremism, and fanaticism—which make it unsuitable as a category of scholarly analysis. Mahmood Mamdani, a scholar on Islam argues that 'fundamentalism' is a cultural phenomenon while terrorism, extremism and militancy are associated with a 'political' phenomenon and therefore does not want to use the two terms—political Islam and Islamic fundamentalism—interchangeably [Mamdani: 2004]. Nevertheless, some scholars have argued that the negative connotations of the term aptly characterize the nature of fundamentalist movements, many of which seek the violent overthrow of national governments and the imposition of particular forms of worship and religious codes of conduct in violation of widely recognized human rights to political self-determination and freedom of worship.

In America, the religious 'Right', comprised of a variety of religious-political organizations and leaders of religious groups, both Christian and other faiths, is the primary expression of fundamentalism. It claims that America was once a Christian nation, and that the concept of separation of church and state is a myth. However, both of these claims are wrong and not all followers of this religious Right are in agreement with this agenda. In the Middle-East, Islamic fundamentalism, of which the Taliban and al-Qaeda are the most well-known products at present, finds expression in many countries, sometimes evidencing itself in terrorist activity. The ultimate goal of Islamic fundamentalism is the reinforcement of current Middle-East theocracies, and the spread of Islamic theocracies to other nations. Although theocracy is viewed favorably by many Muslims, fundamentalist or otherwise, most of them do not condone terrorism. In short, Fundamentalism is about total control ... strict control of those within the conclave, and seeking to forcefully extend that control to those outside the enclave as a way of enlarging the "faith" community.

When the term fundamentalism was first applied to the Protestant Christian movement in America, which believed that every word of Bible is literally true, it was not used in pejorative sense. In the context of Islam, "the word fundamentalism was applied to Islam for the first time by American media when Islamic revolution was taking place in Iran in late 1970s and it was applied in a pejorative sense. Since then the world media has been using it, i.e. Islamic fundamentalism, in a very negative sense. Not only journalists but also academics are using it world-wide pejoratively" [Engineer: 2002]. The Islamic revolution of Iran, led by Ayatollah Khomeini who was very hostile to America and called it a great *Satan,* dealt a great blow to the American interests in the region and hence America began to denounce Islamic revolution of Iran and applied the term "Islamic fundamentalist" revolution. American media used headlines like "militant soldiers of Allah on march". In this way, the term **'Islamic fundamentalism' is basically a political term,** which, according to Engineer, conveys a sense of political hostility rather than religious rigidity, militancy, conservatism or orthodoxy. A rigid dogmatic approach to moral and spiritual questions should not be dubbed as fundamentalism. Fundamentalism, however, **is used in popular parlance today to mean religious rigidity, militancy and extremism as well as use of Islam for political ends** rather than for spiritual and moral development. More than a revival of premodern religious views, fundamentalism is a practical political preference of religious ideologues and political activists who have political agendas and are primarily concerned with political power. It is not confined merely to Islam, though in the current phase of history, it has got explicitly associated with Islam due to Islamists' prevailing anti-Western outlook and their involvement in terrorist activities, specifically in the post-cold war years.

Fundamentalism: A Political Variety: Thus, Islamic fundamentalism is only one variety of a new global phenomenon in world politics. Bassam Tibi, himself a Muslim, is an outspoken critic of much of contemporary Islam and is critical of Islamic fundamentalism as it strongly rejects the spirit of religious pluralism, dismissing it as a heresy threatening the neo-absolutist claim for the dominance of political Islam throughout the world. The intellectual father of Islamic fundamentalism, Sayyid Qutb, who is supposed to have inspired Osama bin-Laden, claimed an Islamic world order to replace the present one. Fundamentalism is not only an intellectual challenge; it is a challenge to security inasmuch as it proposes to topple the existing order. ... The challenge is a very concrete one posed and practiced by ... *jihad*-fighters willing to sacrifice their lives. "By definition, then, this ideology is exclusive in the sense that it attacks opposing options, primarily those secular outlooks that resist the linking of religion and politics. Fundamentalists thus are absolutists, by their very nature, and as we move into the next century, they seem to be placing their imprint on world politics" [Tibi: 1998]. These radical fundamentalists are less concerned with literal substance of Islam, with which they take considerable liberties either deliberately or because of ignorance of orthodox Islamic doctrine. They usually do not have any institutional religious affiliations but tend to be eclectic autodidactic in their knowledge of Islam. al-Qaeda and Taliban fall into this category. They not only glorify the past but apply some more stringent rules more rigorously than the original Islamic community ever did, exercising an arbitrary selectivity that permits them to ignore more egalitarian, progressive and tolerant aspects of Quran and *Sunna,* and inventing some new rules of their own. Though all fundamentalists do not necessarily endorse terrorism, but fundamentalism as whole, be it associated with any religion, is incompatible with the values of a civil society, particularly if it enters the political domain. Thus, Tibi views fundamentalism *per se* as primarily, if not solely, a political phenomenon that is first and foremost "an aggressive politicization of religion undertaken in the pursuit of non-religious ends". In his view, fundamentalism is only secondly and superficially a form of terrorism or extremism.

Islamic Fundamentalism and the difference between Islam, Islamist/Islamic Fundamentalist and Muslim

Though fundamentalism is a term that applies to all religions in so far as they plead for the strict maintenance of ancient or fundamental doctrines of their religion, fundamentalism in Islam is today the most visible and influential of all fundamentalisms. The term is used to describe religious ideologies seen as advocating literal interpretations of the texts of Islam and of *Sharia* law. Muslims believe that the Qoran is the unadulterated word of God as revealed to Muhammad, the Prophet. Islamic fundamentalists, or at least "reformist" fundamentalists, believe that Islam is based on the Qoran, *Hadith* and *Sunnah* and "criticize the tradition, the commentaries, and popular religious practices, the cult of saints, deviations, and superstitions". It aims to return to the founding texts. They hold that the problems of the world stem from secular influences. Further, the path to peace and justice lies in a return to the original message of Islam combined with a scrupulous rejection of all religious innovation and perceived anti-Islamic traditions.

Islam is a religious belief, not an ideology. An ordinary Muslim believes in Islam as a faith and honors its precepts as a source of ethics for humans and of orientation in their conduct. But this is not the way fundamentalists see Islam. Fundamentalisms, of whatever stripe, are not expressions of a renaissance of religion, but rather reflect political ideologies ostensibly drawn from religions in an effort to remake the world, thus clearly for political ends. The true **fundamentalist is basically a political man with a political outlook, and in some cases a political activist with little or no interest in religious ethics and divinities.** It becomes obvious, then, that our concern here is to study Islamic fundamentalism as a political ideology, not to inquire into the religion of Islam itself. The distinction between Islam as a faith and the political ideology of Islamism as a variety of religious fundamentalisms is most important to denying Islamists their claim that they are the true representatives of Islam, which they are not.

Though it has its opponents like Fred Halliday and John Esposito, the term ***Islamism*** has come to denote a set of political ideologies holding that Islam is not only a religion but also a political system and advocating *sharia* (the Islamic law) to be the basis of all statutory law of society under which all Muslims should take refuge and that the western influence in social, cultural, political and economic matters of followers of Islam is un-Islamic. The definition offered by American historian Ira Lapidus distinguishes between mainstream Islamists and Fundamentalists. Although a fundamentalist may also be an Islamist, a **Fundamentalist is "a political individual"** in search of a "more original Islam," while the **Islamist is pursuing a political agenda with which the devout Muslims have nothing to do.** He notes that Islamic fundamentalism "is at best only an umbrella designation for a wide variety of movements, some intolerant and exclusivist, some pluralistic; some favorable to science, some anti-scientific; some primarily devotional and some primarily political; some democratic, some authoritarian; some pacific, some violent" [Lapidus: 2002]. Olivier Roy distinguishes between fundamentalists or neo-fundamentalists and Islamists in describing fundamentalists as more passionate in their opposition to the perceived "corrupting influence of Western culture," avoiding western dress, neckties, laughter, the use of western forms of salutation, handshakes, applause. While Islamists like Maududi didn't hesitate to attend Hindu ceremonies, Khomeini never proposed the status of *dhimmi* (protected) for Iranian Christians or Jews, as provided for in the *sharia*: the Armenians in Iran have remained Iranian citizens, are required to perform military service and to pay the same taxes as Muslims, and have the right to vote (with separate electoral colleges). Similarly, the Afghan *Jamaat*, in its statutes, has declared it legal in the eyes of Islam to employ non-Muslims as experts.

Also called by many as *political Islam*, the *Islamism* represents a form of ideological protest, sometimes against the arbitrary rule and socio-economic injustice also. In Islamic societies, since other western-type democratic channels of opposition are absent, fundamentalism/ extremism and radicalism have been used historically as a religiously sanctioned means to articulate popular dissatisfaction, though today's Islamist terrorists do not seem to be worried about what the popular Muslim opinion is. Bernard Lewis would identify the modern Islamists, whom he prefers to call 'activists', with the 'rebel' Prophet, who before establishing an Islamic state in Mecca and becoming its sovereign, protested against the pagan oligarchy of Mecca, remained in exile, formed a government and returned back triumphantly. Thus, a rebel Prophet can be taken to provide a paradigm of revolution to modern day's radical Islamists [Lewis: 2002, 2003].

The other distinctions between Islamists and fundamentalists are: *Islamists* often talk of revolution and believe that the society will be Islamised only through social and political action: it is necessary to leave the mosque ... Fundamentalists are uninterested in revolution, less interested in modernity or by Western models in politics or economics, and less willing to associate with non-Muslims [Roy: 1994]. As regards *Sharia,* while both Islamists and fundamentalists are committed to implementing *Sharia* law, Islamists tend to consider it more a project than a corpus [Roy: 1994]; on issue of women Islamist generally tend to favor the education of women and their participation in social and political life, Islamist groups include women's associations. On the contrary, the fundamentalist preaches for women to return to the home. **Graham Fuller describes fundamentalism not as distinct from Islamism but as subset,** "the most conservative element among Islamist". Its strictest form includes Wahhabism, sometimes also referred to as *salafiyya.* ... For fundamentalists the law is the most essential component of Islam, leading to an overwhelming emphasis upon jurisprudence, usually narrowly conceived [Fuller: 2006]. Commenting on fundamentalists' approval and endorsement of killing those who do not abide by the *Sharia* and therefore are apostates who can be killed, Bassam Tibi, says that such a killing is unjustified and does not have the approval of Quran. He asserts that the command to slay reasoning Muslims is un-Islamic, an invention of Islamic fundamentalists [Tibi: 2002].

According to the Rand Project Airforce [Rabasa: 2004], the term *Muslim* and *Islamic* are often used interchangeably in discussions on Islam though there are important differences between the two. The

Muslim refers to the religious and cultural reality; *Islamic* denotes political content. Whereas a *Muslim country* is one where majority of the population is Muslim, an *Islamic state* is one that bases its legitimacy on Islam. As in any variant of religious fundamentalism, Islamic fundamentalism implies return to the original foundations of the faith. As long as the fundamentalist adherents of the religion follow the fundamental tenets of their religion, it remains a matter of personal domain, not affecting others. But, when fundamentalist radicals use original tenets of religion to politically mobilize their co-religionists who might be having some real or imaginary discontenments, they might be termed as ***Islamists,*** **such as in the context of politicized Muslims. If these Islamists use unpredictable, premeditated, sporadic and lethal violence in their quest for political power, it becomes an act of terrorism driven by Islamic fundamentalism.** In Muslim world, religion, politics and culture are intertwined in complicated ways and intersect with the geopolitical interests of the world's great power(s), the core problem in post-cold war world.

Ideological Streams within Islam

In fact, though all the Muslims are the followers of Islam, there exist various trends in religious interpretations and behavior amongst nearly 1.2 billion of the world's Muslims scattered over different regions over all the continents, having visible signs of regional influence over these interpretations and behavior. An analysis of certain social, political and cultural indices shows that the Muslims do not constitute a homogenous group having a common pattern throughout the world. Rather, they differ in their perceptions of government and human rights, their social agenda, particularly related to women's status in family and society and their rights and the content of education curricula even for men, their views on covert or overt support to and linkages with international terrorist networks and their propensity of violence. A RAND study by Cheryl Benard clearly summarizes the defining characteristics of the main tendencies in Islam along the different streams of ideological divide, which can broadly be categorized as: ***Fundamentalists*** (Scriptural and Radical), ***Traditionalists*** (Conservatives and Reformists), ***Modernists*** and ***Secularists*** (mainstream and Radical) [Benard: 2003: pp. 8-13]

This classification reveals that though a large mass of the people are oriented in accordance with their falling into a specific category, there is always a small minority within each category whose views on different characteristics might overlap and are identical to the other end of the ideological spectrum. For example, the followers of secularist or modernist group at times might approve of violence. In the case of July 2007 Heathrow and Glasgow airport blasts, the arrested Muslim youth were found to be highly educated and belonged to relatively moderate and highly educated strata of society. All these three moderate turned radicals were suspected to belonging to Tiblighi Jammat (TJ), widely understood to be a missionary organization with no pronounced political agenda. Founded with the objective of reviving Islamic beliefs and practices in 1920s in Deoband, India and having adherents in all the continents, TJ is basically not involved in promoting terrorism or militancy though this does not rule out individuals using it for militant purposes in some isolated cases [see: "Missionary or Militant"—a report in *Hindustan Times*, July 11, 2007, p. 9]. Intelligence reports from India and UK point out that the global terror groups like al-Qaeda seek highly qualified, middle or upper middle-class recruits as they are more suited to handle complicated and nuanced terror operations and due to their well-to-do status, they are expected to be more committed and productive. This example clarifies the possibility of a faction of a moderate/non-violent group drifting into violent terrorist ventures.

In general, however, the religiously moderate groups have greater affinity for democracy and are less inclined to violence while, conversely, radical fundamentalists are both anti-democratic and violent. Sometimes, the radicals and moderates coexist under the same movement or group. Also, there are examples of a radical group evolving into a moderate one and the *vice versa*, thus prohibiting any effort to put a group under one exclusive category. Muslim Brotherhood, for instance, had a history of militancy, conspiracies and assassinations in its early stages, which later on evolved into a still radical but non-violent movement in a number of countries in the Arab world, particularly in Egypt. Even Hamas, the largest and most influential Palestinian militant movement that won the Palestinian Authority's (PA) general legislative elections in January 2006, defeating Fatah, the party of the PA's president, Mahmoud Abbas, has historically the record of doing a lot of social services for Palestinians in the Occupied Territories. But more notoriously, the group has also operated a terrorist wing carrying out suicide bombings and attacks using mortars and short-range rockets. The group has launched attacks both in the Palestinian territories of the West Bank and Gaza Strip, and inside the pre-1967 boundaries of Israel. In Arabic, the word "hamas" means zeal. But it's also an Arabic acronym for "Harakat al-Muqawama al-Islamiya," or Islamic Resistance Movement.

The various streams of Islamist ideologies sanctioning different levels of radicalism/extremism/fundamentalism and terrorism are as follows:

(a) Radical Fundamentalist Ideology has a recent origin and the groups following this ideology have certain common denominators like:

- Total belief in *Sharia* law, the basis to establish an Islamic state;
- Dedication towards fundamentalist religious beliefs;
- Anti-Semitism (hatred for Jews and Zionism and Israel);
- Desire to establish pure, ethnic communities and for imposing strict discipline both on men and women in their day-to-day living and behaviour; and
- Carrying out a revolutionary political program, both violent and non-violent depending on individual group's ideology.

The Islamic fundamentalist' push for *Sharia* and an Islamic state has come into conflict with conceptions of the secular, democratic state, such as the internationally supported Universal Declaration of Human Rights. Among human rights disputed by fundamentalist Muslims are:

- ***Disagreement with western concept of the equality of men and women***: Under *Sharia* law, for example, a "man gets double the share of a woman in inheritance" because "he has much more responsibilities". The Prophet is said to have told early Muslims 'The best woman is she who, ... when you direct her she obeys"[2];
- ***The separation of church and state;***
- ***Freedom of religion***: Muslims who leave Islam, or worse still criticize it, "should be executed", while the right of non-Muslims to convert to Islam is celebrated; and
- ***Freedom from Religious Police.***

As a result of these sharp differences, some would argue that fundamentalist Islam is incompatible with modern liberal democratic states.

Salafi-Wahhabi: An influential strain of Muslim thought came from the Wahhabi movement in Saudi Arabia. The *Wahhabists,* who emerged in the 18th century led by Muhammad ibn Abd al-Wahhab, also believed that it was necessary to live according to the strict dictates of Islam, which they interpreted to mean living in the manner that the prophet Muhammad and his followers had lived in during the 7th century in Medina. Consequently, they were opposed to many religious innovations such as veneration of saints. This ideology represents extreme conservatism and emphasis on outward physical manifestations, such as, growing beards, wearing Islamic garb by men, strict segregation of sexes, wearing of *abaya* and *hijb* by women, five time recitation of *Namaj* (prayer) each day and fasting by all adults in the month of *ramzan* (holy month for Muslims) and so on. *Salafism* is a generic term depicting a Sunni Islamic school of thought that takes the pious ancestors, *Salaf,* of the patristic period of early Islam as exemplary models. *Salafis* view the first three generations of Muslims, who are Muhammad's companions, and the two succeeding generations after them, the Tabi'-in and theTabi'-at Tabi-in, as examples of how Islam should be practiced. The principal tenet of Salafism is that Islam was perfect and complete during the days of Muhammad and his companions, but that undesirable innovations have been added over the later centuries due to materialist and cultural influences. Salafism seeks to revive a practice of Islam that more closely resembles the religion during the time of Muhammad. Salafism has also been described as a simplified version of Islam and is often used interchangeably with "Wahhabism."

Salafism emerged at the at of 19th century to respond to the cultural, religious, political and military challenge posed by the West; sought the 'purification' of Islam by exhorting Muslims to go back to roots and following the teachings of Prophet. *Salafi* movements also have many streams—some are purely religious, non-political, some have evolved modernist tendencies in Islam and yet some others have radicalized Islam and provide support to violent terrorist groups. *Wahhabis* preceded *Salafists* by some 150 years for whom *jihad* is not only permissible, it is obligatory for infidels, idolaters and non-believers. Wahhabism remained marginal amongst Muslims till King Saud established Kingdom of Saudi Arabia in 1932 which contained holy mosques of Mecca and Madina. Since then, extreme doctrines of Islam are imposed not only upon Muslims but also to a great extent on those who are visitors or workers there. Till date, *Wahhabism* has fundamentalism of tribal origins [Rabasa: RAND Project: 2004; Wikipedia search].

Muslim Brotherhood (MB) was established in Egypt in 1928 by Hasan-al-Banna, having a sophisticated organizational structure with access amongst different strata of society and its organizations in almost all the countries of the Arab as well as non-Arab Muslim world. Its sanction to violence depends upon the requirements in the country where its branch is functional. Over all, it has a vision of an Islamic state in its pure form. After Banna, Sayyid Qutb introduced the neo-fundamentalist ideology of seizing power by revolutionary vanguard and openly supported violence against all those who stood in the way of establishing an Islamic state. Today's founders and followers of al-Qaeda can rightly be called the followers of Qutb in their zeal for violence, hatred and terrorism.

Tiblighi Jamaat is a missionary Muslim and revival movement working within Muslim community itself with the objective of bringing a spiritual awakening amongst the Muslims and has no declared political objectives. It aims to work at grass root levels reaching out to Muslims throughout the world across social and economic spectrum. Initially it started its activities in India and then subsequently reaching out to the South Asian states and now has extended its reach within South Asian diaspora in Arab, Southeast Asia, USA and Britain due to a large Asian Muslims community living in these countries. Though a professedly non-political movement, due to its popular stature, many Muslim politicians from Muslim and non-Muslim countries from both right and left ideology associate themselves with *Tiblighi.* It is tolerated even by the intolerant governments such as of Pakistan as its preachings are supposed to be purely spiritual, having nothing to do with politics [see: Wikipedia].

Deobandi movement started in India by Sayyid Abul Ala Maududi, developed as a reaction to British colonialist actions against Muslims and to check the influence of Muslim modernist Sayed Ahmad Khan, who advocated the westernization of Islam. Named after the town of Deoband in Uttar Pradesh in India, where it originated, the movement expanded under the guidance of Maulana Qasim Nanotwi on the traditional methods of *Fiqh* (jurisprudence) and *Aqidah* (theology). Maududi advocated the creation of an Islamic state governed by *sharia,* Islamic law, as interpreted by *shura* councils. To him, because Islam is all-encompassing, the Islamic state should not be limited to just the "homeland of Islam". It is for the world.

However, although Maududi talked about Islamic revolution, he was both less revolutionary and less politically/economically motivated than the later Islamists like Qutb.

Now, as a foremost movement of traditional Islamic thought in the subcontinent, it works for the establishment of thousands of *madarssahs (religious schools)* throughout India, Pakistan and Bangladesh. Deobandi thought is defined foremost by its adherence to the *Hanafi Fiqh* and by its emphasis on *Tasawwuf* (Spiritual reformation and purification). In Pakistan, *Deobandiism* is represented by the Jamat Ulema-e-Islam organization/political party and its splinter groups. The thousands of *madarssahs* these groups established for impoverished Afghan refugees helped spawn the Taliban, a Deobandi-based movement that held power in most of Afghanistan from 1996 to 2001. The Taliban are known for the many restrictions they placed on women, their hosting of Osama-bin-Laden and the execution of their gruesome terrorist acts.

(b) The neo-fundamentalist Ideology: The modern neo-fundamentalists such as represented by some factions of Muslim Brotherhood, Taliban, al-Qaeda, Abu Sayyaf, Jemmah Islammiyah and a number of Palestinian and Pakistani groups, reject western culture and ideas. The rejected category includes Crusaders, Communists, secularists and Jews. The crusadors are the western and American soldiers on the soil of Islamic countries. Communists are rejected for their rejection of God. Secularists also base their state and citizenship on non-religious grounds and thus are guilty of apostasy and there is no place for secularists in the radical Islamists' concept of universe. The Jews have committed the historical blunder of occupying the holy lands of Islam in Palestine and deserve death [see: Rabasa: Rand Project, 2004; Kepel, G.: 2000]. Democracy is condemned as a religion of infidels as voting in elections is prohibited in Islam. Some neo-fundamentalist terrorist groups, such as, Hizb ut-Tahrir (HuT), which now is banned in many countries: the entire Central Asia, in Egypt and also in Germany for its radical agenda, call for reestablishment of Caliphate.

The Jihadi ideology, preached by these neo-fundamentalists is based on the concept of exerting one's utmost efforts in order to attain a goal, the goal of holy war, particularly against the infidels, demands one's supreme sacrifice in order to raise the words of Allah. The aim of *jihad* is to remove oppression and injustice and to eliminate the barriers to the spread of truth. Striving for religious deeds, defending

Islam and community, helping friends and removing treacherous rulers from power, conveying the message of Islam and all these to be pursued in the name of Allah and not for the sake of wealth, goods, fame, glory or power [see: The Jihad Fixation: 2001]. The reasons for the misfortunes and decadence of Muslim societies are attributed to abandoning *jihad* for the love towards this-worldly gain and fearing or running away from martyrdom. The Afghan *jihadis* are elevated to the status of the sons of Islamic movement loved by Allah. Though terrorism and the deliberate killing of non-combatants is forbidden in Koran, the *jihadi* ideology in practice does have no inhibitions in using terror tactics in the name of service to Allah and religion. Killing and dying for Allah are viewed as the highest forms of sacrifice by most of the terrorist groups driven by Islamist ideology and that is the reason why one finds so many suicide operations by a variety of Islamist terrorist groups. Terrorism is only one aspect of the multi-dimensional strategy of engaging the enemy on several fronts, the most challenging strategy being the radicalization and politicization of Muslims all over the world and thus exhorting them to make super sacrifice to book a seat in the heaven. Such an ideological maneuvering succeeds in getting large number of brainwashed and disenchanted Muslim youth as recruits to terrorist groups who happily sacrifice their lives and everything else at the call of their leaders and thus serve as potential sustainers of terrorist violence through religious fundamentalism.

In fact, jihad as a means of warfare was not sanctioned in Quran; it was referred to as a person's inner struggle. But later, in the expansioning period of Muslim state, military struggle was justified in some circumstances, though never as a first resort. Theoretically, even a *jihad* in self-defense prescribed for some normative limits to be observed during fighting an aggressor, such as, using minimum force, protecting the non-combatants and avoiding ambushes and assassinations. However, the modern Islamist radicals use it as an expression of Islamic world revolution and premise it on their own concept of the west, which in their view has victimized the Muslims worldwide and hence turn the Qoranic concept of *jihad* as the main driving force for violence and terrorism. But the modernist/secularists amongst the Muslims do not endorse the view of radical Islamists that the armed jihad is a religious obligation for Muslims. Instead, they emphasize the duty of all Muslims to wage a war against the prevailing evils in Muslim society, such as, poverty, ignorance, social backwardness and orthodoxy.

The radical fundamentalists of all varieties believe that the teachings of Islam are so comprehensive and all-encompassing that they can take care not only of religious but also of political aspects of life and hence, religion and politics in Islam are so intertwined that any effort to separate them is branded as an act of infidelity.

(c) Other Non-Radical Streams in Islam are represented by the traditionalists, *Sufists*, modernists and secularists within Islam. The fact is that they are not as vocal or as imposing as the radical Islamists are, making it believe by a majority in the non-Muslim world that radical/extremist version of Islam is the mainstream version and these being in majority all over the world. However, this is far from truth. ***Traditionalist***, mostly conservative in their beliefs and traditions drawn from the local beliefs and practices adapted over centuries, constitute the majority among Muslims. Differing from Wahhabi severity and intolerance, they offer prayers at the tombs of saints and Faqirs, have belief in spirits and miracles, do not endorse violence and terrorism, do not blame others, particularly the West and US for problems of the Muslims and their activities are largely concentrated in social and cultural spheres like promoting Islamic education and preserving traditional values and norms [see for details: Rabasa: 2004: pp. 63-67]. Thus, traditionalists' views and actions are, by and large, in compliance with secularist, pluralist ethos.

Having a popular following and strong tradition, ***Sufism***, the Islamic mysticism, has always remained popular amongst the Muslims, particularly in Central Asia, North and West Africa, South and Southeast Asia till today, though they are charged for the deviant corruption of Islam and for their pacifistic views that obviously does not suit those Islamists who want to establish Allah's religion on earth by the use of sword, by the *Salafist-Wahhabist* school, modern radical *Sunni* fundamentalists and terrorists. As Anthony Shadid mentions in Boston Globe, wherever the radical Islamist movements have gained power, such as Taliban in Afghanistan in 1996, they tried to suppress *Sufi* practices of singing, dancing, chanting as a way of reaching an ecstatic state, thought by them as a way to be nearing the God [Shadid: 2002]. Nevertheless, the victimization of *Sufis* and traditionalists by *Salafist/Wahhabist* radicalists has made the former a natural ally of the West in their efforts to define the place of Islam in the modern world.

Modernism is another non-radical stream amongst Muslims whose main prank is to free Islam of superstitions propagated by some streams such as traditionalists and to bring it at a level of harmony and compliance with the modern world capable of articulation and defense by reason. The modernist scholars call for Muslim unity and reform not only as a bulwark against the onslaughts of West but also plead to make Islam of practical use in an era of science and technology by adjusting it to modern conditions. Instead of blindly giving into the authority of *ulama*, they invoke the utility of independent reasoning and seek to reconcile Islam with Western concepts of liberal democracy, rule of law and human rights. They believe that the Islamic concept of *Sharia,* consultation with the Islamic community, is quite similar to modern democratic practices. Though, at times, the modernists' support for advocating a government based on *sharia* seems to be in line with what the fundamentalists also advocate, they significantly differ in their views rejecting intolerance, violence and terrorism as a means of implementing it.

A large number of modernists are ***secularists*** too, some believing in western style of secularism in which there is support for separation of religion and politics—religion being a subject of personal domain—and accept the primacy of secular over religious laws, while others belonging to the Marxist secularists who do not believe in western style of democracy but hold religious and secularist values like Arab Socialists and other communist/socialist nationalists of various varieties [see: Mondal: 2002; Cole: 2002].

The non-fundamentalist Muslims, though see a role for religion in politics, they nevertheless believe in the compatibility of Islam and democracy. However, given the current horrifying dominance of fundamentalists over the state of affairs in Islamic world, it may be doubted whether a moderate face of political Islam can really be made to exist and work. In the West too, the religiously-oriented political parties play a role and try to influence political decisions, but only in a secular framework, without dominating the state and politics. But, since there is no history of the process of secularization in Islamic countries that could have tamed religion as a political force, there are serious doubts on the capability of state in Muslim countries, even where moderate Muslims are in dominant position, to accommodate the political power of the religion within a secular framework without getting destabilized itself.

Islamic fundamentalists seek to strengthen the *ummah,* the world-wide Islamic community and condemn the western concept of individual nation-state based on secularism that denies the sovereignty of God. Before Reformation and the rise of nation-state in 16th century, the medieval Christian fundamentalist were also striving for the unity of the western church though they did not have any significant support from the political thinkers or practitioners. Radical fundamentalists in Islam are both religious and political militants/ extremists; scriptural fundamentalists' main focus is on theology, issues of personal behaviour and morality than on politics and political behavior. Islamic fundamentalists have political agenda, have support of a significant number of Muslims also, regional variations notwithstanding and in some countries have the patronage of state and instead of challenging the state, and they are obviously in league with the state's religious rigidity and diktats.

In the Muslim world that stretches from West Africa to the Southern Philippines, as well as the Muslim communities and diasporas scattered throughout the world, the religion, politics and culture are intertwined in complicated ways though significant regional inter and intra-cultural differences are also noticed such as between Shia-Sunni branches of Islam and Arab and non-Arab Muslim worlds. A majority of the world's Muslims are Sunni; the Shiites form only a fifteen percent of the world Muslim population with Iran being the only Shiite majority state. Though both Sunnis and Shiites draw their interpretation of Islam exclusively from the Muslim scriptures, Quran and *hadiths* (the record of the sayings and actions of the Prophet), they differ in that the Shiites add to this the traditions of their Imams. Though the Arabs constitute only a twenty percent of the world total of Muslim population, yet all the interpretations of Islam, political and otherwise, are often filtered through the Arab lens [see: Rabasa: 2004; p. xvii]. As such, as if by default, the Arab issues and grievances are taken as Muslim issues and grievances despite the fact that the non-Arab Muslim world is more inclusive and majority of them having some kind of democratic or partially democratic governance and are being comparatively more secular in their outlook. Looking from the point of view that most of the contemporary innovative work in Islam is being done in the non-Arab part of the world, such as in Indonesia and by Muslim Diaspora in the West, some scholars infer from this that the Islam's centre of gravity is shifting to more dynamic regions of the Muslim world.

Resurgence of Islamist Movements: Character, Factors, History

The character of Islamist movements varies greatly throughout the world. Some Islamists resort to terrorism and some do not. Some espouse leftist political and economic programs, borrowing ideas from Marxism and other varieties of socialism, while others are more conservative. Most Islamists, however, insist on conformity to a code of conduct based on a literal interpretation of sacred scripture. They also insist that religion encompasses all aspects of life and hence religion and politics cannot be separated. Like most fundamentalists, they generally have a Manichaean (dualistic) worldview: they believe that they are engaged in a holy war, or *jihad*, against their evil enemies, whom they often portray as pawns of Jewish and Masonic conspiracies in terms taken directly from the anti-Semitic literature of 20th-century Europe. Messianism, which plays an important role in Christian, Jewish, and Shi'ite Islamic fundamentalism, is less important in the fundamentalism of the Sunni branch of Islam.

The pertinent question here is why and how Islam became all-permeating in the Muslim majority countries encompassing not only religious and cultural but social, economic and political life as well having their impact on the Muslims of other countries and regions too? Why other economic, social and political models were unworkable in countries of Islam? In fact, the last fifty years of twentieth century and the first decade of the twenty-first have been indelibly stamped by the remarkable resurgence of Islam as an international political force. The end of British Empire in the decades following the end of World War II released powerful energies, which had been suppressed by colonialism. The effect of this explosion has been global and profound, benign and sinister. Since the creation of Pakistan in 1947 and the establishment of Israel a year later, Islam became a critical factor in global political events that included mass migration, war and peace, oil blockades, boycotts, political development and political disintegration, famine and plenty, catastrophe and relief, terrorism, ethnic cleansing, hostage-taking, boundary disputes, destruction of mosques, aggression and secession. These events cover the globe: from Palestine to the Philippines, from Kashmir to Kuwait, from Cyprus to Chechnya, from Bangladesh to Bosnia, from the Central Asian Islamic republics to Morocco and the Sahara, from Turkey to Brunei and Yemen. Moreover, the position of Muslim nations and the perception of Islam as an international political force have changed in the last half-century. The 54 predominantly Islamic states constitute one-third of the membership of the United Nations. Muslim minorities, including that of the United States, are becoming influential political force. In fact, the political recognition of Muslims is beginning to equal their bulk as a quarter of the world's people. Some of these issues and factors for Islamic resurgence can be analyzed here:

Puritanical revivalist movements calling for a return to the pristine Islam of the Prophet Muhammad have occurred periodically throughout Islamic history. During the period of European colonial rule in the 19th and 20th centuries, however, these movements began to take on a polemical, apologetic character. Muslim reformists such as Muhammad' Abduh (1849-1905) and Jamal al-Din al-Afghani (1838-97) stressed that a return to the "rationalist" Islam of Muhammad, which was not incompatible, in their view, with science and democracy, was essential if Muslims were to free themselves from European domination. This argument was subsequently adopted by some Islamic fundamentalists, though many others condemned democracy on the grounds that only God's laws are legitimate. Some Jewish and Christian fundamentalists have rejected democracy for the same reason.

Ethnic and Tribal/Clan-based cleavages: One important characteristic of societies in the Muslim world, both Arab and non-Arab, in addition to the ideological differences amongst the *Sunnis* and *Shi'ites* having implications for international relations and world politics, is that most of these suffer from ethnic or tribal cleavages that acts as an important precipitator of Islamic fundamentalism. These elements determine the identity of an individual and how a group or tribe or clan would behave politically. If the components of the modern state are superimposed in such societies, a dysfunctional state is the result, leading to confusion and chaos. Examples of clans and tribes holding a sway over politics, administration, military and intelligence were all too visible in Iraq under Saddam Hussein, in Pushtun region of Afghanistan and in the tribal belts of Pakistan—the areas badly infested with terrorist activities and fertile grounds for terrorist recruitment. The anti-Soviet jihad in 1980s fundamentally altered the internal political balance in the Pushtun region of Afghanistan whereby the religious leaders or *mullahs* gained access to arms and money supplied by international patrons of Afghan *jihad.* Then onwards, these *mullahs* started seeing themselves not merely as the poor villagers leading the prayers but as rightful possessors of political power. These sub-national groups might be detrimental to the growth and strengthening of a nation-state, they are quite popular amongst their groups, not only for their clout in political field but also for the useful social services they provide which the state usually fails to deliver. Many a times, they make it difficult for the state to enforce its writ as is the current state of affairs in Pakistan's Federally Administered Tribal Areas (FATA) including South and North Waziristan where the tribal warlords call all the shots. If the state is not strong enough to submerge these ethnic, tribal and clan identities, the reverse is the case. That is how the failed and failing states are largely in the grip of the warlords, tribes and clans, such as in Somalia too. As the Rand study rightly argues, the convergence of tribalism, radical political Islam and the weak state authority has produced most virulent kind of Islamic extremism and terrorism. Extremist tendencies are very well and quickly received in the segmentary lineal tribal societies. bin-Laden had and still has a very strong appeal in the tribal areas of Pakistan, and Afghanistan. Tribal conservatism and religious extremism make a very good chemistry. One hypothesis is that tribes' leanings with rigid interpretation of Islam are inherent in their lifestyle and history and therefore, many prominent features of radical Islam find their roots in tribal customs that predate Islam [Rabasa: 2004: p. 86].

More or less permanent conditions in the Muslim world, such as the failure of political and economic models in many Arab countries, have fueled anger at the West and US which is held responsible for the failures in Islamic countries. Moreover, the decentralization of religious authority in Sunni Islam has opened the door for extremists with scant religious credentials and they manipulate religion for their own ends. The Islamic resurgence in the Middle-East over the past 30 years and the exportation of Arab ideology and religious practices to the non-Arab Muslim world has increased the support for fundamentalism. Radical Islamic ideology has spread to tribal societies that lack strong central political authority (e.g., Pashtun areas of Pakistan and Afghanistan). Moreover, radical Islamists have succeeded in forming networks that support fundamentalist and even terrorist activities through funding and recruitment. Many of these networks provide social services to Muslim communities, making them difficult to detect and disrupt. Certain *catalytic* events have shifted the political environment in the Muslim world toward radicalism, such as, the Iranian revolution, the

Afghan war with the Soviets, the Gulf War of 1991, and the global war on terrorism after September 11, the Iraq war and the removal of Saddam Hussein. All these have surely had an effect on the Muslim world having long-term implications.

Impact of Modernity and Globalization: Modernity, widespread education and globalization of information technology help making local or regional issues global. Democratic freedoms and the global reach of the media empower the masses, including the fundamentalists, militants and terrorists to use technology, democracy and freedoms to their benefit. Emergence of highly sophisticated means of communication to disseminate information at an unimaginable speed, including private print and electronic media has done wonderful service to the modern terrorist having primordial ethos and violent agendas. Technology developed in the so-called 'infidel' countries has made all this possible for the terrorists. Qatar-based Al Jazeera has widely been used by al-Qaeda radicals to disseminate information, to issue warnings and messages and to exhort people of Islamic faith to rally around what is told to them by these self-appointed crusaders of Islam. It has also been used to expose the anti-Muslim machinations of western media and to counter-attack it adding to the fury and hatred of Muslims against the West and the United States. Since the fundamentalists of any religion, such as *Jihadists* in Islam, have a clear political agenda, they see in the resurgence of Islam a great opportunity to come close to power centres, and hence, try hard to invoke the Islamic concept of *jihad* to provoke the adherents of Islam to practice its most sectarian and extremist version. Through their propaganda machine, they sometimes succeed in maligning great powers like America, which are widely held responsible for creating fundamentalist Islam in various Muslim countries to serve their vested interests.

Modernity and secularism: Since 1970s, the rising wave of modernization resulted in the decay of religion, which disturbed the religious fundamentalists and prompted a sudden expansion of religious groups swearing by *Quran* and calling for *jihad* exhorting the fellow religious followers to establish Islamic states in Muslim countries. Though it was a religious variant of fascism and revivalism of puritan values and ethos, many liberals still saw Islamism as the authentic creed of modern Muslims building an Islamic civilization within the multicultural world. One of Europe's leading authorities on Islamic societies, Gilles Kepel, in his book *The Revenge of God*

[Kepel: 1994] offers a compelling account of resurgence of religious belief in the modern world and reflects on radical movements within Christianity, Judaism and Islam and concludes that each of these movements resists the spirit of modernity and secularism. More than being a reaction to modernity, these are its children. Having an agenda to recreate a society based on a set of values as prescribed in their holy scriptures, they pursue a two-fold strategy: to seize power from above and to evangelize the masses to take control of their lives from below. According to a CRS Study [see: Vaughn: *CRS* Report: 2005], the decline of Islamic power in the wake of European colonial expansion is viewed by the *Traditionalists* as moral laxity and departure from the true path of Islam and by *Reformers* as chronic failure to modernize their societies and institutions. The question before both of them is how to revive Islam so as to develop coexistence between modernization and traditionalism without westernization.

Absence of alternate model of governance: The historical reason for resurgence of Islam lies also in the fact that since the days of western domination of Muslim and Arab world, the western values and political model of governance was discredited without any corresponding effort to evolve or develop an indigenous model that could be non-western and Islamic in a non-autocratic, democratic framework. Consequently, in the post-colonial period, the only model with which these societies were acquainted were the rule of royal families, kingdoms, tribal chiefs or military rules in those countries that tried democracy but failed. In some countries, the influence of socialist and Marxist ideology on the freedom movements drove them towards adopting some sort of socialist model of governance in their post-Independence period, but as is everywhere else, this model also turned into another form of autocracy or dictatorship, such as, Iraq under Saddam Hussein's Baa'th Party which was secular but extremely dictatorial. Pakistan's civil as well as military rulers failed to deliver what was promised on the eve of Partition of India and the resultant creation of Pakistan. Today, the country stands at cross roads presenting an example of a nearly failing state infested with very active terrorist and Islamic fundamentalist networks who are working overtime to initiate a global *jihad* using violence and terrorism. Nasser's Arab socialism, Al Fatah and various splinters groups in Palestine, Sukarno's Indonesia—are but a few examples of such models that were adopted by Muslim and Arab countries, but they could not deliver. At best, these regimes presented a set of corrupt, repressive and unrepresentative regimes incapable of providing a modicum of democracy, individual well-being and social justice. Deplorable condition of women and poor state of other human rights do not need any further substantiation. However, the failures in one way or the other in Muslim countries is not universal. There are exceptions too such as those of the Muslim majority countries in Southeast Asia doing very well economically, socially and politically with a liberal, plural democratic structure. Malaysia, Indonesia after the end of Suharto era and Singapore are some fine examples. Some oil-rich countries of the Gulf are also an exception. But, largely under the strong influence of a rigid Sunni ideology, the confused situation in most of the Muslim majority countries was exploited by Islamists and fundamentalists who found an answer to their problems. The answer was—Islam. The extremist Islamists used their old tendency of externalizing their frustrations borne out of social, economic and political failures on what RAND Project (2004) rightly calls—**structural anti-westernism.** An orthodox, extremist interpretation of Islam facilitates exploitation of these frustrations by radical clerics with political agendas [see: Rabasa: 2004: p. 78]. In short, it can be called resurgence of Islam before and after the cold war.

Oil money and pressures to Arbize: This resurgence appeared in many more ways such as insistence on Muslims, both men and women for maintaining a specific outward appearance everywhere like wearing garb/*abaya*/headscarf, etc. even in schools and universities of the liberal west as an expression of religiosity, segregation of sexes at public places, intolerance for un-Islamic behaviour and on identity-driven politics. Though moderates amongst Muslims would also stress on an internal critique of Islam, those preaching political and religious extremism and violence managed to get heard and followed by masses. The Gulf-money was too attractive to resist and therefore a process or rather competition began in non-Arab Muslim countries to Arabize themselves to 'look' Middle-Eastern. Liberal Saudi funding and export of Wahhabi version of Islam became too permeating since 1970s. It was during this period that a number of student organizations were created to work for radical ideology amongst the youth in different countries. This process got strengthened when in 1973, the oil embargo ushered into an era of steeply high oil prices and the petrodollars started flushing to Gulf states. It was at this point of time that the funding infrastructure for financing the international radical Islamist movements was put in place having long-term repercussions for international politics and relations. Under the garb of aid to war devastated countries, Wahhabism was spread. Not only this, intentionally or unintentionally, these petrodollars reached the hands of Islamic radicals who capitalized on the social dislocations and corruption generated by the oozing-out oil money to implement their radical agendas.

Money Weapon: Others like Hugh Fitzgerald of *Jihad Watch,* who strongly reject the concept of moderate Islam, also attribute the resurgence of Islam to the 'money weapon' provided by the OPEC oil bonanza, a sort of fatalism that prevents Arab and Muslim countries from economic development as they managed to acquire gigantic sums by the accident of geology. Since 1973, the Arab and other Muslim-dominated oil states have received ten trillion dollars from the sale of oil and gas to oil-consuming nations, the greatest transfer of wealth in human history. With this huge amount of money, these countries bought hundreds of billions of dollars worth of western arms, and with those arms, paid a whole network of middlemen, bribes-givers and bribes-takers, and Western hirelings not only working in arms sales, but also in the business of supplying other goods and services to the suddenly rich oil states. The money became the fabled *"wealth" weapon of the Jihad*, by which boycotts, and bribery, and the dangling of profitable contracts contributed to creating a vast and loyal constituency among some very influential and meretricious people in the capitals of the West and the US. This ill-gotten money is spent, among other things, on by and large corrupt royal families and their equally corrupt courtiers, on purchase of arms worth hundreds of billions of dollars by Muslim states who are the biggest buyers of arms since the discovery of oil in Gulf, on building mosques all over the world including the West and US, on madarssahs and terrorist training camps for Taliban in Afghanistan, and on hiring the armies of opinion builders, such as, journalists, public relation officers, academics and useful government officers in West and Muslim/Arab world [Fitzgerald: 2007].

Role of Diaspora Muslim Communities: International networks in Muslim countries as well as within the Diaspora Muslim communities living in the West and US have contributed largely in the Islamic resurgence, their other socio-cultural, humanitarian and other developmental activities notwithstanding. Partly due to the negligence of western governments and partly due to impact of international political events unfolding themselves since the end of cold war and with the overflowing petrodollars, these networks are found to work as nodes in funding and operations of extremists and terrorist groups. Al-Qaeda, Hezbollah, Hamas are a few examples of a large number of such networks providing social services to strengthen their grip and political base amongst the masses [for Al-Qaeda network, see Gunaratana: 2002]. But these networks are clandestinely used to launder money in the service of terrorists who claim to be working for revival of true Islam. Joined by criminal enterprises indulged in kidnapping, extortion and drug-trafficking, these networks use free trade zones, informal *hawala* system of transferring untraceable and unaccountable money from one country to another, and other methods of transferring cash, gold and diamond in the service of extremists.

Palestinian movement in West Asia: Among the Islamist movements that have attracted the closest attention in the west is the Palestinian movement, largely nationalistic in character, carried out by Hamas, which was founded in 1987. Its name, which means "zeal" in Arabic, is an acronym of the name Harakat al-Muqawamah al-Islamiyyah ("Islamic Resistance Movement"). Hamas was created primarily to resist what most Palestinians viewed as the occupation of their land by Israel. There is thus a clearly nationalist dimension to this movement, though it is also committed to the creation of a strictly Islamic state. Hamas opposed the idea of a Palestinian state in the West Bank and Gaza and insisted on fighting a *jihad* to expel the Israelis from all of Palestine, from the Jordan River to the Mediterranean and from Lebanon to Egypt. It justified its terrorist attacks on Israelis as legitimate acts of war against an occupying power. Like some other Islamist movements in the Middle-East, Hamas provides basic social services—including schools, clinics, and food for the unemployed that are not provided, or are inadequately provided, by local authorities. These charitable activities are an important source of its appeal among the Palestinian population. In January 2006 Hamas was the victor by a wide margin in elections to the Palestinian Legislative Council, and it was asked to form a government. This development led to much speculation among political observers about whether Hamas could evolve into a moderate non-violent political party, as many other terrorist groups have done, (e.g., Irgun Zyai Leumi in Israel and the Irish Republican Army (IRA) in Ireland ["Islamic Fundamentalism", *Encyclopedia Britannica online*].

In short, Palestine-Israel conflict is responsible not only for resurgence of Islam and *jihadi* fervor, it has negatively impacted the peace and security environment of the Middle-Eastern region and of the world. In fact, this chronic condition has shaped the political discourses in and of the Middle-East and the allies of the two rivals for over half a century. Due to the primacy of this territorial-*cum*-political dispute, the region has badly suffered in terms of development and most of the areas in Occupied Territories—West Bank and Gaza Stripe—look like a devastated heap of mortar and concrete. A lot of Zionist and Islamist propaganda on both sides has created an army of suicide-bombers, hate-mongers, and religious zealots. Therefore, merely the usage of the word—Islamist fundamentalism—is incomplete and does not define the whole truth. In the context of this Palestine-Israel conflict, the term—religious fundamentalism—would be more suitable to the existing conditions as the fundamentalism and terrorism of the one is paid back in the same coin by the other. Given the American hegemony in the world affairs and its multidimensional support to the Israelis, the Arab cause remains neglected and suffers from lack of support at various regional and international forums. The dominant and sharp western media also projects the issue in a slanted image thus putting most of the blame for violence and terrorism on Islamic radicalism. Much of anti-US anger in a common Muslim's perception owes its origin to pro-Israel policies of the US and the West. From this, it can be deduced that if this outstanding issue is resolved, much of Islamist terrorism might reduce to a great extent. But the question remains how to achieve this? In any event of the solution being just round the corner, violence, sabotage and disintegrative acts from both the sides complicates the task of finding a permanent solution to the problem, at least in near future. To a great extent, the Palestinian issue has served as a mechanism by some Arab and non-Arab Muslim countries to externalize their internal or domestic failures. In Majority Muslim view, the crux of the problem is America. The Palestinian agony was emotionally explained when after 9/11, a Palestinian journalist, who was asked to comment on the terrorist attacks on US, replied, "America, we feel your pain. Isn't it time you felt ours ?"

The defeat of Arabs by Israel in 1967 marked the end of Nasserism and Pan-Arab socialism and the beginning of a surge in current forms of Islamist extremism. Rise of oil prices in 1970s, use of petrodollars to export fundamentalism by Saudi Arabia, Iranian

Revolution and Soviet occupation of Afghanistan accelerated the forces of radicalization. The Israeli occupation of the previously Arab-held territories of West Bank and Gaza has given much impetus to the Palestinian cause and served as a rallying point not only for the Arab public opinion but also a justifying logic to the violent activities of the rising wave of radical Islamism.

Influence of West on Islamic states: Islamist movements, both violent and non-violent, have been politically significant in most Muslim countries primarily because they articulate political and social grievances better than do the established secular parties, some of which, such as the leftist parties were discredited following the collapse of communism in Eastern Europe and the Soviet Union in 1990-91. Despite their claims for conforming strictly to Islamic law, the Middle-Eastern autocracies continue to face internal opposition from Islamist movements for their pro-Western political and economic policies, the extreme concentration of their countries' wealth in the hands of the ruling families, and, in the Islamists' view, the rulers' immoral lifestyles.

To some extent, the Islamists' hostility toward the west is symptomatic of the rejection of modernity attributed to all fundamentalist movements, since much of what is modern is derived from the West. It should be noted, however, that Islamists do not reject modern technology. It would be a mistake to reduce all such hostility to a reactionary rejection of all that is new; it would also be a mistake to attribute it entirely to xenophobia, though this is certainly an influence. The Islamists' main resentment is against western political and economic domination of the Middle-East. This is well illustrated by the writings and pronouncements of Osama bin-Laden, the founder and leader of al-Qaeda, who repeatedly condemned the United States for enabling the dispossession of the Palestinians, for orchestrating international sanctions on Iraq that contributed to the deaths of hundreds of thousands of Iraqi citizens in the 1990s, and for maintaining a military "occupation" of Saudi Arabia during the Persian Gulf War (1990-91). Bin Laden also condemned the Saudi regime and most other governments of the Middle-East for serving the interests of the United States rather than those of the Islamic world. Thus, the fundamentalist dimension of bin Laden's worldview is interwoven with resentment of western domination.

Islamic Revolution in Iran and Soviet Invasion of Afghanistan: Three events in 1979 initiated a turning point adding more fervor to Islamist zeal and these were: Iranian Revolution, march of Soviet troops in Afghanistan and taking over of Grand Mosque in Mecca by Saudi dissidents. Whereas Islamist Revolution in Iran set-off a furious competition between Riyadh and Tehran for influence over the Muslims throughout the world, seizure of Mosque in Mecca had potential to shake the foundations of monarchy in Saudi Arabia and therefore, the Saudi rulers handled the situation with a more fundamentalist brand of Islam. While the observance of religious rituals was strictly tightened at home, Khomeinis were handled by stepped up funding of Mosques and *madarssahs* and of social welfare organizations. The countries like Pakistan, where state institutions were already non-existent, benefited largely monetarily as it reciprocated this Saudi generosity by managing a boom of *madarssahs* that prepared the later Taliban and other religious fundamentalists and terrorists to take on its all-time rival—India. These developments directly impacted the already existing conflicts, such as those of Palestinians and of Lebanon's militants. Emergence of Lebanese Hezbollah is taken as a proxy of an Islamic Iran. A number of fundamentalist groups in Middle-East, Persian Gulf, North Africa and South-East Asia owe their origin and sustenance to Iranian Shiite mullahs.

The Soviet invasion of Afghanistan in 1979 occupied the central stage in political and academic analyses over resurgence of Islamist terrorism especially after the September 11, 2001 attacks on the United States by al-Qaeda, an international Islamist terrorist network. The withdrawal of the mighty Soviet troops from Afghanistan in 1989 and the eventual collapse of the Communism itself afterwards emboldened the *Jihadi-Salafist* Islamists who arrogated to themselves and to their world-wide Muslim followers, who had flocked to Afghanistan and took training in *Jihadi* techniques in the training camps of Pakistan, the credit for the defeat of Soviet infidels. This euphemism of their victory with the weapon of Islamic fundamentalism and terrorist tactics encouraged them to implement the ambitious *jihadi* project elsewhere in those locations where the Muslims were the real or perceived victims. Now, they thought, they could defend *Dar al-Islam*, the realm of Islam everywhere, could go ahead to recreate a Caliphate by uniting all Muslims around the world into a central entity and could continue to fight on behalf of oppressed Muslims who in their opinion were being persecuted for their religious and political beliefs. Such a self-serving thought-process precipitated the emergence of one of the most deadly terrorist groups—al-Qaeda—and over one hundred other contemporary Islamist movements in the Middle-East, Asia, Africa, Caucasus, Balkans and also in Western Europe. United States and the west were their main targets [Gunaratana: 2002].

Thus, the anti-Soviet jihad in Afghanistan, having massive support of Saudi subsidies channeled through Pakistan's Inter-Services Intelligence (ISI), sophisticated arms from the United States—the arc rival of Soviet Union in Cold War politics and terror-training in Pakistani *madarssahs*, attracted large number of Muslims easily susceptible to be influenced by a *jihadi* ideology. This situation can be taken as most combustible in terms of having generated much of Islamic resurgence in last three decades. Al-Qaeda and Taliban and the subsequent rise of numerous terror networks regionally and globally are the product of this event that consciously or unconsciously had brought in to play all important world powers in such a way that the situation slipped out of everybody's hands and placed radicalists at the helm of affairs. Their program did not finish after Soviet withdrawal; rather it was taken further to different conflict situations in which the Muslims, rightly or wrongly, were a party to the conflict. Kashmir and Chechnya in the immediate neighborhood were not the only next targets; strategists like Osama had an agenda of destruction of pro-west Arab nations as well as United States. Many of the returning Afghan-Arabs played a key role in the Algerian Islamist Insurgency, in Islamist terrorist campaign in 1990s, and in Bosnia, Chechnya, Kashmir and Southeast Asia [Rabasa: 2004: p. 89-90].

A serious factor that led to embolden the *jihadi* agenda and spirits was that after the cold war's 'evil' empire came to an end and the US retreated from this location, no effort was made to strip the Mujahideen of their weapons and funds that were made available to these anti-Soviet fighters since 1979. These powerful resources were diverted into other regional conflict situations where Islamist guerrilla were involved, thus forging links between several far flung Islamist terrorist groups and strengthening and binding them together ideologically and strategically. Both international media and intelligence failed to assess the dangerous implications of such growing terrorist linkages when they were in the process of being forged whose impact was tragically felt in the years to come by a number of states though most vulnerable to terrorists' targets remained western liberal democracies and US. Their culture of tolerance and freedom of association, enshrined in their constitutions, were fully exploited by the terrorists to their advantage

Gulf-War of 1991 that made Saudi regime to call for entry of 'infidel' US troops on the Saudi soil created a sort of disillusionment amongst the radical elements in Muslim world, particularly amongst the religious *ulamas* of Saudi Arabia who were so assiduously cultivated by the Royal family. This Saudi alignment with the west in the war by one Muslim state against another Muslim state provided a justification to Islamists to launch another alternative Islamist movement against the now labeled as 'infidel' Saudi Royal rulers and their 'official' clerics who had the patronage of the Saudi government. This phenomenon gave Osama and other religious fundamentalists a rallying point to make coalition not only against US but also against the Saudi Royal family for being false Muslims and tried to win the support of other dissidents whom he had cultivated earlier. This was followed by attacks on British and American nationals in Saudi Arabia and the subsequent crackdown by the regime on terrorists and their cells. Sensing trouble, he relocated to Sudan from where Al-Qaeda networks spread worldwide, unprecedented communication networks established and Afghan veterans sent to various locations where Muslims were involved in conflicts with their regimes, thus 'creating a free-floating pool of Arab *mujahidins* extending from New York via Algeria to the Philippines' [Gunaratana: 2002: p. 35]. It was the time after the end of cold war when the militancy in Kashmir also started picking up as the trained Afghan *mujahidins* were used by Pakistan and were sent to unleash violence in the valley so as to fulfil Pakistani dream of winning over Kashmir by inducing its secession from India. In fact, the Kashmir issue has been skillfully externalized by Pakistan's ruling elite since 1947 to divert the public attention from real internal issues that have never been addressed seriously. Other than Islam, Kashmir is the main unifying force in Pakistan. Successive Pakistani governments have pursued a proxy war in Kashmir, to which they have subordinated the other purposes for which the state of Pakistan was created. This dynamic has dramatically changed the fabric of Pakistan's domestic politics by empowering extremist movements and their sponsors in the Pakistani security services [Rabasa: 2004: pp. 90-91].

Revival of ethnicity, religion and culture in Balkans: At the end of cold war and disintegration of the communist world, there was a sense of triumph in the west, as it was perceived as the major source of international tensions. But this euphoria soon evaporated as so many local and regional conflicts appeared as never before and on a much more deadly scale. As the restraining power of bipolarity could no longer maintain a system of checks and balances in international relations, the conflictual factors such as ethnicity, religion and culture, that so far lay dormant appeared with full strength and were perhaps in a hurry to make their impact felt. One important example of this revival appeared in Yugoslavia where Christian fundamentalists butchered Bosnian Muslims in 1993-4; the Muslims were slaughtered, mercilessly killed or forced into exile. The chastity of their women was infringed for the simple reason that they were Muslims and the women were gang-raped by Christian soldiers in uniform as a part of the state policy. This was perhaps a demonstration of an age-old hatred against the community as a whole. All this went deep into the psyche of Islamic fundamentalists and sharpened the divide along religious lines. There were retaliatory killings in Balkans by Arab *mujahidins*. In many important ways, the war in Balkans triggered the proliferation of many Islamic terrorist groups and also Islamist NGOs moving into conflict zones where Muslims were perceived to be under threat. Saudi Arabia, Kuwait and many oil-rich countries of the Gulf were the major donors with Saudis alone giving over $600 million to Bosnia [see: Gunaratana: 2002: p. 132].

Post-Cold War world and Islamic Fundamentalism

Though the process of the repoliticization of Islam began in the Arab part of the Middle-East as a response to the repercussions of the Six-Day War in 1967, it soon spilled over beyond the 'Abode of Islam' to its periphery and overshadowed the Pan-Arab nationalism of the post World-War II. It appeared as a new and challenging global phenomenon after the end of cold war. In fact, Islam as a world religion and as a major civilization, embracing one-fifth of the population living on this planet, is not at all a threat to any world order, but Islam as a political ideology—when translated into 'fundamentalism' by its radical followers—might pose a serious threat to world peace and security.

September 11, 2001 attacks on United States and War on Terror

September 11, 2001 terrorist attacks on America, most spectacular terrorist event at the dawn of 21st century will remain a defining paradigm in all discussions about the causes and consequences of terrorism and efforts to combat it though there certainly are more pressing issues of poverty and development that also require immediate attention. The attacks were neither the first terrorist event nor the last; not even the losses in terms of man and material were the topmost. These attacks became a defining paradigm in that there were **many 'firsts'** in it, such as,

(a) These attacks were on the soil of the world's unipolar power for the first time whose vulnerability was revealed to them and to the rest of the world for the first time;
(b) The event polarized the world into those with terrorists and those with US-led coalition on war against terrorism; and
(c) Also, America realized for the first time how and where the shoe pinches, which till 9/11 was either the problem of the underdeveloped countries of the Third World or a phenomenon acutely related with Communism.

9/11 not only turned the world upside down in American perception, but also brought to the surface the issue of the resurgence of radical Islam as in most of the terrorist events, Muslim fundamentalists were found to be involved. In some Islamist circles, the attacks were described or rather celebrated by remarks such as— 'America deserved it'—though the attacks were condemned by the mainstream Muslim community around the world. As Booth and Dunne have brought out in their book, the collision between the United States and Islam (a preferred word should have been Islamists) raised two parallel questions seeking answers: Why the United States is hated in so many parts of the world? And why is Islam so feared? The answers to these disturbing questions require a ruthless introspection, but unfortunately what is done by both sides is self-justification [Booth and Dunne: 2002: pp. 1-2]. Some in Muslim world interpreted 9/11 as a consequence of what America has visited upon Muslims. Condemnation was a sort of mixed reaction and was conditional. The moderates among Muslims called it a criminal act on the part of the radical fundamentalists; it was also described as a result of American foreign policy. Different conspiracy theories were doing the rounds such as: it was the handiwork of American government itself to malign the Muslims and Islam (Micheal Moore's film *Fahrenheit 9/11* touches upon this issue); yet others called it the work of Mossad, the Israel's intelligence service and so on.

One very unusual development was that though only a few Muslim states were on the side of terrorists, *most of the Muslim countries declared not only to be with United States in its War on Terrorism but also strengthened their own counter-terrorism activities.* Pakistani President Pervez Musharraf, who had been such an important player in escalation of radical brand of Islam through the flourishing *madarssahs* on his country's territory and in the making of Taliban, offered his services and support to be a bulwark against Islamic extremism. 9/11 made countries across the world adjust their foreign policies in accordance with the rapidly changing global political realities. Even US reviewed its relations with most of the countries. India, after its nuclear tests in 1998 which it justified on the basis of its own domestic and foreign policy compulsions, had to face American wrath and limited sanctions. A new strategic configuration after 9/11 made US understand India's security concerns and nuclear power requirements. The short-term objective of global war on terror was to eradicate al-Qaeda, Taliban and other terrorist networks, the long-term goals were to strengthen relations with friendly and moderate governments in the Muslim world to advance democratic and pluralistic values that are conducive to the peaceful co-existence of the countries despite vast differences of culture and religion. *Whereas global war on terror became a preeminent US interest, a quantum change was noticed in Muslim attitudes.*

One of the main objectives of Islamists for 9/11 attacks on America was a provocation of gigantic proportion, to bring America into a repression of the similar magnitude which would forge universal solidarity among Muslims in reaction to victimization and suffering of their Afghan brothers, particularly if US and its allies failed to carry their offensive with surgical precision, a situation the terrorist actors wanted to capitalize in order to become the catalysts of a mass movement in the name of *jihad* against the 'invaders' of Islamic land who massacre innocent Muslims. Had US and its allies not planned to minimize civilian casualties, it would have been the best opportunity for terrorists to seize to rouse the Muslim world in solidarity against American offensive and sweep Islamists to power in Muslim countries throughout the world, an objective they had failed to achieve so far despite all the efforts.

Islamist fundamentalists perceived it as 'war against Islam': Islamist fundamentalists belonging to all varieties of ideological streams and a large number even amongst the moderates perceived 'War on Terror' as a 'war against Islam'. A feeling of satisfaction in some quarters and of celebration in others was all too visible. Anti-American attitudes got solidified further after Operation Enduring Freedom, carried out in Afghanistan by US-led coalition, and other counter-terror measures implemented against the countries held responsible for sponsoring terrorism. Military assistance and training was extended to the countries complying with US anti-terror measures. Such measures further polarized the radical Muslims who were working either independently or as terror networks resulting in the intensification of militant activities in the form of planning and executing terror acts in future. *Ulamas* in countries like Pakistan and Saudi Arabia expressed their anger over their governments' decision to aligning with an infidel US and the West that have historically been anti-Muslim and anti-Islam. *Fatwas* were issues against the 'apostate' Muslims such as the Saudi Royal family. There were a number of mainstream Muslim organizations that though condemned attacks on America nevertheless did not approve US action in Afghanistan and later in Iraq too.

Islamism after US invasion of Iraq: A preemptive war on Iraq without sufficient grounds for substantiating the alleged acquisition of the weapons of mass destruction (WMD) by Saddam Hussein's regime and his pre-planned execution in the way that defies all legal justification and moral international standards, raised Muslim anxieties not only provoking radical Islamists to intensify their terrorist attacks on US and Western interests internationally and forcing moderates to rethink their alliance with US, but further deepened the dividing lines between, what Samuel P. Huntington would call, the civilizations, i.e., Christianity and Islam. Though the Bush administration accepted its mistakes by assigning the responsibility to go to war in Iraq on Intelligence failure, it was on defensive and tried to justify its Iraq Project on flimsy grounds. It tried to obscure huge civilian casualties and destruction of infrastructure by highlighting the atrocities committed by Saddam regime three decades ago, in 1980s, by conducting guided opinion polls and surveys whose almost 'predetermined' results show an aversion to the return of Ba'ath Party, or of the Islamic government. Terrorism is as old as civilization itself and many political movements have been forced to resort to it in varying degrees, especially in their early stages of struggle. However, powerful and established political powers have regularly resorted to state terrorism, known euphemistically as war. **Thus the "war on terror" is in fact fought with state terror.** Even the most heinous war is always rationalized with high moral justification, such as the US justification for war in Iraq. Moreover, Islamists perceive it as an unevenly matched conflict between a powerful state military machine and clandestine cells engaged in asymmetrical warfare. To them, the "war on terrorism", in its most diabolical manifestation, has also been called a religious war between a faith-based Christian nation and Islamic extremists, both groups controlled by their respective fundamentalists.

The problem in Iraq during occupation was not the popular support for extremists, but the real apprehension was that the radical elements might move into the state institutions to fill the vacuum created by the weak state and non-existent civil society institutions. The Saddam loyalists, the nationalists who opposed foreign rule in Iraq and the Islamists—all the three gave a stiff opposition and challenge to coalition forces and the peace and normalization in the country looked to befragile. So was the case with pull out of troops of the coalition partners. This confused situation gave much fodder to the Islamists' grinding mill that used this mess in their favor to sustain radical Islam and to keep alive the anti-American propaganda. Also, as Rabasa points out, [see Rand Project: 2004: p. 97] the rift between the three major players in Iraqi politics—the Shiites, the Sunnis and the Kurds—might strengthen Islamic fundamentalist forces, both Sunni and Shiia, and ran the risk of manipulation of Shiite movements by Iran, using pro-Iranian Iraqi cadres as proxies and of radicalization of Sunnis by the Wahhabis of Saudi Arabia. The electronic and print media's footage of violent resistance to coalition authorities in Sunni area of Iraq, the figures of civilian casualties and the well-publicized abuse of prisoners at Abu Gharaib prison not only developed and intensified insurgency in Iraq, it somehow helped in the resurgence of Islam in some quarters at least.

A serious fall out of 9/11 was—***intensification of radicalism*** in the most fanatically Islamic country: Saudi Arabia. It felt the pinch when on May 12, 2003 Riyadh was attacked by the terrorists and a series of other such attacks followed. In fact, the intolerance and fanaticism fostered by Wahhabism has spread in the Kingdom through religious police, school curricula and mosque sermons. However, there were certain positive developments also that went in favor of secular politics in some Muslim countries as the moderate Muslim leaders made efforts to carefully project and popularize the thesis that war in Afghanistan and Iraq should not be taken as directed against Islam. This thesis was reinforced after Bali bombings in Indonesia in October 2002 as people in general refused to endorse such unnecessary violence in the name of Islam.

US policies in the Middle-East and other Muslim countries after 9/11: In addition with the more than a century-old persistence of the Palestine-Israel conflict, which has overwhelmingly topped the regional and global political agenda as a potential source of increasing tension in the region and the revealingly pro-Israel stance of US in it, the post-9/11 turn of events has added to much of the already escalating anti-Americanism not only in the Muslim World but in other parts of the world as well, resulting in escalation of more intense Islamist radicalism. An analysis of the American foreign policy in the region reveals (see Chapter on Political Fundamentalism in this book) that it is in fact not as much about civilizational values as about the U.S policy in the Middle-East. The United States is criticized especially for its alleged unbalanced, pro-Israeli policy in the Palestinian-Israeli conflict and because of its cooperation with authoritarian-repressive regimes of the Middle-East. It is especially related to the Bush administration, whose 'axis of evil' rhetoric has attracted serious criticisms not only from the Muslim world but also from elsewhere. Furthermore, US hegemony and unilateralism, and not the 'Western-Christian values' can be considered as cause of growing anti-American sentiments in the Muslim world. In this respect, ignoring the different political perspectives within the 'Western' civilization and talking about 'clash of civilizations' is considerably questionable.

Thus, at the dawn of the 21st century, political Islam continues to be a major force for order and disorder in global politics, one that participates in the political process as well as in acts of terrorism, a force that poses a challenge to the Muslim world and to the West. Understanding the nature of political Islam today remains critical and important for governments, policy-makers and students of international relations. The contemporary Islamic revivalism that started with emergence of Islamic republics (Iran, Sudan, Afghanistan), proliferation of Islamic movements, and confrontational politics of violent radical extremists has put the 'Muslim' and 'Islam' in question, a theme of intense scrutiny. While Iran offered a visible and sustained critique of the West embodying rejectionist anti-Westernism, the fear of the threat of Westernization and its cultural penetration have been pervasive themes of Islamic resurgence. Modernization, viewed as a progressive Westernization and secularization, is seen as a form of neocolonialism exported by the West and imposed by their local clients, undermining non-Western cultural identity and values and replacing them with imported foreign values and models of development. International issues involving Muslims, such as, Palestine-Israel conflict and the issue of liberation of Jerusalem, chain of foreign occupations of Afghanistan, events of anti-Muslim violence in Bosnia, conflict in Kashmir, Muslim uprising in Chechnya, Western eyes on the Gulf-oil—all these have confirmed the Muslim fears and hence the resurgence of Islam.

Such developments and the spectacular nature of events like 9/11 and a preemptive war on Iraq may have lent plausibility to a common but mistaken belief in the West that Islam and Islamic fundamentalism are closely connected, if not identical. In fact, however, not all Muslims believe that the Qur'an is the literal and inerrant word of God, nor do all of them believe that Islam requires strict conformity to all the religious and moral precepts in the Qur'an. More important, contrary to Islamic fundamentalists' belief, most Muslims are not ideologically committed to the idea of a state and society based on Islamic religious law.

Agenda/Objectives of Islamic Fundamentalists

Justification of Violence

Ideological Islamist groups and leaders have a clearly defined ideology of violence. The common denominator among all these groups is that they all offer a systematic set of ideas that justifies the use of violence which is highly regulated and controlled from above and is carefully adjusted in accordance with the social and political requirements of the environment in which an operation has to be executed. Having no rational political approach, these terrorist groups do not resort to peace dialogue or negotiations; rather, they try to sabotage if there were any. Engaging in catastrophic acts of terrorism resulting in mass-casualties, they consider all this as divinely ordained. Some of these groups, such as al-Qaeda, though have Asian origins, but they subsequently grew to have global networks to bring global *jihad* and campaign on a common shia-sunni platform. Thus, despite being puritanical in ideology, some of these are highly pragmatic in their approach and keep refining their ideology in order to be operationally effective.

Use of Non-state Sector for Preaching a Jihadist Ideology

The terrorist groups use strategic methods to get their recruits from *madarssahs*, universities, cultural centres, mosques, charities, health and welfare NGOs and organizations and also from other social structures such as families that are economically and socially marginalised. The motivation to join terrorist groups might range from sheer frustration from the very depressing conditions of life, self or induced desire to die for the 'cause'—in the service of Allah, peer pressure, economic incentives from recruiters, to get counted as somebody great in their community or as a channel to get rid of a life of boredom particularly in the case of unemployed youth. Collective prayers are occasions when young potential recruits are incited to join a group. Many a times, particularly when the political climate of an area is communally charged, it is very common even in a pluralist democracy like India to hear instigating speeches delivered from mosques over the loudspeakers and amplifiers. In Islamic countries, delivering such speeches is not at all a taboo; rather, it may be done with the governmental connivance who might silently be praising the 'bravery' of their 'boys' for joining armed *jihad* which is elevated to the level of five essential canons of the faith, namely—total belief in faith, five-times-a-day formal prayers, fasting during the holy month of *Ramdaan, zakat* (alms-giving) and *hajj* (pilgrimage to Mecca and Medina).

Out of Context/Fundamentalist Interpretations of Quran and Selective Interpretation of History and Culture

The eccentric and violent ideas of bin-Laden drew largely on the selective interpretation of history, culture and body of beliefs which he used to spread his message amongst Muslims—with his call for indiscriminate killings of Americans anywhere—the message of restoring the pride of people who consider themselves the victim of successive foreign masters. Drawing upon multiple sources—Islam, history and region's political and economic malaise—he appealed to the people disoriented by cyclonic changes coming due to confrontation with modernity and globalization. His points of reference were: American troops on Saudi soil, suffering of Iraqis due to sanctions imposed after first Gulf War, and US support for Israel. The contemporary Islamists are engaged in an unprecedented exercise of corrupting and misinterpreting the words of Quran to achieve their hatred-generating political mission by threatening even the simple, God-fearing Muslims making them surrender to their will.

The Establishment of Islamic State

Establishment of Islamic state in predominantly Muslim societies on the basis of most authoritative Islamic scriptures, Quran and *Sunna is their declared objective*, though these two prescribe only the general principals of governance with no specific blueprint. Moreover, those countries that brought Islamic states into being, such as Iran, Saudi Arabia, Pakistan and Sudan differ significantly in their Islamic thrust. Whereas the political model of predominantly Shia Iran displays the principle of cleric supremacy over political authority, the Sunni Saudi Arabia and Sudan are models of governance by consultation with clerics without subordinating to them. Pakistan represents an Islamic state where a combination of unstable military-civilian stints have existed with military remaining an integral part of governance and where some efforts of westernization at governance are also tried every now and then since it's coming into existence. In Taliban type of Islamic State such as Afghanistan, particularly between 1996-2001, a terrorist/fundamentalist model was implemented where only the terrorists ruled the roost with no regard for either Quran or Sunna. In most of the other states which look like Islamic states because of their predominantly Muslim population, such as Indonesia and Malaysia in Southeast Asia, the western democratic principles are largely followed where the sovereignty is derived not from God, but from the people and secular laws—not the *sharia*—are the highest legal authority of the land defined by a nationalist constitution.

Attacking the West/US and their Policies in Muslim Countries

The short-term strategy of the Islamic fundamentalists, particularly before 9/11 was withdrawal of US troops from Saudi soil and creation of Caliphate there. The mid-term goal has been to oust the corrupt rulers of Muslim states in the Middle-East and setting up true Islamic states everywhere. The long-term strategy is to build formidable Islamic states with nuclear capacity and wage war against the United States and its allies—the"Satans" of Zionist-Crusader alliance. For this to achieve, terrorism based on religious justification is used, either through direct assault or through guerrilla tactics against both the

Muslim as well as non-Muslim states where the Muslims are perceived as victims.

The objective of Osama bin Laden and his cohorts was to drive the United States out of Muslim lands, topple the region's current rulers, and establish Islamic authority under a new caliphate. The path to this goal, they have made clear, is to "provoke and bait" the United States into "bleeding wars" on Muslim lands. Since Americans, the argument goes, do not have the stomach for a long and bloody fight, and they will eventually give up and leave the Middle-East to its fate. Once the autocratic regimes responsible for the humiliation of the Muslim world have been removed, it will be possible to return it to the idealized state of Arabia at the time of the Prophet Muhammad. A caliphate will be established from Morocco to Central Asia, Sharia rule will prevail, Israel will be destroyed, oil prices will skyrocket, and the United States will recoil in humiliation and possibly even collapse—just as the Soviet Union did after the *Mujahideen* defeated it in Afghanistan [see: Gordon Philip H.: 2007]. Bin Laden's version of the end of the war on terror was based on an exaggeration of his role in bringing down the Soviet Union, a failure to appreciate the long-term strength and adaptability of US society, and an underestimation of Muslim resistance to his extremist views.

De-westernization

Though the global standards of science and technology also create a globalization of civilization that made Leslie Lipson comment that "future does not belong to Khomeinis", the realities of many of those Islamic and even some non-Islamic civilizations belie this optimism where science and technology are adopted just as means of comfort or in the service of terrorists and fundamentalists, without any regard to the broader outlook and scientific and secular views of global civilization. Distinguishing between compassionate Islam and totalitarian Islamism, Tibi advocates combining the tradition of enlightenment in Europe and in Islam. To him, it is possible to reconcile Islam with modernity and democracy only if Muslims are willing to do this job. **Islamism is the Anti-Semitism of the Middle-East** and has become part of political ideology in most parts of the Muslim world. With Islamic migration to Europe, it can be seen to exist among the Muslim diaspora within Islamist networks. After 9/11, there is an Islamophobia in Europe, but one must distinguish between Muslims of Islam and *Jihadists*/Islamists, the former being few and poor and latter resourceful and powerful knowing full well how to make themselves heard. Agreeing with Bernard Lewi's comments that Europe would be Islamic by the end of this century 'at the very latest', Tibi says that the current situation warrants that either Islam gets Europeanized or Europe gets Islamized. Islamic fundamentalists, like some Christians of the West, believe in the universal applicability of their faith. The critics of West and US would argue that many of the American and west European scholars are enlisted to promote westernization under the garb of modernization and therefore, the agenda of Islamist fundamentalists is 'de-westernization'.

Hatred for Secularism

Secularism is a much-hated theory among the Islamists and they very often appeal to Muslims to recognize and elevate the place of Islam in politics and society in those countries, which have suffered from imposition of secularism and western paradigm of democracy. Instead of believing in the concept of sovereignty of state, Islamists call for establishment of a revolutionary vanguard of true believers to establish Islamic State.

Invocation of Regional and Civilizational Awareness

Many a times, it is seen that regional or civilizational awareness is invoked by politically motivating the local cultures. In such situation, religious fundamentalism easily moves to centre stage as an expression of political ideologies and provide potential ground for drawing faultlines between competing civilizations leading to conflict and violence. There are circumstances where people perceive themselves as members of ethnic, communal or sectarian groups overstepping their state citizenship. For example, a large number of Muslims or Arabs would like to see themselves as Muslims or Arabs, at times at least, irrespective of their living in any part of the world. Religious fundamentalists easily exploit such sense of ethnic belonging to their advantage by mobilizing their co-ethnic people in the service of a 'cause'. For example, a large number of Arabs and Muslims, drawn from different parts of the world, fought together as *mujahids* to fight against the Soviet troops. However, after the Soviet withdrawal, that ethnic unity diluted and now what we see in Afghanistan is that those once united *mujahids* have split into various ethnic factions fighting and slaughtering one another in an endless war. On this basis, it can be inferred that civilizational identity and framework facilitates Islamic fundamentalists to call on the entire Islamic *umma* of 1.3 billion Muslims globally whenever a major civilizational or external threat is perceived. But the sectarian and ethnic strife amongst this *umma* at local levels weakens the fundamentalist movements. Arguably, and happily, this is the reason why so far no religious fundamentalist movement has brought the proclaimed 'divine' order.

Modus Operandi to Execute Fundamentalist Agenda

- Ideological penetration of Muslim communities by fully exploiting the religious affinity and identity;
- Infiltrating the already existing Islamic NGOs and other 'charities' both in Islamic and non-Islamic societies;
- Radicalization and mobilization of the Islamic diaspora through a carefully crafted religious propaganda so that they either join terrorist ranks or support them financially, politically and logistically;
- Training the Islamist fighters in such a way that they carry on the 'mission' unflinchingly, willing to lay down their lives without any second thoughts as this sacrifice in the service of 'Allah' is the best kind of martyrdom, ensuring a seat in the paradise—al-Qaeda institutionalized these techniques of suicide terrorism;
- Glorifying martyrdom through suicide bombing is now fully embedded in the collective psyche of Islamist terrorist fighters, be it in Afghanistan, Iraq or Palestine. Suicide bombing is a technique by which an indoctrinated bomber aims to inflict maximum damage on the enemy target by fearlessly striking it and in the process also destroying himself [Gunaratana: 2002: p. 7]. Such a technique has been adopted in terror attacks on US Embassies in East Africa in 1998, USS *Cole*, Yemen in 2000 and in September 11, 2001 attacks on World Trade Centre and Pentagon on US territory; and
- Issuing *Fatwas* to dissuade Muslims who dare to challenge fundamentalists' agenda and programs and to terrorize simple, gullible Muslims to submit to their diktats.

Al-Qaeda, the most highly organized terror group based in Afghanistan but having networks all over the world and responsible for 9/11 attacks on US, is known for its extensive planning right from the preparatory stages, choosing high profile and symbolic targets, indoctrinating Muslims cutting across *shia-sunni* divide through a puritanical ideology, exploiting up-to-date technology and projecting West and US—the Zionist infidels—as the real enemy with whom a long-term and continuous conflict was imminent. Taliban used religious weapon to create a perfect Islamic state and to achieve this, it inflicted untold miseries on a nation that had already suffered the deaths of millions of its people and seen 1.5 million injured and 6 million refugees since the Soviet occupation of Afghanistan [Gunaratana: 2002: p.43]. Though all sections of society had to encounter the Taliban's religious frenzy, it came heavily on women. With the increasingly mounting global criticism of Taliban's ever new and gruesome religious diktats, Afghanistan got isolated from the rest of the world except a few that had recognized this regime, such as, Pakistan, Saudi Arabia and United Arab Emirates (UAE). Taliban's religious fervor went to the extent of putting conditions on certain NGOs, the only conduit to get foreign funding, putting the country and the people in serious financial troubles. Osama's *fatwas* were mainly targeted against United States for the reasons of latter's military presence in Saudi Arabia, for plundering its resources against the will of the people, for inflicting devastation and massacre on the Iraqis through a Crusader-Zionist alliance with the ultimate aim to serve the petty state of Jews despite their illegal occupation of the territories of Muslims. All these crimes committed by US or at its behest was construed as a declaration of war on God, on His Messenger and on Muslims and hence the call of *jihad* by the Muslims all over the world as a sacred duty in the service of religion and God. It was the sacred duty of all Muslims, announced Osama, to comply with God's order by killing Americans and their allies, both civilian and military, irrespective of location.

Both state and non-state groups joined the anti-American coalition on call from Al-Qaeda. Prominent among the states was Sudan, Pakistan, Afghanistan, Iran and Iraq. In addition, a number of Islamic groups, including some of those Muslims living in the West but responding to the call of *jihad,* were among the tactical and strategic partners in terrorist activities and their plans. Taliban fundamentalists committed the gruesome acts of destroying the age-old statues of Bamiyan Buddha in March 2000, in retaliation for sanctions being imposed on Afghanistan, inviting international outrage. But the fundamentalists remained undeterred in implementation of their terrorist acts and as a prelude to September

2001 attacks on US, went ahead with assassination of Ahmed Shah Masood, the commander of Northern Alliance, the symbol of resistance to Taliban and Al-Qaeda since 1994. Though CIA was also trying to execute plans to kill Osama but its failure went only in favor of Osama bringing him to the centre-stage of anti-American terrorism, making him a hero in the eyes of supporters of Islamist fundamentalism.

Looking at some Relevant Controversial Issues

Clash of Civilizations[3] *Thesis?*

Though as mentioned above, the September 11 events brought to the surface the 'Clash of Civilizations' debate in some academic circles and media, the ensuing developments go in favor of discrediting this thesis. Of course, the event reinforced a rethinking on relations of Islam *vis-a-vis* West. Many in the West, including President Bush himself in his public utterances, reiterated that Islam is a religion of peace and al-Qaeda cannot be considered as representative of Islam. *Islam* vs. *terror* debate has also frequently come into discussions and not unexpectedly, the Western media looked at 'Islamic roots' of the terrible attacks. Thereafter, 'Islam', 'Islamism', 'political Islam' and 'Islamic fundamentalism' became the most frequently used terms in the media. But as Booth and Dunne reiterate in their *Worlds in Collision*, "it would be an error of historic proportions to exaggerate the incompatibility between the thought-worlds of the so-called West and the so-called rest. We, in the West may not be able to understand the thinking of the mass murderers of September 11, but few in the rest of the world could comprehend their motives either. In the weeks after the attack, a noted Muslim writer, Ziauddin Sardar, wrote: Islam cannot explain the actions of the suicide hijackers, just as Christianity cannot explain the gas chambers, Catholicism the bombing of the Omagh. They are acts beyond belief by the people who long ago abandoned the path of Islam" [Booth and Dunne: 2002: p. 7].

Bernard Lewis has reflected deeply on issues of religious fundamentalism and a continuous clash between Christianity and Islam since the rise of Islam in 7th century Arabia, the two major religions and civilizations of the world, though he also dismisses the much publicized thesis of 'clash of civilizations' propounded by Samuel P. Huntington [see: Lewis: 2002]. Both of these religions and civilizations believe their truth to be universal, as is the case with other local and regional civilizations too such as Egyptian, Babylonian, Persian, Indian or Chinese, both claiming to be the possessor of God's final revelation to humanity and entrusted by God with a duty to carry it forward to the rest of the world. Both call each other infidels and use basically the same theological language. All this seems to make the conflict between these two almost inevitable.

Taking a journey into the history of the clash between Islam and Christianity, Lewis would argue that for most of the fourteen centuries of their conflict, Islam in the earlier period was, by far, the most advanced, creative, original and powerful religion, spreading to different parts of the world with an extraordinary speed posing a challenge to its rival—Christianity—which was poor, primitive and confined substantially to Europe though there were small and insignificant groups of Christians outside Europe. In most of the encounters between Christianity and Islam, it was Islam that was triumphant till late 17th century when in 1699 the Treaty of Karlowitz was imposed upon Ottomans, the last and the most enduring of all Islamic Empires, by Christian 'enemies' who replaced it by establishing the Holy Roman Empire that was representing Christianity. Thence onward, it was the turn of Christianity to register victory for centuries, leading ultimately to the defeat of the last of the great Islamic empires by occupation of Istanbul in 1918 and the partition of its territories between Christian allies. So, Lewis raises the question as to 'what went wrong' that turned the tables so completely, initiating the defeat of Islam at the hands of Christians in most of the future battles. Was it better military, a written constitution and elected assemblies in the Christian world, or were it the shabby and insulting treatment meted out to half the humanity—women—in Muslim countries, establishment of the self-centered and frightening tyrannies with horrifying methods of repression, or the religious fanaticism practiced to the level of controlling each and every aspect of personal and political life by the religion or is it Islam itself? The answers to these questions are many. Some explanations have their origins either in conspiracy theories while the others are grounded in the blame game, such as, the Mongol invasions of 13th century destroyed the great Islamic civilization of the Middle Ages; West European imperialism did it and so on. But these explanations, Lewis would call, merely as looking at symptoms rather than cause of the disease. The two other arguments carry some weight. *One* is the argument of *'modernity'*, arguing that the **Islamic countries were left behind because they failed to live up to the expectations of modernity** whereas those latecomers in Southeast Asia who started the process of modernization late, are doing wonderfully well. They succeeded because they experimented with two prerequisites of modernity, namely: separation of religion from government and democracy and did it successfully. Except Turkey, most of the Muslim countries failed at this front. The *second* argument given by many Islamists is: the ills of the Islamic world are **because *Muslims abandoned Islam*,** its heritage and traditions, God-given faith and started aping the ways and life-styles and values of 'infidels'. And the remedy they suggest is: go back to your roots, return to true, authentic Islam. And it is exactly this argument that is gaining ground in most of the Islamic countries after Khomeini's theology took roots in Iran in 1979 after Islamic revolution. Taliban's, Osama's and al-Qaeda's fundamentalism is the product of the same breed.

In fact, the 'Clash of Civilizations' thesis was first expounded by Samuel P. Huntington's controversial article in *Foreign Affairs* in 1993 in which it was argued that the post-cold war conflicts would occur more frequently and violently along the cultural as opposed to ideological lines. With the disintegration of communist bloc, ideological rivalry between the west and the communist bloc disappeared bringing in its place the civilizational rivalries and divisions amongst what he identified as eight or nine major civilizations, namely: Western, Latin American, Islamic, Sinic, Hindu, Orthodox, Buddhist, Japanese and African. The future locus of war would be culture and not the state. Though the nation-states will remain the most powerful actors in world affairs, but the principal conflicts of global politics will occur between nations and groups of different civilizations. The clash of civilizations will dominate global politics. The fault lines between civilizations will be the battle lines of the future. This was in contrast to the theories of some post-cold war analysts such as Francis Fukuyama who argued in his *End of History* that the defeat of communism has proved the supremacy and validity of the liberal democracy and Western values that have become the only remaining ideological alternative for nations in the post-Cold War world. Specifically, Francis Fukuyama argued that the world had reached the 'end of history' in a Hegelian sense.

Wars such as those following the break up of Yugoslavia, in Chechnya, and between India and Pakistan in the subcontinent were cited as evidence of inter-civilizational conflicts. The war on terror is seen as its largest manifestation. Huntington also argues that the widespread Western belief in the universality of the West's values and political systems is naive and that continued insistence on democratization and such "universal" norms will only further antagonize other civilizations. Huntington sees the West as reluctant to accept this because it built the international system, wrote its laws, and gave it substance in the form of the United Nations. Huntington identifies a major shift of economic, military, and political power from the West to the other civilizations of the world, most significantly to what he identifies as the two "challenger civilizations", —Sinic and Islam.

Huntington argues that the Islamic **civilization has experienced a massive population explosion** which is fueling instability both on the borders of Islam and in its interior, where fundamentalist movements are becoming increasingly popular. Manifestations of what he terms the "Islamic Resurgence" include the 1979 Iranian Revolution, the War on Terror, and extremely widespread Islamic opposition to the United States during both Gulf Wars. Perhaps the most controversial statement Huntington made in the *Foreign Affairs* article was that "Islam has bloody borders", which he believed to be a real consequence because of several factors, including the Muslim youth bulge and population growth and Islamic proximity to many civilizations including Sinic, Orthodox, Western, and African.

Like Bernard Lewis, Huntington also believes that civilizational conflicts are "particularly prevalent between Muslims and non-Muslims", identifying the "bloody borders" between Islamic and non-Islamic civilizations. He believes that the current global war on terror between the West and Islam is not a modern consequence of a few crazed radicals, but rather reflects a millennium-plus history of conflict between the two civilizations. This conflict dates as far back as the initial thrust of Islam into Europe, its eventual expulsion in the Spanish reconquest, the attacks of the Ottoman Turks on Eastern Europe and Vienna, and the European imperial division of the Islamic nations in the 1800s and 1900s. Some *common factors* contributing to this conflict between Christianity, upon which Western civilization is based, and Islam are: both are Missionary religions, seeking conversion by others; both display universalistic pretensions, strongly believing in "all-or-nothing" religions in the sense that both sides believe that 'only their faith' is the correct one, and both are teleological religions, meaning that their values and beliefs represent the goals and purposes of human existence.

In fact, it is the ambitious thesis of western universalism that infuriates Islamic fundamentalists and hence their warning to the west about the inherent dangers in their universalistic agenda. Western and American social scientists and policy-makers since then have systematically campaigned against the clash of civilizations thesis. However, Huntington's warning and observation that the Western-Islamic clash would represent the bloodiest conflicts of the early 21st century and the subsequent September 11, 2001 attacks and the events that followed including the Afghanistan and Iraq wars have been widely viewed in some academic circles as support for the Clash theory.

Rejection and opposition to this thesis notwithstanding, Huntington seems to be correct at least in his argument that modernization should not mean westernization which his critics refuge to accept. The examples of a modern Japan, a resurgent China on the path of modernization, South Korea, Singapore, Malaysia prove that modernization is very much possible without being westernized. The critics of Clash theory, both Muslim and non-Muslim, would also say that more than and prior to Islamic countries' despotism and violent conflicts, West itself was rife with despotism and fundamentalism for most of its history. Some also see Huntington's thesis as creating a self-fulfilling Prophecy and reasserting differences between civilizations. Edward Said issued a response to Huntington's thesis in his own essay titled "The Clash of Ignorance." Said argued that Huntington's categorization of the world's fixed "civilizations" omits the dynamic interdependency and interaction of culture. According to Said, it is an example of an imagined geography, where the presentation of the world in a certain way legitimates certain politics.

Whatever the criticism and the inherent dangers of such a controversial analysis of the emerging world order after the cold war, the chain of events after 9/11 leading to the US invasion of Afghanistan and the long continued mess that followed, 2003 invasion of Iraq and the global criticism against US actions there, the 2005 Danish cartoon crisis, the Iranian nuclear issue, the 2006 Israel-Lebanon conflict and the Pope Benedict XVI—Islam Controversy somehow fueled the perception that Huntington's *Clash* is well underway and seems to be valid at least for the time being.

Pippa Noris and Ronalds Inglehart in their comparative study of 75 societies around the globe, including many Islamic and Western states, argue in their working Paper that the political events between 1995-2001 provide such evidence that confirm the first claim in Huntington's thesis: culture does matter, and indeed matters a lot, so that religious legacies leave their distinct imprint on contemporary values. But Huntington is essentially mistaken in assuming that the core clash between the West and Islamic worlds concerns democracy, as the evidence suggests striking similarities in the political values held in these societies. It remains true that Islamic nations differ from the West on issues of religious leadership, but this is not a simple dichotomous clash, as many countries around the globe display similar attitudes to Islam. Moreover, the original thesis fails to identify the primary cultural fault line between the West and Islam, concerning the social issues of gender equality and sexual liberalization. The values separating Islam and the West revolve far more centrally around Eros than Demos. [Norris and Inglehart: 2002]

In a similar vein, Amartya Sen highlights the dangers inherent in the stereotyping of Muslims about their religious beliefs and affiliations. To give an automatic priority to the Islamic identity of a Muslim person in order to understand his or her role in the civil society, or in the literary world, or in creative work in arts and science, can result in profound misunderstanding. Criticizing such a use of imagined singularity, he terms a civilizational clash as conceptually parasitic. In partitioning the population of the world into those belonging to "the Islamic world," "the Western world," "the Hindu world," "the Buddhist world," the divisive power of classificatory priority is implicitly used to place people firmly inside a unique set of rigid boxes. Other divisions say, between the rich and the poor, between members of different classes and occupations, between people of different politics, between distinct nationalities and residential locations, between language groups, etc. are all submerged by this allegedly primal way of seeing the differences between people. The thesis somehow seems to presume that the relations between different human beings can be seen only in terms of relations between different civilizations. Such a reductionist view does damage not only by way of helping both the Western chauvinists and Islamic fundamentalists, it can also be counterproductive to "dialogue among civilizations" [Sen: 2006]. To focus just on the grand religious classification is not only to miss other significant concerns and ideas that move people, it also has the effect of generally magnifying the voice of religious authority.

Edward Said, in his critique of the Orientalist School of Thought describes the historically rooted Christian hatred and aversion towards Muslims and Islamists in his *Covering Islam* (1981) and says that for most of the Middle Ages and during the early part of the Renaissance in Europe, Islam was believed to be demonic religion of apostasy, blasphemy, and obscurity. It did not seem to matter that Muslims considered Mohammed a prophet and not a god; what mattered to Christians was that Mohammed was a false prophet who sowed the seeds of discord, a sensualist, a hypocrite, an agent of the devil... Real events in the real world made of Islam a considerable political force. For hundreds of years, great Islamic armies and navies threatened Europe, destroyed its outposts, colonized its domains... Even when the world of Islam entered a period of decline and Europe a period of ascendancy, fear of 'Mohammedanism' persisted. Closer to Europe than any of the other non-Christian religions, the Islamic world by its very adjacency evoked memories of its encroachments on Europe, and always, of its latent power again and again to disturb the West. Other great civilizations of the East—India and China among them—could be thought of as defeated and distant and hence not a constant worry. Only Islam seemed never to have submitted completely to the West; and when, after the dramatic oil-price rises of the early 1970s, the Muslim world seemed once more on the verge of repeating its early conquests, the whole West seemed to shudder [Said: 1981].

Western concept of popular sovereignty and the Islamists' view of sovereignty of God is another important point of departure between the West and the Islamist thought. The international order of the sovereign states that emerged as a result of Treaty of Westphalia in 1648, which shaped the state system of the entire world, recognizing nation-state as what Giddens calls a "bordered power-container", where sovereign states interact peacefully with one-another, honoring each-other's boundaries and in which each state is a 'secular' and not 'divine' entity. This is the point of departure from what is perceived by fundamentalists who take these Western, secular views as a challenge to the sovereignty of the God, the Allah. In other words, the collision between popular sovereignty and the sovereignty of God is one of the most important points of departure between western universalism and the universalism of Islam, the former signifying the 'end of God's rule' and latter determined to establish it anyway. Add to this the abolition of Islamic caliphate in Turkey in 1924 after which many Muslim states adopted the idea of popular sovereignty underpinning the system of nation-state—an idea that quite never took roots and the current crisis of nation-state has led to the politicization of the classic Islamic doctrine of sovereignty of God, under the new confines of Islamic fundamentalism. Since long, ideology of political Islam has floated the idea that the institution of nation-state denies Islam's claim of a universal Islamic order based on *Sharia,* i.e. the Islamic law and hence the Islamists' challenge to the secular, western-style world order. Moreover, the loss of bipolarity in the aftermath of cold war has brought to the surface a number of irregular wars and regional conflicts having the spill-over potential also. Some such examples are Balkans, Caucasus, Central Asia, Iraq, Afghanistan, and Somalia, in all of which Islam remains an important factor making the prophets of doom declare a 'clash of civilizations'.

Islam's confrontation with Modernity: Islam's confrontation with modernity and the question of maintaining a compatibility between the two in the face of the continuous and rapid change taking place in the world that has put pressure of maintaining the legacy of the past in Islam's religious tradition and integrate change into society and in the lives of its followers without disrupting societies and dislocating values are some very pertinent questions that have surfaced more intensely after the catastrophe of 9/11 due to involvement of Islamists in it. There are various responses to this problem, depending on one's point of view. An economist, a sociologist, a secularist and a fundamentalist might have different answers and responses. Whether one likes it or not, modernity has virtually permeated all aspects of life, including those of Muslims, by way of technological innovations and comforts. But modernity, as a way of thinking and accepting change in political and cultural processes by integrating new ideas into society, may not always be welcome. There are cases when one is using modern technologies and modern communication system, but maintains a primordial and closed mindset and resists new ideas of modernity such as democracy or pluralism. The opposite can be true; one may lack modern facilities and live traditionally but adopt the attitudes of modernity. These people assume an attitude of inquiry into how people make choices, be they moral, personal, economic, or political. This problem of rational choice is central to modernity. Choice, query and doubt—which imply rationality, debate, discussion and disagreement—are part and parcel of modern mindset [Cooper: 2000: pp. 2-3].

To understand the Muslims' reaction to modernity, one must look at the interpretations of Islam by the reformist/modernist who are devout, knowledgeable Muslims whose mission is **threefold:** ***first,*** to define Islam by bringing out the fundamentals in a rational and liberal manner; ***second,*** to emphasize the teaching of Islam in such a way as to bring out its dynamic character in the context of the intellectual and scientific progress of the modern world. The modernists sincerely endeavor to reconcile differences between traditional religious doctrine and secular scientific rationalism, between unquestioning faith and reasoned logic and between continuity of Islamic tradition and modernity [Husain, 1995: p. 95]. Jamal al-Din al-Afghani (1838-97) can be called the father of Islamic modernism who stressed the need for independent judgment and interpretation, the so-called *ijtihad,* a necessity and the duty to apply the principles of the Qur'an afresh to the problems of the time. He was extremely critical of traditional *ulama* (religious scholars) who discouraged the basic ideals of Islamic brotherhood, tolerance, and social justice; and *third,* to bring upon the simple, religion following Muslim the ill-effects of medieval mentality that is primarily responsible for the decline of Muslim power and influence in the world. Like Europeans Muslims should also integrate change and must do it in their own way by becoming better Muslims. Europeans had modernized because they were no longer really Christian; and Muslims, conversely, were weak because they were not really Muslims. Afghani strongly recommended acquiring Western knowledge, technology, and services, as long as borrowing from the West was selective and served the basic needs and aspirations of the Muslim people and struggled to initiate an Islamic reformation similar to the successful Christian Reformation sparked by Martin Luther. This tradition of modernity in Islam was further sustained by Afghani's most prominent Egyptian student and ardent follower, Muhammad Abduh (1849-1905), who insisted that Muslims could improve their lives and their society only by carefully studying the Qur'an in the light of reason and rationality. He taught that the Qur'an gives all Muslims the right to differ even with the *ulama,* if the latter were unreasonable or irrational. Many others in this tradition believe in the convergence of Islamic and universal ethics and are eager to introduce them into Muslim societies. They welcome non-Islamic ideas and practices that they consider beneficial to the progress and prosperity of Muslim societies.

On the contrary, the fundamentalists' views on the subject of Islam and modernity are steeped into superficiality and reductionism. As their understanding of religion is limited merely to superfluous and rigid reading of religious texts, they literally fail to capture the deeper spiritual dimension of the religion. Fundamentalist thinking is no more relevant in the currently fast changing world because human problems are so complex and diverse. Religious texts need to be reinterpreted by putting at the forefront the goal of the religion (*maqasid al-syari'ah)*. Ideologically, fundamentalism is marked by several characteristics. First, it reacts against marginalization of the religion. Fundamentalist movements form in reaction to, and in defense against, the processes and consequences of secularization and modernization that have penetrated the larger religious community. Protestants, Catholics, Muslim, Jews, Hindus, Sikhs and Buddhists are losing their members to the secular world outright or to relativism—the assumption that any given religion is culture-bound and thus relatively true or false. Second, fundamentalists demonstrate moral Manichaeanism, a dualistic worldview which uncompromisingly divides the world into two: the light (the world of the spirit and of the good) and the darkness (the evil). Third, they are selective. For example, they accept much of the modern science and modern technology such as radio, television, computer, and so on but refuse the concepts arising out of modernity such as democracy. ***Fourth***, fundamentalists are absolutist and inerrant. They steadfastly believe in the infallibility of certain religious interpretation and oppose hermeneutic methods developed by secularized philosophers or critics.

Modernity is the common denominator of the outside forces, which is often viewed as an external threat by the fundamentalists. The unbalanced pace of modernization and development has led many Muslim countries into developmental crises. The rapid changes through a process of technological, economic, political, social, and cultural innovation, however, have not been followed by the development of their people. In contrast to modernization, development denotes the relative welfare of a nation's population. In most Muslim countries, appropriate development has not happened. Modernization and development have become paradoxical.

Modernization has occurred rapidly, while appropriate development has not. In the West, modernization accompanied the growth of a middle class. Because of its relative success in the West, modernization has become identified with Westernization and secularization. The unhappy predicament of the nation-building, modernization, and secularization of the Muslim world has given rise to a number of crises, afflicting the fragile nation-states of developing world. Of the five developmental crises—identity, legitimacy, penetration, distribution, participation—the identity crises is often the precipitating crisis, triggering political chaos and national catastrophe. Rapid modernization has broken the familiarity of traditional society, uprooting people from their traditional communities and moving them to new social environments where they oft become victims of the development. These conditions are fertile grounds for the breeding of fundamentalism.

In fact, modernity is a complex and multidimensional, rather than a unified and coherent, phenomenon. Modernity swept through Europe in the 18th and 19th centuries with its resultant outcomes, such as: shifting of power from religious leaders to a growing civil service, a deliberate separation of church and state, respect for science that led to respect for scientific ideals such as rationality and empiricism. Since these developments were western in origin, modernity by traditionalist Muslims and Islamists came to be associated with west, taken as something west-centric in which secularism replaces or displaces religious values from most aspects of daily lives, reason and enlightenment pose a challenge to the unquestionable authority and sovereignty of God and individualism becomes central in social, economic and political debates.

But such a clichéd vision of life in the West as seen on cable television and films—a world of sex, violence and desperate loneliness—that overlooks the role religion, custom and tradition play in the lives of most people in Western countries is only a partial or flawed understanding and interpretation of modernity. **Most Muslims do not really think of Modernity in terms of a break with the Past.** To them, it also means a *renewal* along with keeping pace with the Past. In fact, there has been a modernist tradition in Islam which got a knee-jerk in the aftermath of World War I that resulted in the fall of Ottoman Empire and the domination of the Middle-East by European powers such as Britain and France. As the intellectual historian, Peter Watson, suggests that World War I marks the end of the main Islamic modernist movements, and that this is the point where many Muslims "lost faith with the culture of science and materialism" [Watson: 2001]. However, in some parts of the world, the project of Islamic modernity continued from the same trajectory before the Great War as in the case in the new Republic of Turkey under Mustafa Kamal Ataturk. The liberal movements within Islam have continued to attempt to reconcile Islamic Sharia law with modern concepts like feminism and human right norms of international law.

According to Bassam Tibi, 'Islamism', in contrast to 'Islam', poses a serious challenge to world politics, security and stability. Political Islam is a particular synthesis of religion and politics and Islamic fundamentalism is the result of Islam's confrontation with modernity, of western variety, and not only—as is widely believed—of economic adversity. It also marks the inability of Islamic nation-state to integrate into the 'New World' secular order [see: Tibi: 1998, 2002]. Though on the basis of their own imagination and brand of interpretation of Islam, the Islamic fundamentalists dream of supplanting the much 'discredited' western world order and its secularism by establishing a new one based on Islamic tenets, but anybody who is closely watching the recent developments in world politics since the end of cold war can easily infer how weak, frightening and divided these so-called Islamic fundamentalist movements have been. At the most, these fundamentalists can and have engineered the frightening levels of terrorism and turmoil making life a hell for their own ethnic and religious fellow being and the imagined alternative of a new world order is nowhere in sight. And argues Tibi, fundamentalism is much more than extremism or terrorism; it is a powerful challenge to the existing order of the international system of secular nation-states, a western institution. In other words, Islamic fundamentalism that has appeared since 1990s is a challenge to West and to its secular values and institutions [Tibi: 2002, 2003]. The fundamentalist revolt against the West is not merely a revolt against its political hegemony; more than that it is a revolt against cultural imperialism, meaning primarily a revolt against western norms and values. As the earlier revolt of the non-western people in the colonized world was in the course of decolonization for the purpose of getting political independence, the current revolt, particularly in the form of Islamic fundamentalism, is for the reassertion of indigenous cultures by challenging the political and cultural fundamentalism of American hegemony—a continuation of European hegemony after World War II.

The US foreign policy in the oil-rich countries of the Middle-East and Persian Gulf is a significant precursor of Islamic fundamentalism. The 1979 Islamic revolution in Iran disturbed the American strategies in this region as it lost the 'valuable' support of at least some of the Middle-Eastern regimes it had for its oil politics. Before the Islamic Revolution, the Shah of Iran not only supported American interests in the region but also supported Israel and Israeli policies, which strengthened American cause further. This turned the US policy establishment to harbor a hostile perception against those militants and extremists who tried to expose US policies in the region. In 1952 when Mosaddeg had captured power in Iran and nationalized the oil companies the CIA had managed to stage a coup with the help of some religious leaders. But, the Islamic revolution in Iran gave a basis to the Muslims of the region to come together against US, heightening the threat perceptions not only of America but also of the other supporters of American policies in the region, such as, the Saudi monarchy or Kuwaiti sheikdom. Instead of countering the radical Iranian Islam with democratic secularism, which US professes, it used *wahabi* Saudi Islam to condemn Iranian brand of religious fundamentalism.

The West's projection of Islam through media is also problematic: A perception amongst the Islamists about the mainstream media all through the world, particularly in the West is that it is severely hostile to Islam and Muslims, projecting them as intolerant, violent, bearded or *Burqa* clad men and women unwilling to think and act logically and rationally and are fit for being governed by the dictators and autocratic royal families. In the age of the media, of the sound bite, of television images, Muslims have not yet found a way of expressing themselves adequately, they feel. Their leaders—of sorts of Gaddafi or/and Saddam—rely too much on mob oratory and controlled television appearances which appear artificial and ineffective in the West. Besides, the vernacular translates poorly. Muslim leaders either appear as military dictators ordering the chopping of hands and the whipping of the poor, such as Zia in Pakistan and Nimeiri in Sudan, or as tribal tyrants like Arab rulers or as socialist dictators of the ilk's of Saddam in Iraq and Assad in Syria. Those English language Dailies and Weeklies in Muslim countries that are edited by the journalists in UK or USA often echo western views and jargons. Therefore, it remains a dilemma for Muslim countries how to project their culture and ideas in right perspective. Whatever media has come in recent years in Muslim societies is largely the replica of AL Jazeera that has gone to precipitate Islamist, radical ideology in the service of resurgence of Islam. Such media propagate stories with anti-American, anti-Western and anti-Semitic content, carry out terrorists' propaganda and occasionally attack Saudi and Jordanian monarchies.

After September 11th, the Western media has done a great service to terrorists and Islamists by its overemphasis and focus on personalities such as Osama bin-Laden for condemnation, and exaggerate what are often unknown terrorists into forerunners of "Islamic jihad." This causes the creation of stereotypes of Muslims in the Middle-East and moreover, results in the grants of prominence to Islamic fundamentalists who might otherwise have been insignificant political characters, and legitimizes extremist opinions and views which might otherwise have been shunned by mainstream Muslims. However, as John Esposito notes: The tendency to judge the actions of Muslims in splendid isolation, to generalize from the actions of the few to the many, to disregard similar excesses committed in the name of other religions and ideologies...is not new. Yet the number of militant Islamic movements calling for an Islamic state and the end of Western influence is relatively small. Nevertheless, these groups are causing great fear among people in the Middle-East and in the West. Finding a solution to this problem of fear, will depend not only on how Islam deals with concepts such as modernity, but also on how the West deals with Islam.

Is Islam Compatible with democracy? To this question there are conflicting answers and opinions expressed on both sides of the spectrum. Robin Wright [Wright: 1996] says that out of the more than four dozen predominantly Muslim states, only a few, such as, Turkey, Lebanon, Bangla Desh, Iraq, made some strides, but the secular rulers of most of these countries refused to share power with others and also the largest single blocs of Muslim countries: Middle-East and North Africa are against the global trends towards political pluralism and democracy. Though neither Islam nor its culture is the major obstacle to political modernity, it is the undemocratic rulers who sometimes use Islam as their excuse. In Saudi Arabia, for instance, the ruling House of Saud relied on Wahhabism, a puritanical brand of Sunni Islam, first to unite the tribes of the Arabian Peninsula and then to justify dynastic rule. Like other monotheistic religions, Islam offers wide-ranging and sometimes contradictory instructions. In Saudi Arabia, Islam's tenets have been selectively shaped to sustain an authoritarian monarchy. Also, in the largest and poorest Muslim countries, problems common to developing states, from illiteracy and disease to poverty, make simple survival a priority and render democratic politics a seeming luxury. Moreover, like their non-Muslim neighbors in Asia and Africa, most Muslim societies have no local history of democracy on which to draw. As democracy has blossomed in Western states over the past three centuries, Muslim societies have usually lived under colonial rulers, kings, or tribal and clan leaders. The constitution of the Islamic Republic in Iran in 1979, the first of its kind, created structures and positions unknown to Islam in the past. But Islam in principle, which acknowledges Judaism and Christianity as its forerunners in a single religious tradition of revelation-based monotheism, also preaches equality, justice, and human dignity—ideals that played a role in developments as diverse as the Christian Reformation of the sixteenth century, the American and French revolutions of the eighteenth century, and even the "liberation theology" of the twentieth century. Thus, Islam is not lacking in tenets and practices that are compatible with pluralism. Among these are the traditions of *ijtihad* (interpretation), *ijma* (consensus), and *shura* (consultation).

However, one should also take into account the fact that the implementation of Sharia is fraught with problems and controversies. In Islamic countries, the justice based on *sharia* is characterized by highly objectionable methods of meting out physical punishments such as stoning to death for adultery and in most of the cases, even of rape, women are adjudged to be guilty. Amputations, blood for blood and numerous restrictions on women's rights are prevalent. . In Saudi Arabia, women must go out only with a male, like, husband, brother or father, they should not drive, should cover themselves from head to toe with *abaya,* all eligible adults should observe fasting during the holy month of *Ramdaan,* all shopping centres close for prayers five times a day, no cinema-halls allowed, magazines and other media material is painted black if not in conformity with the concepts of religious police and a host of other restrictive and prohibitive practices which are sufficient to make women the private property of men, having no independent will or desire of their own. Add to this the practice of polygamy reducing women virtually to the life of a hell. Due to non-compliance with international human rights standards, these countries become centre of criticism world-wide including many Muslim countries too which have restricted the use of *sharia* only in some family matters and social practices such as adoption, inheritance and marriage. The inhuman punishments to both men and women are severer where radical fundamentalists hold the sway over political power by silent consent of the government or by force. In fact, as RAND Study (2004) comments, the oppressive social and anti-women practices ascribed to Islam have less to do with Islam or Islamic law than with tribal laws and customs and even if no one in government should be bothered about some of these practices, they are adhered more as a matter of habit than out of fear of punishment.

Focusing on three important and controversial themes of inquiry—namely, Islam and Governance, Islam and Gender and Islam and modernity, the project *Living Islam* [Living Islam: 2002] reflects that answers to these questions can be found by looking back into the founding years of Islam and on this basis conclude that the difference between the radical Islamists and liberals is that while the former take Islam as a total social and political ideology, the latter use the principles of the *past* to forge a new synthesis that takes into account the contemporary realities. The problem is that the words and action of the Prophet, who was also a human-being, though for most Muslims *He* represents a model of *Perfect Man,* have become so central to everyday living of a Muslim down from a highly intellectual to a lay Muslim in the street that they do not want to give any latitude or scope for varied interpretations and interpolations that might have crept-in during the gap of more than a century between when Mohammed lived and revealed Quran and when these were given a written form by way of *Hadiths.* Passed down over the centuries by word of mouth, through sermons, in folk-tales and folk wisdom, these *hadiths* directly inform the thinking and behavior of millions of ordinary Muslims till date. For centuries, *mullahs* or *Imams* (prayer-leaders in the mosque) have claimed, whether they understood Arabic or not, a monopoly of their interpretation and continue to deny ordinary Muslims the capacity and right to reach their own individual understanding and judgement of either *Quran* or *Hadith* in accordance with their own or generally changing circumstances. Some of these clerics assert that only those steeped in *ilm* or knowledge, who are trained in religious science, are qualified to decipher the true meaning of these texts. A direct understanding of God without intermediaries is simply denied. Both the *Qur'an* and *Hadith* were transmitted, first orally, then in written form by human beings. This fact makes them open to historical and subjective interpretation.

Context, in fact, is often the key to understanding *Qur'anic* texts. Many of the revelations from the later Medina period of the *Quran* are in fact answers from God to specific dilemmas facing the Muslims. Many *suras (*chapters), for example, are addressed solely to men. One of the Prophet's wives complains. Almost immediately, God answers (Sura 33 Ayat 35), specifically addressing both men and women as equals, both emotionally and intellectually. This verse was then and is once again today the basis of feminist interpretations of the Qur'an. But unfortunately, its unequivocal message is also contradicted by other verses from the same *surah.* Islam is stereotyped as inherently and irredeemably hostile to women completely ignoring the comparatively more egalitarian thrust of Quran in the context of women. The 7th century Arabia in Muhammad's own lifetime was more egalitarian if looked backwards but a shift from matrilineal to patrilineal culture ushered into an era of more restrictive future for women. During Prophet's own lifetime, women worshipped with men in mosques, *purdah* (veil) was introduced simply to protect the privacy of the Prophet's bedroom from the prurient gaze of male guests who stayed too long after dinner (Sura 33.54). Women imams also led the prayers for their households. Women fought on the battlefield and were entitled to own property in their name. But all this deteriorated within the space of one or two generations of the Prophet's death in 632 AD as Islam spread and came into contact with well-established, patriarchal cultures to its North—Judaism, Christianity, Zoroastrianism. Veil came to be applied on all women though Quran enjoins upon both men and women not to make a shameless display of their bodies. Women were relegated out of public life and back into the seclusion of the home. Men arranged marriages. Women lost most of their rights to divorce or to remarry. Their testimony and their worth were literally devalued. How all this happened? Some ascribe it to mutually contradictory and basic ambiguity towards women's equality within Quran itself; others believe that perhaps Arabian Islam could not swim against the rising tide of fast creeping-in patriarchy. However, had Quran been clearer and more unequivocal in its statements about gender equality, perhaps the process of the patriarchal influence on gender could have been delayed and the Islamist fundamentalists would have been contained in their atrocities against women which they ascribe to as Quranic injunctions.

As for ***governance***, though the Islamists claim Islam to be self-contained and all-encompassing ideology and argue that the way Muslims governed themselves in Mecca and Medina fourteen hundred years ago provides clear and adequate models for a specifically Islamic political theory and for Islamic political institutions, the fact is that there is little serious rethinking about Islam as a political theory. The first thirty odd years after Mohammad's death (632 CE), when the first four Caliphs were the *Chosen Companions of Prophet*, the problem of how Muslims were to govern themselves became the stuff of murder, intrigue and military conquest. Even the theory that a Caliph having the virtue of governing in accordance with *sharia* should be chosen offered no institutional model for peaceful political change and therefore what followed was Muslim communities' preference to accept authoritarian rule cloaked in the legitimacy of the *Caliphate* rather than contemplate democracy. The 20th century-Egyptian scholar Ali Abd al-Raziq's argument that the Prophet did not in fact establish for all times a model for an Islamic state and the various texts did not set out any definite system of government, merely general principles and therefore Muslims must themselves invent effective forms of government on the basis of consent of the governed, not implementation of the *Shariah*, has remained not only intriguing and controversial view but strictly a minority view also, easy to discredit with the label of *westernization*. The thoughts expressed by people like Hasan al-Banna of Muslim Brotherhood that Muslims and societies must be made virtuous, of Maulana Maududi (1979) of Pakistan rejecting the concept of sovereignty of *ummah* as being blasphemous and of Ayatollah Khomeini of Iran who propagated the traditional Shi'i doctrine of *Velayat-i-faqih*, the Guardian who stands in for God—all might seem attractive, but are dismally impractical. In short, they negate the very idea of democracy based on the Sovereignty of the People.

With globalization's rising tide, the non-western people are increasingly exposed to a global matrix that somehow makes one believe that western culture is the dominant, universal culture. Sure, modern technology has shrunk the globe, bringing an unprecedented degree of awareness about different cultures and civilizations. But this does not necessarily mean the emergence of a unity of outlook. In a way, it is a world more unified, but also more fragmented on many levels and grounds leading to a decline in international consensus.

Hence, the merits of globalization notwithstanding, some of its merits have become demerits when they serve the vested interests of fundamentalists and terrorists and create divisions that are detrimental to world peace and security. This is also the reason why we see the religious fundamentalists calling upon their fellow beings to go back to their roots, meaning—returning to local cultures as against the global one. If we accept this premise, it becomes understandable that the religious fundamentalism is neither a mere political phenomenon, nor a cultural revival only, nor fundamentalists are totally traditionalists as they themselves are the products of modernity and their projection of anti-modernity is merely a rhetoric.

It needs to be clearly understood that fundamentalism is the mainstay not only of Islamists. It has a political variety as well. The presence and use of Islamic fundamentalism demonstrates that some basic structural changes are taking place both in Islamic countries as well as in international system. Few fundamentalists are in power not only in Islamic countries, but in non-Islamic world as well. It is not the clash of civilizations that is taking place; rather, it is the clash of two types of Fundamentalisms: *political fundamentalism* on the one hand, practiced by the hegemon of the world to implement its project of secular nation-state, democracy, popular sovereignty and westernization and *religious fundamentalism* on the other unleashed by those radical fanatics who politicize religion for implementing their agenda universally but lacking wherewithal, they are weak enough to achieve this goal. The result is—terrorism, violence and rejection of everything that brings stability and peace.

Separation of church and state is the political and legal concept that pleads that the government and religion should be separate, and not interferes in each other's affairs. Since religious institutions and their adherents are a part of civil society, private religious practices often come into conflict with broad legislation not intending to target any particular religion or religious group. Hence, there emerged the logic of separation of the church and state. Beliefs about the proper relationship between religion and government cover a wide spectrum, from the state atheism through secular government to varying degrees of theocracy.

In the United States, the "Separation of Church and State" is generally discussed as a political and legal principle derived from the First Amendment of the United States Constitution, which reads, "Congress shall make no law respecting an establishment of religion, or prohibiting the free exercise thereof ...." The concept of separation is commonly credited to the combination of the two clauses: the *establishment clause*, generally interpreted as preventing the government from establishing a national religion, providing tax money in support of religion, or otherwise favoring any single religion or religion generally; and the *free exercise clause*, ensuring that private religious practices are not restricted by the government. The effect of prohibiting direct connections between religious and governmental institutions while protecting private religious freedom and autonomy has been termed the "separation of church and state." Nevertheless, issues of free exercise are also implicated by the extent to which laws are permitted to impinge upon private religious practice. In the United State, the state laws can prohibit practices such as bigamy, sex with children, human and occasionally animal sacrifice, use of drugs, or other criminal acts, even if citizens claim the practices are part of their religious belief system. As such, the federal courts give close scrutiny to any state or local laws that impinge upon the *bona fide* exercise of religious practices. The courts ensure that genuine and important religious rights are not impeded, and that questionable practices are limited only to the extent necessary. The courts usually demand that any laws restricting religious practices must demonstrate a fundamental or "compelling" state interest such as protecting citizens from bodily harm.

Ever since, Mustafa Kemal Ataturk abolished the Caliphate and separated Church from state, a great debate has raged in the Islamic world over this issue. Most Islamists consider the Western concept of separation of Church and State to be rebellion against God's law. Secularism cannot be a solution for countries with a Muslim majority or even a sizeable minority, for it requires people to replace their God-given beliefs with an entirely different set of man-made beliefs. Separation of religion and state is not an option for Muslims because it requires abandoning Allah's decree for that of a man. Based on the premise that religion interferes with, and possibly upsets matters of state, it is possible for a secular state to take a neutral stand toward all religions. However, religions, feel the supporters of unity between state and religion, do not only deal with collections of beliefs, rituals and individual behaviors that do not affect the society. Most of the well known religions—Judaism, Christianity and Islam—have laws that regulate relationships between people; whether on an individual basis, among the family, or with the society at large that cannot be separated from the business of the state. The basic belief in Islam is that the Qur'an is one hundred percent the word of Allah, and the Sunna was also as a result of the guidance of Allah to the Prophet *sallallahu allayhe wasalam*. Islam cannot be separated from the state because it guides through every detail of running the state and a Muslim's day-to-day lives. Muslims have no choice but to reject secularism for it excludes the law of Allah. To them, the western concept of the separation of state and religion was a reaction against the cruel hold of the medieval church, and is not applicable to Islam, which blends both the church and state in great harmony. To the argument of secularists that the values of one religion cannot be imposed on members of other different religions, the followers of Islam, including some moderate fundamentalists, reply that under the law of Islam, other religions are not prohibited. At the same time, non-Muslim people are provided with doctrines for legislation and running of state that will protect people of all faiths living in the Islamic state [Separation of Church and State: *Wikipedia search; islaam.com*].

However, in the contemporary debate in Islam whether obedience to Islamic law is ultimately compatible with the Western secular pattern, one finds an increasing number of moderate and liberal Muslims in India, Indonesia, Turkey and the Arab world who support such a separation between state and religion. In Europe and North America, a number of Muslim organizations have the demand for secular democracy in their mission statements. The Westerners according to them know only a single Islamic term, it is likely to be *jihad*, the Arabic word for holy war, in which the image of Islam as an inherently aggressive and xenophobic religion has long prevailed and at times appear to be substantiated by current events. L. Carl Brown challenges this conventional wisdom with a fascinating historical overview of the relationship between religious and political life in the Muslim world ranging from Islam's early centuries to the present day [Brown: 2000]. He examines the commonplace notion—held by both radical Muslim ideologues and various Western observers alike—that in Islam there is no separation between religion and politics and shows that both the modern-day fundamentalists and their critics have it wrong when they posit an eternally militant, unchanging Islam outside of history. As the historical record shows, mainstream Muslim political thought in pre-modern times tended toward political quietism. In order to illuminate the distinguishing characteristics of Islam in relation to politics, Brown compares this religion with its two Semitic sisters, Judaism and Christianity, drawing striking comparisons between Islam today and Christianity during the Reformation. With a wealth of evidence, he recreates a tradition of Islamic diversity every bit as rich as that of Judaism and Christianity. According to Eric Bobsbawn, as has been quoted by Bassam Tibi, the interpretation of the unity of state and religion (*din wa dawla*) is no more and no less than a fundamentalist invention of a tradition [Tibi: 1998: 165]. In fact, Islam knows no institution in the sense of a 'church'—a gathering or a council—which can be convened in case of a controversy or conflict. There are historical examples of how the Muslim scholars have collaborated with the state and also of how the Muslim rulers were confronted or involved in the overthrowing of regimes. But there have been no institutionalized confrontations between the church and the state comparable to those as we have in European history. In Ottoman Turkish Empire, the scholars were incorporated into the state apparatuses and authorities had great regard for religious scholars. The Egyptian scholar Ali Abd al-Raziq argued clearly for the separation of the church and the state. A present-day Egyptian scholar, Nasr Abu Zayd argues along the similar lines and says that people do not need the state to fulfil their religious obligations. The faith is a personal message and is not intended for the state. He mentions how the explanations of unity between church and state came about. According to him, this is a fundamentalist interpretation who thinks state to be necessary for consolidation of religion and therefore no belief is possible without an Islamic state. But fundamentalists are not interested in history. In history, there are examples of both separation and integration between the state and religion. Contrary to what has been often thought and claimed, a survey of the history of fourteen centuries of Islam makes it clear that there is a *de facto* separation of state and religious community. Rather than recognizing a divine right to rule, Islam sanctioned a divinely-sanctioned need to rule [Brown: 2000].

Wesseles argues that no doubt politics has played an important role in the history of Islam and it still does. But Islam does not proclaim any dogma for the indissoluble connection between religion and politics or religion and state. Quran does not prescribe any specific form of government, nor does it contain any blueprint for politics in the state. Only a small number of verses in Quran contain rules for public legislation. Certainly, there have been movements which put forward a view of such a link between the two but these movements are of fundamentalist nature unsupported by the sources and by average Islamic thought [Wesseles: 2006]. In fact, it is for the Muslims themselves to think dispassionately of the issue and take a stand that is good for the entire Muslim community around the globe.

Freedom of religion as a legal concept is related to, but not identical with, religious toleration, separation of church and the state, or *a secular state.* It refers to the freedom of an individual or community, in public or private, to manifest religion or belief in teaching, practice, worship, and observance. It is generally recognized to also include the freedom to change religion or to not follow any religion. Freedom of religion is considered by many in many nations and people to be a fundamental human right.

In a country with a *state religion* freedom of religion is generally considered to mean that the government permits religious practices of other sects besides the state religion, and does not persecute believers in other faiths. The Universal Declaration Human Rights adopted by the fifty-eight Member-States of the United Nations General Assembly on December 10, 1948 defines freedom of religion and belief as follows: "Everyone has the right to freedom of thought, conscience and religion; this right includes freedom to change his religion or belief, and freedom, either alone or in community with others and in public or private, to manifest his religion or belief in teaching, practice, worship, and observance." Historically *freedom of religion* has been used to refer to the tolerance of different theological systems of belief, while *freedom of worship* was defined as freedom of individual action. Each of these has existed to varying degrees. While many countries have accepted some form of religious freedom, this has also often been limited in practice through punitive taxation, repressive social legislation, and political disenfranchisement.

In Islam, the contemporary idea of religious freedom as a human right remains a contested topic. Some Islamic theologians quote the Quran—"There is no compulsion in religion," (Sura 2:257)—to show scriptural support for religious freedom and which can rightly be called "a charter of freedom of conscience unparalleled in the religious annals of mankind." However, it contrasts with the hard-line Islamist understanding, derived from other Qur'anic verses (9:12; 29; 36; 73; 123): "Fight the unbelievers" is the message of all these verses. The term "unbelievers" is popularly and among hardliners understood to mean all non-Muslims. Hardliners also quote other verses and the *Hadith* that mandate severe treatment for unbelievers, which is reflected in the high levels of intolerance shown in many past and contemporary Islamic societies, such as in Saudi Arabia and Iran. In Iran, the constitution recognizes four religions whose status is formally protected: Zoroastrianism, Judaism, Christianity, and Islam, whereas the constitution prohibits Bahai'sm who have been subjected to arrests, beatings, executions, confiscation and destruction of property, and the denial of civil rights and liberties, and the denial of access to higher education. Likewise, according to the moderates and the modernists, the term *ridda,* or apostasy, should not be understood in terms of inner religious conviction, but rather in terms of political and military treason. However, there are moderate Islamic scholars like Dr. Yusuf Al-Qaradawi who believes that Islam does not compel people to join it nor does it force anybody to accept or to leave any other religion, but it places great importance upon conviction for those who embrace it. Dr. Muhammad Salim Al-'Awwa believe that the punishment for apostasy is discretionary and is left entirely to the concerned authorities in the Muslim state to decide.

Wherever Sharia is adopted as the basis of national or regional law, or even where adherence to Sharia is the expected norm in a sub-culture, the whole idea of religious freedom as described in Article 18 of the Universal Declaration of Human Rights is problematic. Sharia as traditionally understood runs counter to the ideas expressed in Article 18. There are countries traditionally belonging to the Muslim world, however, where a long history of religious tolerance has made it easier to accept the ideas of Article 18, even among Muslims. The best example is Indonesia, but even in Indonesia the legislative process is at present strongly influenced by a conservative Islamist agenda.

Among the most contentious areas of religious freedom is the ***"Right to Change" one's religion*** which relates to restricting certain kinds of missionary activity by religions. Many Islamic states, and others such as China, severely restrict missionary activities of other religions. Religious practice may also conflict with secular law creating debates on religious freedom. For instance, even though polygamy is permitted in Islam it is prohibited in secular law in many Western countries. Does prohibiting polygamy then curtail the religious freedom of Muslims? The USA and India, for instance, have taken two different views of this. In India, polygamy is permitted, but only for Muslims, under Muslim Personal Law. In the USA, polygamy is prohibited for all. [Religious Freedom under Islam: *Wikipedia Search; Dhimmi Watch*, January 20, 2004; for Apostasy and Freedom of Religion, see: Islamonline.net, April 13, 2006; Pipes, Daniel: "The Issue of Compulsion in Religion: Islam is What its Followers Make It": New York Sun, September 28, 2004, visit also: DanielPipes.org].

Freedom from Religious Police: The *Mutaween* or as a variant English spellings would call it: *mutawwain, muttawa, mutawallees, mutawa'ah, mutawi', mutawwa* are the government-authorized or recognized **religious police** or clerical police, mainly existing in Saudi Arabia. More recently the term has gained use as an umbrella term indicating any religious-policing organization in an Islamic nation with at least some government recognition or deference, which enforce varied interpretations of *Sharia* Law. It originally referred solely to Saudi Arabia's infrastructure of proselytization and enforcement of Wahhabist tenets under the Committee for the Propagation of Virtue and the Prevention of Vice. However, the term has gained increasing use as a generic term for any religious-policing organization in a Muslim nation such as the Islamic Revolutionary Guards Corp in Iran. Recently "mutaween" has appeared to describe the enforcement of *Sharia* by autonomous groups within Muslim enclaves located inside secular nations and has also entered the lexicon of media as sarcastic pejorative describing politicized, non-Islamic religious groups.

Secular versus Theocratic debate: A highly relevant theme of this study is to analyze the role of religion, particularly its fundamentalist version, in political domain so as to ascertain if religious fundamentalism is responsible for the current sway of terrorism in the international sphere. Based on the "secularization theory," which holds that religion will wither as modernity advances, most countries in the secular West and US have formulated and shaped their foreign policy without making a serious space for religion in it. From the inception of international relations as a discrete discipline, its approach has been defined by the seventeenth-century Westphalian subordination of religion to the state. Consequently, as the international relations scholar Daniel Philpott has observed, most in the field have simply "assumed the absence of religion among the factors that influence states." But the world today is, as the sociologist Peter Berger puts it, "as furiously religious as it ever was, and in some places more so than ever." Those scholars, such as Peter Berger, who challenge the 'secularization theory', argue that contrary to the belief of secularists, the role and intervention of faith is on the rise, particularly since last several decades even as modernization has proceeded apace. Iran's Shiite revolution in 1979, the Catholic Church's role in the "third wave" of democratization, the 9/11 attacks—all illustrated just how important a global force religion has become. For the most part, however, analysts and policy-makers have remained either ignorant or baffled. Scholars are now scrambling to reexamine the question of faith in international affairs—its "return from exile," as one study puts it [Petito: 2003]. Unfortunately, policy-makers are lagging even further behind, and the implications for the international security are troubling. Instead of condemning religion, its surge needs to be reconciled with international society. To see today's religious movements as part of a fundamentalist-driven clash of civilizations misses their role in a broader struggle to find alternative paths to modernity. In fact, religion is as much an opportunity as it is a threat. Rather than being inimical to the advance of freedom, as many secularists assume, religious ideas and actors can buttress and expand ordered liberty. For much of the world, the religious quest lies at the heart of human dignity. History suggests that protecting religious freedom and harnessing it for the common good are vital if democracy is to endure. Social science data show strong co-relations between religious freedom and social, economic, and political goods. Accordingly, state diplomacy should move resolutely to make the defense and expansion of religious freedom a core component of its foreign policy. Doing so would give the states, United States in particular, a powerful new tool for advancing ordered liberty and for undermining religion-based extremism at a time when other strategies have proved inadequate.

Having said all this, one important question remains as to ***why the Muslims hate US and the West*** so much to which something more that the usual, familiar answers, such as, they hate US because it stands for freedom which they hate, it is rich and they envy, and that US is strong and they resent this is required. These answers might be true but then there are billions of poor and weak and oppressed people around the world who do not resort to acts of terrorism because of their lesser position. "There is something stronger at work here than deprivation and jealousy. Something that can move men to kill but also to die" says Fareed Zakaria [*Newsweek*: Oct. 2001]. Osama bin Laden's answer might be 'religion', 'jihad' or the holy war against the west. But it is nothing more than terrorists' bid to place their own twisted morality above mankind's. Despite the fact that every Islamic country in the world has condemned the attacks of September 11, they come out of a culture that reinforces their hostility, distrust and hatred of the West—and of America in particular. This culture does not condone terrorism but fuels the fanaticism that is at its heart. Every religion is compatible with the best and the worst of humankind. Through its long history, Christianity has supported inquisitions and anti-Semitism, but also human rights and social welfare. A quest for better life has weakened the forces of religion in most parts of the world with the exception of the Arab world, some variations notwithstanding. So the question that needs to be probed is: ***what has gone wrong in the world of Islam*** that explains not the conquest of Constantinople in 1453 or the siege of Vienna of 1683 but Sept. 11, 2001 also?

If we look at the large Islamic world, we find that many of the largest Muslim countries in the world show little of anti-American rage. The biggest, Indonesia, had, until the recent Asian economic crisis, been diligently following Washington's advice on economics, with impressive results. The second and third most populous Muslim countries, Pakistan and Bangladesh, have mixed Islam and modernity with some success. While both countries are impoverished, both have voted a woman into power as prime minister, before most Western countries have done so. Next is Turkey, the sixth largest Muslim country in the world, a flawed but functioning secular democracy and a close ally of the West (being a member of NATO). Only in the context of the Middle-East, the Islamic countries represent an example of religious rigidity that goes to precipitate forces of extremism and fundamentalism. In Iran, Egypt, Syria, Iraq, Jordan, the occupied territories and the Persian Gulf, the resurgence of Islamic fundamentalism is virulent, and an anti-West/anti-American feeling is widespread. And this anti-West rage is merely three decades old. In the late 1950s as Gamal Abdel Nasser consolidated power in Egypt, he spoke for the entire Arab world and his language expressed modern ideas like self-determination, socialism and Arab unity. And before oil money turned the Gulf States into golden geese, Egypt was the undisputed leader of the Middle-East. The Middle-East desperately wanted to become modern. But, eventually these ideas could not take roots in that region. The republics calcified into dictatorships. Arab unity cracked and crumbled as countries discovered their own national interests and opportunities. Worst of all, Israel humiliated the Arabs in the wars of 1967 and 1973. When Saddam Hussein invaded Kuwait in 1990, he destroyed the last remnants of the Arab idea. The rest was done by the West and American eyes on Gulf oil and the disillusionment with the West came to become the central problem of the Arab world. In this background, the political progress is fraught with difficulty. Modernization is now taken to mean, inevitably, westernization or even worse, Americanization. This fear has paralyzed Arab psyche.

The new age of globalization has hit the Arab world in a very strange way. Its societies are open enough to be disrupted by modernity, but not so open that they can ride the wave. They see the television shows, the fast foods and the fizzy drinks. But they don't see genuine liberalization in the society, with increased opportunities and greater openness. For the regimes it is an unsettling, dangerous phenomenon. As a result, the people they rule can look at globalization but for the most part not touch it. Disoriented young men, with one foot in the old world and another in the new, now look for a purer, simpler alternative. Fundamentalism searches for such people everywhere; it, too, has been globalized. Mohamed Atta, the Hamburg-educated engineer who drove the first plane into the World Trade Center on 9/11 was surely a globalized youth.

Arab societies are going through a massive ***youth bulge***, with more than half of most countries' populations under the age of 25. When they go out to the new world for jobs and other opportunities, they see great disparities of wealth and the disorienting effects of modernity. A huge influx of restless youth when exposed even to a small economic and social change lands up into a new politics of protest. In the past, societies in these circumstances have fallen prey to a search for revolutionary solutions. (France went through a youth bulge just before the French Revolution, as did Iran before its 1979 revolution.) In the case of the Arab world, this revolution has taken the form of an Islamic resurgence.

In political domain, the ***Arab world is a political desert*** with no real political parties, no free press, few pathways for dissent. As a result, the mosque turned into the place to discuss politics. And fundamentalist organizations have done more than talk. From the Muslim Brotherhood to Hamas to Hizbullah, they actively provide social services, medical assistance, counseling and temporary housing. For those who treasure civil society, it is disturbing to see that in the

Middle-East these illiberal groups are civil society. When the state or political parties fail to provide a sense of legitimacy or purpose or basic services, other organizations have often been able to step into the void. In Islamic countries there is a ready-made source of legitimacy in the religion. So it's not surprising that this is the foundation on which these groups have flourished. The particular form—Islamic fundamentalism—is specific to this region, but the basic dynamic is similar to the rise of Nazism, fascism and even populism in the United States. If there is one great cause of the rise of Islamic fundamentalism, it is the total failure of political institutions in the Arab world. Muslim elites have averted their eyes from this reality. As the moderate majority looks the other way, Islam is being taken over by a small poisonous element, people who advocate cruel attitudes toward women, education, the economy and modern life in general. Muslims should come forward to rescue their religion and themselves from medievalists, fundamentalists and terrorists.

While the Arab world has long felt betrayed by Europe's colonial powers, its disillusionment with America begins most importantly with the creation of Israel in 1948. Arabs see it as a state in which foreign people are being imposed on a region with western backing. The anger deepened in the wake of America's support for Israel during the wars of 1967 and 1973. They look at American policy in the region as cynically geared to America's oil interests, supporting thugs and tyrants without any hesitation. Finally, the bombing and isolation of Iraq have become fodder for attacks on the United States. While many in the Arab world did not like Saddam Hussein, they believed that the United States chose a particularly inhuman method of fighting him—a method that starved an entire nation.

America must now devise a ***three-fold strategy to deal with this form of religious terrorism*** that would include: **military, political and cultural** strategy. On the ***military*** front, the global terrorist networks need to be destroyed, their operations to be disrupted, finances drained, hideouts destroyed. The ***political strategy*** must include multilateralism, working in cooperation with institutions like the United Nations Security Council and respecting international law. On Israel the US should make a clear distinction between Israel's right to exist and its occupation of the West Bank and Gaza. Israel cannot remain a democracy and continue to occupy and militarily rule 3 million people against their wishes. The third, vital component to war on terror is a ***Cultural Strategy***. The United States must help Islam enter the modern world. The whole world today faces a dire security threat that will not be resolved unless the political, economic and cultural collapse is arrested that lies at the roots of Arab rage. To achieve this, the moderate Arab states should be genuinely helped by involving them and their willing cooperation to embrace moderation. International efforts are needed to make the case to the Arab people that Islam is compatible with modern society, that it does allow women to work, that it encourages education and that it has welcomed people of other faiths and creeds. Sept. 11 has been a wake-up call for many. The Saudi regime denounced and broke its ties to the Taliban in Afghanistan. The moderate Muslim groups and scholars should be encouraged to spread a fresh thinking across the Arab world, all aimed at breaking the power of the fundamentalists.

It is important that Islamic fundamentalism still does not speak to the majority of the Muslim people. In Pakistan, fundamentalist parties have yet to get more than 10 percent of the vote. In Iran, having experienced the brutal Puritanism of the mullahs, people are yearning for normalcy. In Egypt, for all the repression, the fundamentalists are a potent force but so far not dominant. If the West can help Islam enter modernity in dignity and peace, it will have done more than achieved security. [Fareedzakaria.com]. The onset of Arab Spring since early 2011 and the departure of many autocrats from their long repressive rules is a good sign in favor of achieving success in war on terror.

Notes and References

1. The research involved in Martin and Appleby's *The Fundamentalism Project* (1991-95), five volumes, is unparalleled, and the conclusions are the most authoritative offered about religious fundamentalisms. The authors examine Fundamentalisms from statistical, religious, sociological, cultural, historical, political and other dimensions. At 8000 pages, it is a very thorough treatment of the subject. [see: Fundamentalism Index: brucegourley.com]
2. To the one very popular and common attack and criticism of the west against Islam regarding the status of women in which the Islam is charged with denying all rights to women, that women are oppressed and are considered inferior to men, the Islamists counterattack such charges and say that Islam has given the highest position to womankind on earth, according them a position of honour, respect, safety and love, that has not yet been matched to this day, let alone

being superceded and excelled. What Europe and the Western civilization have given to women in the name of freedom and human rights are Prostitution, Escort Agencies, Massage Parlours, Lesbianism, Illegal Mistresses, Nudity and Shamelessness, making women the cheapest commodity on earth whose nude or semi-nude body is sold with almost every saleable product as if her body is the property of one and all to which every lusty and lecherous man is at full liberty to cast his filthy and dirty gazes and commit everything evil and profanity in his mind and heart. West has despised women beyond description. On the contrary, the woman in Islam is a "precious jewel" not to be viewed by all and sundry except for the only person who truly appreciates and loves her—her husband. Islamists believe that whenever and wherever woman was lured, in the name of social and economic rights, to neglect her domestic life by the glitter and glamour of the outside world and indulged in economic and other pursuits, her home and children was ruined. In countries that have given women the so-called rights of freedom and equality and left them free to do as they wish are the degeneration and disintegration of their societies. Their women being economically and socially independent, are no longer faithful and dedicated daughters, wives, sisters and mothers. Living-in relationships and companionship, which is not binding upon the man or woman, are replacing the institution of marriage. Children of such parents become delinquents and drug addicts. The whole society is decaying and disintegrating so fast that they have reached a point of no return. Sharia indulges in details on the rights of the wife and makes it obligatory for men to care for his wife and family, to respect his mother before he gives respects to father and also speaks about women's right to inheritance [allahuakbar.net].

3. The *Clash of Civilizations* is a theory, proposed by Harvard Political Scientist —Samuel P. Huntington—who argues that people's cultural and religious identities will be the primary source of conflict in the post-Cold War world. The theory was originally formulated in 1993 in a *Foreign Affairs* article titled "The Clash of Civilizations?" as a reaction to Francis Fukuyama's 1992 book, *The End of History and the Last Man* in which Fukuyama had argued that human rights, liberal democracy and capitalist free market economy had become the only remaining ideological alternative for nations in the post-Cold War world. With the end of Communism, the world had reached the 'end of history' in a Hegelian sense. Huntington later expanded his thesis in a 1996 book *The Clash of Civilizations and the Remaking of World Order*. The term itself was first used by Bernard Lewis in an article in the September 1990 issue of *The Atlantic Monthly* titled *The Roots of Muslim Rage* [Huntington: 1996, 2002; Fukuyama: 1992]. Huntington believed that while the age of ideology had ended, the world had only reverted to a normal state of affairs characterized by cultural conflict. In his thesis, he argued that the primary axis of conflict in the future would be along cultural and religious lines. As an extension, he posits that the concept of different civilizations, as the highest rank of cultural identity, will become increasingly useful in analyzing the potential for conflict.

REFERENCES

Ahmed Akbar S., *Islam Under Siege,* New Delhi: Vistaar, 2003.

Almond, Gabriel A., R. Scott Appleby (eds.), *Strong Religion: The Rise of Fundamentalism Around the World,* Chicago and London: University of Chicago Press, 2003.

Antoun, Richard T., *Understanding Fundamentalism: Christian, Islamic and Jewish Movements,* New York, Oxford: Alta Mira Publishers, 2001.

Appleby, R. Scott, Gabriel Abraham Almond, and Emmanuel Sivan, *Strong Religion,* Chicago: University of Chicago Press, 2003.

Bayat, Asef, *Making Islam Democratic: Social Movements and the Post-Islamist Turn,* Stanford University Press, 2007.

Benard, Cheryl, *Civil Democratic Islam: Partners, Resources and Strategies,* Santa Monica,: RAND, 2003.

Berger, Peter, *The Sacred Canopy: Elements of a Sociological Theory of Religion,* Anchor Books, 1990.

____, *Questions of Faith: A Skeptical Affirmation of Christianity,* Blackwell Publishing, 2003.

Berke, Jason, *Al-Qaeda: Casting a Shadow of Terror,* NY: I.B. Tauris, 2003.

Booth, Ken and Tim Dunne, *Worlds in Collision: Terror and the Future of Global Order,* NY: Palgrave Macmillan, 2002.

Bostom, Andrew (ed.), *Legacy of Jihad: Islamic Holy War and Fate of Non-Muslims,* Prometheus Boos, 2005.

Brill, E.J., *Encyclopaedia of Islam (EI),* Brill Publishers, 1913-36; Reprinted, 1987 (it is the standard encyclopedia of the academic discipline of Islamic Studies and not a Muslim or an Islamic encyclopedia).

Brown, L. Carl, *Religion and State: The Muslim Approach to Politics,* Columbia University Press, 2000.

Bruce, Lawrance, *Shattering the Myth: Islam Beyond Violence,* Princeton: Princeton University Press, 1998, 2000.

Bunt, Gary R., *Islam in the Digital Age: E-Jihad, On-Line Fatwas and Cyber Islamic Environments,* London: Pluto, 2003.

Cole, Juan, *Sacred Space and Holy War: The Politics, Culture and History of Shiite Islam,* London and New York: I.B.Tauris, 2002.

Cooper, John, *et al.,* (eds.), *Islam and Modernity: Muslim Intellectuals Respond,* London: I.B. Tauris, 2000.

Daniel, Benjamin, *The Age of Sacred Terror,* NY: Random House, 2002.

Dixon, A.C., *The Fundamentals: A Testimony To The Truth,* Bible Institute of Los Angeles, 1910-15.

Ehrehfeld, Rachel, *Funding Evil: How Terrorism is Financed and How to Stop It,* Chicago: Bonus Books, 2003.

Engineer, Ashgar Ali, "Islam, Globalization and Fundamentalisms" in *Islam and Modern Age,* August 2002.

Esposito, John L., *Islam. The Straight Path,* Oxford, 1988.

____, *Oxford Encyclopedia of the Modern Islamic World* (4 volumes), 1995.

____, *Islam and Civil Society,* European Univ. Inst., 2000.

Farr, Thomas F., "Diplomacy in an Age of Faith: Religious Freedom and National Security", *Foreign Affairs*: March/April 2008.

Fitzgerald, Hugh, "Understanding the Resurgence of Islam", *New English Review*, July 2007.

Fuller, Chris, *Renewal of Priesthood*, Princeton: Princeton University Press, 2003.

Fuller, Graham E., *The Future of Political Islam*, NY, Palgrave Macmillan, 2006.

Gordon, Philip H., *Winning the Right War: The Path to Security for America and the World* : Brookings Institution Press, 2007.

Gunaratana, Rohan, *Inside Al-Qaeda: Global Network of Terror*, New Delhi: Roli Books, 2002.

Hilladey, Fred, *Islam and the Myth of Confrontation*, NY; I.B. Tauris, 2003.

____, *100 Myths About the Middle-East*, University of California Press, 2005.

Hinde, Robert A., *Why Gods Persist: A Scientific Approach to Religion*, Routledge, 1999.

Hiro, Dilip, *War Without End: The Rise of Islamic Terrorism and Global Response*, Delhi: Roli Books, 2002.

Husain, Mir Zohair, *Global Islamic Politic*, New York: Harper Collins College Publishers, 1995.

Jarisha 'Ali M. and Muhammad Sh. Zaibaq, *Methods of Intellectual Invasion of the Islamic World*, second printing (Cairo: Dar al-I'tisam, 1978).

Juergensmeyer, Marc, *Terror in the Mind of God: The Global Rise of Religious Violence*, University of California Press.

Kaplan, Esther, *With God on Their Side: How Christian Fundamentalists Trampled Science, Policy and Democracy in George Bush's White House*, NY: Free Press, 2004.

Kepel, Gilles, *Jihad: The Trail of Political Islam*, London: I.B. Tauris, 2000.

____, *Revenge of God: Resurgence of Islam, Christianity and Judaism in the Modern World*, Cambridge: Polity: 1994.

Kohut, Andrew *et al*, *Diminishing Divide: Religion's Changing Role in American Politics*, Washington D.C.: Brookings Institution Press, 2000.

Lapidus, Ira M., *A History of Islamic Societies*, Cambridge University Press, 2002.

____, *Islam, Politics and Social Movements*, 1988.

Larsson, J.P., *Understanding Religious Violence: A New Framework for Conflict*, Ashgate Publishers, 2004.

Lewis Bernard, *What Went Wrong? Western Impact and Middle-Eastern Response*, Oxford Univ. Press, 2002.

Lewis, Bernard, *The Crisis of Islam: The Holy War and Unholy Terror*, NY; Modern Library, 2003.

Lincoln, Bruce, *Holy Terrors: Thinking About Religion After September 11*, Chicago: Chicago University Press, 2003.

Lining Islam, A Project of Independent Broadcasting Associates Inc. (IBA), USA, 2002 (Funded by Ford Foundation).

Maley, William, *The Afghanistan Wars*, London and NY: Palgrave, 2002.

Mamdani, Mahmood, *Good Muslim, Bad Muslim: America, The Cold War and the Roots of Terror*, Pantheon Books, 2004

Marty, Martin E. and R. Scott Appleyby (eds.), *The Fundamentalism Project*, (Five Volumes), University of Chicago, 1991-95.

Mondal, Anshuman A., "Liberal Islam?", *Prospect Magazine*, January 2000.

Nassar, Jamal, *Globalization and Terrorism,* New York: Rowman & Littlefield, 2005.

Norris, Pippa and Inglehart, Ronald, "Islam & the West: Testing the Clash of Civilizations Thesis", *KSG Working Paper No. RWP02-015,* April 2002.

Petito, Fabio and Pavlos Hatzopoulos (ed.), *Religion in International Relations: The Return From Exile,* Palgrave Macmillan, 2003.

Phares, Walid, *Future Jihad: Terrorist Strategies Against the West,* Palgrave Macmillan, 2006.

Phillips, Kevin, *American Theocracy,* NY: Viking, 2006.

Philpott Daniel, *Revolutions in Sovereignty: How Ideas Shaped Modern International Relations,* Princeton University Press, 2001.

_____, *The Politics of Past Evil: Religion, Reconciliation, and Transitional Justice,* Notre Dame, 2006.

Pipes, Daniel, In the Path of God: Islam and Political Power, 1983.

_____, The Long Shadow: Culture and Politics in the Middle-East, 1990.

_____, Militant Islam Reaches America, Norton, 2002.

Prince of Wales, *Islam and the West: a lecture given in the Sheldonian Theatre, Oxford on 27 October 1993* (Oxford: Oxford Centre for Islamic Studies, 1993).

Pryce-Jones, David, The Closed Circle: An Interpretation of the Arabs, 1989.

Qutb, Syed, Social Justice in Islam, NJ: Islamic Publications International, 1949 (Tr: Hamid Algar).

Rabasa Angel M. *et al, Muslim World After 9/11,* Rand Project Air Force, 2004

Ramadan, Tariq, *In the Footsteps of the Prophet: Lessons From the Life of Muhammad*: Oxford University Press, 2007, p. 242.

Rashid, Ahmed, *Jihad: The Rise of Militant Islam in Central Asia,* Yale University Press, 2002.

_____, *Taliban: Militant Islam, Oil and Fundamentalism in Central Asia,* Yale University Press, 2000.

_____, *Al-Qaeda in 2007: Striving to Regain the Initiative,* 2006.

_____, *Taliban: The Story of Afghan Warlords,* London: Pan Books, 2001.

Roy, Olivier, *The Failure of Political Islam,* Cambridge, Harvard University Press, 1994, 2001.

_____, *Globalized Islam: The Search for a New Ummah,* CERI Series in Comparative Politics and International Studies, 2004.

Rubenstein, Richard E., *Alchemists of Revolution. Terrorists in the Modern World,* (New York, 1987).

_____, *Aristotle's Children. How Christians, Muslims, & Jews Rediscovered Ancient Wisdom and Illuminated the Dark Ages,* Orlando, 2003.

Rubin, Berry and Judith Rubin, *Anti-American Terrorism and the Middle-East,* NY: Oxford, 2002.

Sachedina, Abdulaziz, The Islamic Roots of Democratic Pluralism, Oxford University Press, 2001.

Saha, Santosh C. (ed.). *Religious Fundamentalism in the Contemporary World: Critical, Social and Political Issues,* New York: Lexington Books, 2004.

Said, Edward, *Orientalism,* Vintage Books, New York, 1979.

Said, Edward W., *Covering Islam: How the Media and the Experts Determine, How We See the Rest of the World,* New York: Vintage Books/Random House, 1981, rev.ed. 1997.

Schacht, Joseph, *An Introduction to Islamic Law*, London, OUP, 1964.

Sen, Amartya, *Identity and Violence: The Illusion of Destiny*, W.W. Norton & Company, Inc, 2006.

Shadid, Anthony, "Reconstruction: For the Sufis, Taliban's Fall Means a Revival", *Boston Globe,* January 23, 2002.

Shaffer, Brenda (ed.), *Limits of Culture: Islam and Foreign Policy*, MIT Press, 2006.

Skaine, Rosemarie, *The Women of Afghanistan Under Taliban*, London: McFarland & Co., 2002.

Stern, Jessica, *Terror in the Name of God,* Harper Collins, 2003.

Sutton, Fillip W., Stephan Vertigaus, *Resurgent Islam: A Sociological Approach,* Blackwell, 2005.

Takeyh, Ray, Hidden Iran: Paradox and Power in the Islamic Republic.

Tibi, Bassam, *The Challenge of Fundamentalism: Political Islam and the New World Disorder*, Berkeley, University of California Press, 1998, updated edition, 2002.

____, *Islam Between Cultures and Politics*, Palgrave, 2005.

Vaughn, Bruce, Islam in South and South-East Asia, CRS Report for Congress, February, 2005.

Watson, Peter, *Modern Mind: An intellectual history of the 20th century,* 2001.

Wessels, Antonie, *Muslims and the West: Can They Be Integrated?*, Peters Publishers, 2006 (Translated by John Bowden).

Wordsmith Compilation: *The Jihad Fixation: Agenda, Strategy, Portents,* Delhi: Wordsmith Compilation, 2001.

Wright, Robin, Islam and Liberal Democracy: Two Visions Of Reformation, *Journal of Democracy*, 7.2 64-75, John Hopkins University Press, 1996.

Zakaria, Fareed, "*The Politics of Rage: Why Do They Hate Us?*" *Newsweek*, 14.10.2001. (see various Newsweek articles by Farid Zakaria on this issue.)

Chapter

3

Political Fundamentalism: Mapping Fundamentalist Trends in US Foreign Policy 1991-2001 and Onwards

What difference does it make to the dead, the orphans, and the homeless, whether the mad destruction is wrought under the name of totalitarianism or the holy name of liberty and democracy?
—*Mahatma Gandhi*

Great is the guilt of an unnecessary war. —*John Adams*

Political Fundamentalism Defined

Though 'Fundamentalism' is a religious construct signifying religious orthodoxy placing a high priority on the doctrinal conformity of its adherents with the fundamentals of that particular religion taking only its doctrinal beliefs to be right to the exclusion of others, it has political orientations and connotations as well. The term 'Political Fundamentalism' in this Chapter stands for the unilateralist tendencies displayed through the foreign policy pronouncements and behaviour of a government or state, the government of United States in the present context, since the end of Cold War in general and after the terrorist attacks of September 11, 2001 on US in particular. Like its religious counterpart, be it in Christianity or Islam or any other major religion of the world, political fundamentalism chooses to take orthodox, rigid positions on matters related to the behaviour of a state in its domestic as well as foreign policy matters. In an international system of nation states where a very limited few possess and exercise the asymmetrical power in comparison to others, political fundamentalism is the privilege of the most powerful, such as that of the unipolarist United States after the collapse of its cold war rival USSR. Like all fundamentalists, the political fundamentalists also insist that only their way of thinking and doing is right, they seem to have a divinely ordained duty to set the things right in the existing world and to bring changes in other political systems too in with their beliefs and way of thinking, irrespective of the cultural, religious and regional variations existing in other parts of the world. Absolute conformity with their doctrines and policy prescriptions is the requirement. A political fundamentalist will not hesitate even to intervene in the political process of other states to ensure that society in those states conform to the behaviors their world view requires. The belief that they are right, without any question, justifies, in their own minds, taking upon themselves the right to impose their point of view on others, by force if necessary.

In the context of the current US-led War on Terrorism, David Domke in his book: *God Willing? Political Fundamentalism in the White House, the 'War on Terror,' and the Echoing Press* [Domke: 2004] underlines the Bush administration's use of rigid and unilateralist language and concludes that it is to justify a policy that America would like to see implemented even in a culturally different Middle-East, such as policies of democracy promotion, rights and liberties in the way they are pursued in the west. The use of the language after 9/11 attacks: either you are with us or against us, echo a fundamentalist rigidity. The way President Bush talked about liberty and freedom sounded just like the way the Bible talks about the Christian Gospel, feels Domke. The Jewish Lobby, Christian Right and the hardliner neoconservatives in Bush administration had the virtual sway over the Pentagon, White House and other political processes, driving American foreign policy in a direction that compelled President to adopt rigid policies pertaining to the oil-rich Gulf countries of the Middle-East. The preemption, regime change, democracy promotion were some examples of following a rigid, almost fundamentalist policy by United States leading to raising the anti-American and anti-West sentiments in the Muslim world. However, it may also be mentioned here that the kind of terrorist catastrophe brought on US could perhaps be responded in the way American did. With this explanation of the term 'political fundamentalism' it is essential to critically examine the US Foreign policy since the end of the cold war to test the hypothesis that a fundamentalist trend in it is responsible for much of the international terrorism that the world faces today.

Unipolarist Tendencies in American Foreign Policy

Any attempt to understand the causes of the rising wave of international terrorism at any given point of time in history requires the critical analysis of the complex play of politics of different nations, particularly of the powerful ones, that itself is the outcome of different motivations, aims and objectives framed on the basis of the concept of national interest that might ensure their security and well-being in the best possible ways. United States of America, since the close of World War II, particularly in the post-Cold War unipolar world, invites attention, interest, curiosity and criticism and even hatred, depending on one's perceptions of its policies, actions and motivations that have a significant impact on the course of events taking place on the world stage. This study on international terrorism and world politics sees a co-relation between the politics of US foreign policy and the rise of militant religious movements that seems to be one of the main causes of international terrorism today. This chapter examines the ways in which the US political fundamentalism displayed and practiced through its post-cold war unilateralist tendencies and behaviour having the support and backing of neoconservative foreign policy elite in Congress, Pentagon and Oval Office, with or without the widespread public support, became a powerful instrument in exacerbating Islamic resurgence and its most dangerous outcome: terrorism.

A social science approach is useful in identifying a number of contributing factors for international terrorism and violence, such as, rapid modernization and globalization resulting not only in the intentional or unintentional expansion of westernization, secularization, democratization, consumerism and the growth of market capitalism, but also in the inevitable growth of social and economic injustice in the developing world resulting in the spread of extremist ideologies and religious fanaticism. In the globalization process, an enhanced ethnic and religious identity emerges and strengthens and comes in direct clash with forces of modernization and globalization that are held responsible for repression, domination and occupation of the territories, resources and self-respect of the peoples of specific ethnic and religious origin [Griest and Mahan,

2003; Cronin, 2002; Crenshaw, 1995]. In this context, the best way to understand the post-cold war dynamics of international terrorism is to examine the clash between unipolarist America's unilateralism and Islamic fundamentalism—the one using its military, economic and political power to have its way anyway and the other resisting such a hegemonic approach by way of calling their co-religionists to start a world-wide *jihad* to arrest the perceived onslaught on their religion, culture and resources. Munthe (2005) argues that radicalization, Islamism and Jihadism in Muslim countries at present are a response to three kinds of oppression and occupation: occupation of domestic, political and cultural sphere, occupation of global sphere of influence and occupation of territory. Much of Islamist terrorism today is avowedly against international economic and political system led by United States of America, particularly against its foreign policy in the oil-rich Middle-East and Persian Gulf region. In the perception of most of the Muslims around the world, while democracy promotion, human rights, war on terrorism and development are the declared objectives of US foreign policy-makers, the fact is that American eye on Middle-Eastern oil and undue favor towards Israel are the driving forces behind its policies and actions, be it Afghanistan or Iraq or its next potential target—Iran. To counter this kind of political fundamentalism, the religiously motivated political terrorism of Islamists is chosen as the answer leading to a complex and unending spiral of violence and terrorism on the one hand and violence and counter-terrorism on the other. As such, violence remains a key component in both terrorism as well as counter-terrorism.

Non-Interventionism to Neo-interventionism

Throughout the twentieth century, except for the last decade, the list of world's great powers was small comprising of only a few countries, such as, United States, Soviet Union, Japan and some countries of north-western Europe. In 21st century, the list is smaller, consisting by and large of only one superpower—United Sates of America—though some would disagree with this kind of opinion and would like to include a number of great powers previously existing and also the newly emerging ones, such as, China and India and plead for the incorporation of these countries into the great power framework in the emerging multipolar era in world politics. But the recent unipolar impulses vividly displayed by American behaviour in Iraq and setbacks in its war on terrorism raises questions about the validity of its multipolar principle and invites criticism for US foreign policy in the post-cold war world, particularly after 9/11, to the extent of overshadowing some of its genuine efforts at multilateralism and at creation of a new world order which are interpreted by strategic analysts as merely the efforts to further US goals. Since the early years of cold war, the US was engaged in a "struggle for the world" with the Soviet Union, and have since rewritten that model to overcome every rival that has come along, right down to Osama bin Laden in his cave [see: Drezner, Daniel W., "The New World Order", *Foreign Affairs*, March/April, 2007]. Due to this kind of the journey of US foreign policy down from a non-interventionist policy to a neo-interventionist one, it is argued that the US behaviour confirms the assumption that Washington believes that it has a universal responsibility for the world affairs and therefore some foreign policy analysts, particularly the right-wing analysts, would call for a new security strategy that would address all threats to human life, whether they stem from terrorism or from environmental degradation, disease, natural disasters or from global poverty. It is this kind of unipolarist impulse that invites indignation or even the hatred of some rivals leading to acts of violence and terrorism against the 'self-appointed' hegemon of the world.

End of Cold War and Emerging New Trends in International Politics: The Role of US

With the ascendancy of Mikhail Gorbachev to power in Soviet Union in 1985, the rapidly collapsing Soviet economy forced the new leader to announce an agenda for rapid economic reforms and openness, popularly known as *perestroika* and *glasnost*, and the country's resources were redirected from costly Cold War commitments to more profitable areas in civilian sector. Major concessions were offered to the United States on the levels of conventional forces, nuclear weapons, and policy in Eastern Europe. In response, US President Reagan agreed to renew talks on economic issues and the scaling-back of the arms race. The East-West tensions that had reached intense new heights earlier in the decade rapidly subsided through the mid-to-late 1980s. In 1988, the Soviets officially declared that they would no longer intervene in the affairs of allied states in Eastern Europe—the so-called Sinatra Doctrine. In 1989, Soviet forces withdrew from Afghanistan. In December 1989, Gorbachev and

George H.W. Bush declared the Cold War officially over at a summit meeting in Malta. Warsaw Pact, the Soviet military alliance system, which was already on the brink of collapse, came to an end. By February 1990, the Communist Party was forced to surrender its 73-year old monopoly on state power. By December 1991, the union-state also dissolved, breaking the USSR up into fifteen separate independent states.

If we look at the legacy of this cold war, which met a relatively bloodless end, we find that this nearly forty-five years long war had heavy economic, military and strategic costs. It did cost the US up to $8 trillion in military expenditures, and the lives of nearly 100,000 Americans in Korea and Vietnam only. It did cost the Soviets an even higher share of their gross national product. In Southeast Asia, local civil wars were intensified by superpower rivalry, leaving millions dead. More importantly than these economic costs, the Cold War still continues to structure world affairs. Since the balance-of-power system no longer held validity, a severe power vacuum was created. The newly emerged unipolar world almost institutionalized the role of the United States in the post-cold war global economic and political system and has a great deal of responsibility for Islamic resurgence and international terrorism.

To understand the developments leading to the end of cold war and afterwards and to grasp how American politics and Islamic radicalism got an inverse co-relation, it is essential to look at the events that have their origin in the year 1979. In 1979, Iran's Islamic Revolution and the Soviet invasion of Afghanistan set the stage for a rise in Islamic radicalism. Historians consider 1979 a watershed for the United States and the Middle-East. It was the year that the US-supported Shah of Iran was forced to flee his nation and the Ayatollah Ruhollah Khomeini, a spiritual leader revered in Iran but virtually unknown in the West, returned to his country after 15 years in exile. Later that year, Islamic militants seized the US embassy in Tehran and took 66 diplomats hostage in a crisis that would last for more than a year. Following the U.S.S.R.'s invasion of Afghanistan, the United States funded, armed and organized an anti-Soviet guerrilla war there. The move "would have horrifying unforeseen consequences many years later when Osama bin-Laden, a veteran of the war in Afghanistan, would turn his anger and terrorist agents against the United States" [Shuster, Mike: 2004].

The Iran-Iraq war, from 1980-88, was another conflict that

would have an impact on the United States. The United States shared battlefield intelligence with Saddam Hussein's Iraq and helped to re-supply its weapons stockpiles. But a few years later, Saddam invaded Kuwait and "another erstwhile friend of the US in the Middle-East had become a deadly enemy," Shuster says. Though a US-led coalition swiftly ousted Iraqi forces from Kuwait in the 1991 Persian Gulf War, problems for the United States in the region would only multiply. Led by Bin Laden's al-Qaeda network, terrorism against America and its allies erupted with murderous determination in 1993, with the first bombing of the World Trade Center in New York, then bombings in Saudi Arabia in 1996, against the US embassies in Kenya and Tanzania in 1998, culminating in the catastrophic events of 11 September, 2001. In the chain of events that thus unfolded since the Soviet occupation of Afghanistan in 1979 till the September 2001 terror acts against US, one can witness a direct or indirect role of America in most of the major, and even minor, political events taking place anywhere in the world. In other words, the US foreign policy, political, economic and military acts, role of its strategic institutions like Central Intelligence Agency (CIA) and of neoconservative think-tanks are understood to be greatly responsible for inciting Islamists' violent reaction and terrorist outbursts against itself as would be explained in the following passages.

1979-91 Role of CIA-ISI Nexus in Fostering Future Terrorists

The anti-Soviet *jihad* that was being spearheaded by Osama bin Laden, who left Saudi Arabia and reached Afghanistan within a month of Soviet invasion in 1979, and by some other Afghan leaders, had the financial, military and logistical support of a multinational coalition organized by the US Central Intelligence Agency (CIA) and comprised of United States, Britain, Saudi Arabia, Pakistan, China and several other countries. This coalition was also utilizing fully and in significant ways the Pakistani co-operation as Afghan and Arab fighters were being trained in its training camps and its Inter Service Intelligence (ISI) was serving as CIA's conduit to transfer sophisticated weapons to *mujahidins*. In these early years of Afghan *jihad,* US was greatly in league with Osama who as part of his field work to raise funds and to lecture Muslims was touring the West and US quite often [Gunaratana: 2002: p. 19]. Like Pakistan and US, Saudi Arabia was also a front runner in providing all sort of financial help in this *jihadi* project under the cover of philanthropy and relief work. In fact, al-Qaeda had a good start as it inherited a full-fledged training and operational infrastructure that had been funded by the US, European and Saudi Arabian and other governments throughout the 1980s. For recruitment, it drew on the vast *mujahidin* database originally created by Osama for tracking martyred or missing *mujahidins* in the later stages of the anti-Soviet *jihad*. American failing also lies in the fact that it failed to realize the threat that this growing monster was about to pose soon even after attacks on US interests at different places. America was complacent in paying attention to constant complaints from the governments that had been victims of terror, such as, India where terrorism in Kashmir escalated exactly at the time of departure of Soviet troops as the availability of trained terrorists and operative infrastructure came handy to the hostile governments such as that of Pakistan. Failing to see a co-relation proved costly in 2001.

Selig Harrison from the Woodrow Wilson International Centre for Scholars and an expert on South Asian affairs said that the CIA made a historic mistake in encouraging Islamic groups from all over the world to come to Afghanistan. The US provided $3 billion for building up these Islamic groups, and it accepted Pakistan's demand that they should decide how this money should be spent. Harrison, who had meetings with CIA leaders at the time when Islamic forces were being strengthened in Afghanistan, told these facts before the Taliban assault on the Buddha statues was launched in early 2001. "They told me these people were fanatical, and the more fierce they were the more fiercely they would fight the Soviets," he said. "I warned them that we were creating a monster" [http://www.timesofindia.com/today/07euro1.htm]. With the help of the CIA and the US Armed Forces intelligence services, Bin Laden began to organize in the early 1980s and created network to raise money and to recruit fighters for the Afghan *mujahideens* that were fighting the Soviets. He did this from the city of Peshawar in Pakistan, bordering Afghanistan. Part of these activities was financed with the production and sale of morphine, the base of heroin. This was the beginning of today's al-Qaeda (the base) network led by Bin Laden. Steve Coll, [Coll, Steve: 1992] also confirms the views of Harrison and said that "In all, the United States funneled more than $ 2 billion in guns and money to the mujaheddin during the 1980s. It was the largest covert action program since World War II." William J. Casey, the CIA Director between 1981-87 who played a significant role in shaping

President Reagan's foreign policy, particularly pertaining to Soviet Union, visited various terrorist training camps in Pakistan and startled his Pakistani hosts by proposing that they take the Afghan war into enemy territory—into the Soviet Union itself. He wanted to ship a subversive propaganda through Afghanistan to the Soviet Union's predominantly Muslim southern republics. The Pakistanis agreed, and the CIA soon supplied thousands of Korans, as well as books on Soviet atrocities in Uzbekistan and tracts on historical heroes of Uzbek nationalism, according to Pakistani and Western officials. Casey's visit was a prelude to a secret Reagan administration decision in March 1985 to sharply escalate US covert action in Afghanistan and to secretly let loose on the Afghan battlefield an array of US high technology and military expertise in an effort to hit and demoralize Soviet commanders and soldiers. Casey saw it as a prime opportunity to strike at an overextended, potentially vulnerable Soviet empire. Eight years after Casey's visit to Pakistan, the Soviet Union collapsed and Afghanistan had fallen to the heavily armed, fratricidal *Mujahideen* rebels. The Afghans, who themselves did the fighting and dying—and ultimately won their war against the Soviets—owe much to CIA's overt and covert help, though not all of them laud the CIA's role in their victory. But even some sharp critics of the CIA agree that in military terms, its secret 1985 escalation of covert support to the *mujahideen* made a major difference in Afghanistan, the last battlefield of the long Cold War. [see: Yousaf, M: 2001; Prados, John, 2002]. By all accounts of the horrendous situation in Afghanistan after Soviet invasion in 1979, with all its implications in terms of the terrorist events later in 1990s and in first decade of the 21st century, the role of the CIA and Pakistani intelligence in the creation of what became the Taliban is undeniable. Pakistan and the USA covertly controlled the largest guerrilla war of the 20th Century, dealing to the Soviet Russia's presence in Afghanistan a military defeat that has come to be called 'Russia's Vietnam'. In a *Review* of the Yousaf's Bear Trap, Eliot A. Cohen, [*Foreign Affairs*, March/April 2002] says that despite abundant US military aid, the Pakistani Inter-Services Intelligence Directorate viewed the United States as calculating and untrustworthy—a mixture of innocent abroad and Machiavellian superpower. Indeed, in Yousaf's view, the Americans came to Afghanistan only to avenge Vietnam; after they succeeded, they abandoned those who had helped them.

John Prados takes us back to the year 1978 when, during the

Carter administration, America's CIA tried to augment a resistance movement against Communists which was started by Muslim fundamentalists and tribal groupings [Prados, John: 2002]. Drawing a parallel between the political machinations of local clients that deepened the American stakes in South Vietnam in 1960s and the Communist coups in Kabul in mid-1970s that committed Soviet Union more deeply in Afghanistan. Prados argues that Washington perceived this as an opportunity to turn the Afghan commitment into a running sore for Moscow—the severest cold war rival. He goes as far as saying that CIA project in Afghanistan preceded the Soviet intervention with three motorized and airborne divisions and other units reaching Kabul in December 1979 under President Carter. Therefore, Prados belives, the US covert operations in Afghanistan were not in direct response to Soviet invasion, which nevertheless was perceived by US as a Soviet thrust towards the Persian Gulf and hence the stage was set for the **Carter Doctrine** whose wordings were framed on the advice of Carter's National Security Advisor, Zbigniew Brzezinski. It said: Any attempt by any major power to seize control of the Persian Gulf will be opposed by the United States by any means necessary, including the use of force. Carter also proclaimed that Soviet troops in Afghanistan posed a grave threat to the free movement of Middle-Eastern oil. On the basis of this policy, Carter increased the impetus for CIA's Afghan project, which transformed from a spoiling operation into a crash program by 1980. Since United States had no organizational means to funnel arms and equipment into the country through Kabul, Americans dealt through local allies: Pakistan and its Inter-Service Intelligence (ISI) Agency as they already had links with Afghan Muslim guerillas and knew the players and their networks well. Despite having a huge budget, the largest for any CIA covert program till date, an enlarged CIA station and a major rebel army, the CIA's Afghan project was run through Pakistan, which created severe problems soon for its mentor—United States.

With CIA acting through Pakistan's ISI, the more and more fundamentalist groups and individuals were selected by ISI to receive weapons and cash aid. One of the prime recipients of this aid, labeled "Freedom Fighters" by President Reagan, and were also known as the *mujahideen*, was Gulbuddin Hekmatyar, an Afghan guerilla warlord who led the US proxy forces against the Soviet Union. Alfred McCoy, in his well-documented work *The Politics of Heroin [1991]*, details the nefarious nature of Hekmatyar, an Islamic fundamentalist who not only commanded the largest guerilla army, but also used it—with the full support of ISI and the tacit tolerance of the CIA—to become Afghanistan's leading drug lord. Such a trend foreclosed the chances of the emergence of more moderate political groups, an evolution the United States would have preferred. But the success of these fundamentalist guerillas confirmed the correctness of ISI's choice. This encouraged the agency to shift the support in early 1990s to an even more virulent strain of fundamentalism in Afghanistan, that later came to emerge as 'Taliban', which the United States felt it necessary to combat after September 11. One more dangerous brand of fundamentalist operations emerged during this Afghan guerilla war in 1980s as the more a group adopted fundamentalist postures, the more they were successful in private fund-raising and thus opened the doors for the individuals and groups to influence the larger conflict. Osama bin-Laden's evolution from a wealthy Saudi Arabian to a mastermind of 9/11 attacks is an example of this pattern which is feared to be followed elsewhere also. During CIA's secret war, Afghanistan became the world's leading producer of heroin. This was not only a blow to Reagan administration's declared war on drugs, it created conflict between US agencies working in Afghan guerilla war and Drug Enforcement Administration. US government could not resolve this internal cleavage in 1980s as it had accepted the interests of the Afghan local groups, including those of drug-traffickers, as a condition of their participation from American side against Soviet forces. This entailed problem of a very conspicuous nature as many a times, these local allies refused to co-ordinate with CIA officials and sometimes worked at cross-purposes with US policies. To happily trade in loyalties has been a norm with Afghan fundamentalists.

Chalmer Johnson draws the attention towards how CIA's Afghan campaign in 1980s continued to have and still has it impact on the United State's current War on Terror (WOT) as the phenomenon of 'blowback' has a tendency to rebound. Osama bin-Laden was one such example who was rebel fighter during CIA's secret war in Afghanistan and was suddenly at the central stage of new terrorism that erupted after September 2001. The Afghan rebels under CIA's 1980 project were popularly called 'Afghan Arabs', and thousands of these rebels went on to other locations such as in Bosnian civil war in 1990s and served on the Muslim side. These CIA-products also participated in the civil war in Azerbaijan and in the struggle against

Russia in Chechnya. Several of the terrorists involved in the 1993 attempt to car bomb New York's World Trade Centre were also Afghan veterans [Johnson, 2002]. No one in US administration could foresee that these very rebels or '*mujahideens*' would one day come on the US soil and would give it a body blow.

An important and fatal aspect related to the period after Soviet withdrawal from Afghanistan is, as mentions Gunaratana, that the Western governments started no weapons collection or buy-back program. The weapons supplied by CIA to its clients in ISI, and then from ISI to Taliban were later passed-on to al-Qaeda, which already had a worldwide network for procurement of weapons. After intense bureaucratic battle within the US government, CIA had supplied to the Muslim rebels between 1986 and 1988 approximately 1,000 of the Stinger missiles of which nearly 340 were expended in anti-Soviet Afghan operation. Either the US government was quiescent to get them back or the Afghan leaders were unwilling to part with these missiles as they needed them for their future projects. 1991 onwards, CIA started a buy-back program. Over next several years, the agency spent more than $ 65 million, but could get back only an estimated 200 missiles. The rest of them were believed to be in the hands of Muslim radicals and might have been used in 2001 against their producers [see: *Washington Post*, September 25, 2001, p. A-15]. The CIA experience in its covert war in Afghanistan should have been a useful guide to the American foreign policy-makers, which if taken seriously, could have avoided the catastrophe of 2001. This, however, was not to be.

Also, the financial network of al-Qaeda flourished on finances provided not only from the Islamic philanthropists and foundations working in developing countries, particularly of Middle-East and Gulf but also from the overseas Muslims in US and West which later surprised even the CIA analysts. In terms of resources, al-Qaeda was a 'state within state' so much so that instead a host state controlling a terrorist organization, the later was controlling its host state. American government did very little to monitor the extremists within US who provided regular funding and personnel for *jihadi* campaigns. US, instead, has remained one of the very important centres of procurement of sophisticated weaponry and other technological devices such as aircraft and satellite phones which later were used by terrorists against US targets at home and abroad. US intelligence failed dismally to correctly and timely assess the dangerous potential of such activities. Its misplaced sense of security that by insulating itself from the rest of the world, American security will be ensured proved disastrous.

Reflecting on the state of ideological and physical penetration of Islamist groups in US, one of the leaders of Islamic Council of America said in 1999, "We can say that Islamists took over eighty percent of the mosques in US. There are more than 3000.... This means that the ideology of extremism has been spread to 80 percent of the Muslim population, mostly the youth and the new generation." [Kabbani: 1999]. However, a little before the 9/11, FBI started infiltrating many Muslim organizations but al-Qaeda was aware of this monitoring and relocated its team for 9/11 attacks away from Muslim strongholds. Use of American aircraft and crashing them into the symbols of American supremacy splits open the shortfalls of American intelligence and excellence of terrorist planning in complete secrecy. A strong intelligence could have avoided not only the success of the novel tactic of using commercial passenger aircraft as improvised guided missiles by terrorists but it could also have tracked down much earlier their training in hijacking and aircraft flying in the American institutes itself.

The 25 years of war in Afghanistan with foreign interference killed over 1.5 million people and devastated the country. The Soviet Union poured some US$ 5 billion a year into Afghanistan to subdue the *mujahideen* and lost. The US committed some 4 to 5 billion dollars between 1980 to 1992 in aid to *mujahideen* and created Taliban and al-Qaeda, the future challengers to US power and status by carrying terrorism to its own soil. Most of this aid was in the form of modern lethal weaponry, distributed by Pakistan's ISI, who tended to treat Kandahar as a backwater. It was given to simple agricultural farmers who used these with devastating results [see: Rashid: 2002].

1991-2001 Creation of Taliban, Osama and al-Qaeda

Between 1979-91, CIA-ISI nexus was working overtime to recruit radical Muslims from around the world to form *mujahideens* who could fight the Soviet forces in Afghanistan. Osama bin Laden was originally brought to Pakistan to help with this effort. Although he was under contract to the CIA, the CIA gave Osama free rein in Afghanistan, as did Pakistan's intelligence generals and bin-Laden used that free rein and his accumulated wealth to begin organizing al-Qaeda in 1985. In the late 1980s, Pakistan's then President Benazir

Bhutto, seeing how strong the *Mujahideen* movement was becoming, told President Bush: "You are creating a Frankenstein" [*Newsweek*, October 1, 2001]. Highlighting the close relationship between CIA and its counterpart in Pakistan, ISI, Chossudovsky argues that just as without ISI there could have been no Taliban government in Kabul, without unbending support of US there couldn't have been a powerful military-Intelligence apparatus in Pakistan [see: Chossudovsky: 2002]. United States pressurized Sudan, where Osama had shifted after Soviet withdrawal from Afghanistan, to expel him after his involvement in attempts to kill Egyptian president, Hosni Mubarak in Ethiopia in June 1995 and when Sudan seemed unwilling to yield to international pressure, US stepped up military assistance to its hostile neighbors Uganda, Eritrea and Ethiopia with intentions of containing Sudan. Under these unrelenting pressures, Osama relocated his group to Afghanistan in May 1996 with his main office in Peshawar in Pakistan and by this time a new force called Taliban (the students or the seekers of knowledge) had not only risen, it had seized control of two-thirds of Afghanistan and was aided and abetted by Pakistan's Inter service Intelligence (ISI).

The CIA ended its aid in 1992, the Russians sometime later, and the pro-Russian government in Kabul fell. In the final stages of that struggle the Taliban began to emerge as a major force in Afghan politics and it subsequently drove the Northern Alliance from Kabul, confining the remnants of the original rebel alliance to a small enclave in the north-eastern part of the country. The fundamentalist leader Osama bin-Laden, though getting his start in the CIA-funded war of the 1970s and 80s, did not become a prominent fugitive in Afghanistan until he returned to the country as the Taliban's guest in 1996. As far as the role of US in creation of Taliban is concerned, Ahmed Rashid in his much renowned book *Taliban* [Rashid: 2000] draws attention towards US interest in oil and natural gas in the neighboring Persian Gulf region and therefore considered Taliban to be a force to be reckoned with as it would provide sufficient stability for America's oil project that planned to use Afghanistan as a potential route for oil and natural gas exports from Central Asia to the Arabian Sea. Hence, the pipeline project was central to this support. Impressed by the ruthlessness and willingness of the then-emerging Taliban to cut a pipeline deal, the State Department and Pakistan's ISI agreed to funnel arms and funding to the Taliban. When the Taliban, with this financial support from Saudi Arabia and the CIA funneled through the ISI, conquered Kabul in 1996, Unocal was hopeful that it would provide enough stability to allow its pipelines to be built and protected. Indeed, it was reported, "preliminary agreement [on the pipeline project] was reached between the [Taliban and Unocal] long before the fall of Kabul" [*Telegraph*: October 11, 1996]. Unocal even reportedly provided some of the financial support for the Taliban [Stobdan P: "The Afghan Conflict and Regional Security", *Strategic Analysis*, 719:47]. The fact that the Taliban continued to serve the purposes of the ISI is illustrated by the fact that when Taliban troops were about to conquer the major city in northern Afghanistan in 1998, an ISI officer sent a message saying: "My boys and I are riding into Mazar-i-Sharif". In any case, after the Taliban conquered this city, it had control of most of Afghanistan, including the entire pipeline route. CentGas then announced that it was "ready to proceed" [*Telegraph:* August 13, 1998]. Later that year, however, Unocal, having become dubious about the Talibans ability to provide sufficient stability, pulled out of CentGas. From then on, says Rashid, the US grew progressively more hostile toward the Taliban, and began exploring other possibilities to secure its regional supremacy, while maintaining basic ties with the regime, to negotiate a non-military solution, which reportedly occurred at a four-day meeting in Berlin in July of 2001. The Bush administration tried to get the Taliban to share power, thereby creating a joint government of "national unity." Taliban were warned by US representative either to behave as they ought to...or we will use another option...a military operation" [Griffin: New Pearl Harbor: 2004]. According to Niaz Naik, the Pakistani representative at Berlin meeting in July 2001, military action against Afghanistan was planned by mid-October, 2001, but it came one month earlier. Those believing in conspiracy theory, such as Niaz Naik, would argue that even if bin-Laden was to be surrendered by Taliban, Washington wouldn't have dropped its attack on Afghanistan plan due to wider 'oil' interests and go as far as saying that bin-Laden was an American agent [This statement from the Israeli newspaper *Ma'ariv* was quoted in the *Chicago Tribune,* February 18, 2002, which is in turn quoted in "Timeline," February 14, 2002.]

Sajit Gandhi, a research associate at the National Security Archives' (George Washington University) in the *'September 11th Sourcebook Series'* 7th Volume (declassified and released on 11th September, 2003) says: "The Taliban Files" details the rise of the Taliban from its meager start in Kandahar to a full-fledged military force and ultimate control of the country. The documents discuss Pakistan's support for the Taliban, US dealings with the Taliban, post-9/11 thinking on military strategy in the War on Terror, and the relationship between the assassination of the Northern Alliance Commander Ahmad Shah Masoud and the terrorist attacks of September 11. Formed in 1994, the Taliban began with only a few followers, mostly religious students who fought with the *Mujahideen* in the war against the Soviets and who were schooled in Islamic seminaries (*madrasas*) in Pakistan. These students, or seekers, as they are referred to in the documents, wanted to rid Afghanistan of the instability, violence, and warlordism that had been plaguing the country since the defeat and withdrawal of the Soviets in 1989.

The departure of the Soviets, while welcomed by Afghans and the United States, left a political vacuum in Afghanistan. The resulting chaos and civil war led to the involvement of the United Nations, which tried unsuccessfully to bring about political transition through the mission led by Special Representative Mahmoud Mestiri. Despite the UN's efforts, and those of the international community, the various factions, as well as the Kabul government led by Barnahuddin Rabbani and Ahmad Shah Masoud, in addition to other outside parties, made a definitive peaceful or military solution difficult. As a result, the civil war continued with Rabbani and Masoud attempting to fill the government role, while the other warlord remnants of the Afghan resistance, such as the Uzbek commander Abdul Rashid Dostum, Pakistani-backed Gulbuddin Hekmatyar, and Ismail Khan, remained unwilling to cede any power or make concessions that could have resulted in a peaceful solution. Consequently, outside forces saw instability in Afghanistan as an opportunity to press their own security and political agendas.

However, the American version of the rise of Taliban and the role of Osama and his al-Qaeda in 11 September, 2001 attacks on US is somewhat different, putting America into a positive image. The '*September 11th Sourcebook Series*' prepared by National Security Archives (7th, 16th, 17th and 18th documents) show that the US made tremendous efforts to obtain a political solution for Afghanistan, not just because of the desire for American companies to take advantage of business opportunities with the Trans-Afghan gas Pipeline (TAP), but also due to other key concerns: human rights, narcotics, and terrorism. In many instances, American officials pressed the Taliban on their counternarcotics strategy, their treatment of women, and on allowing Afghanistan to be used as a base for terrorist operations and home for Osama bin-Laden. The cable traffic shows the difficulty the US had negotiating with Taliban representatives in all these areas. Cultural and political miscommunication was rampant. It soon became clear that Taliban rule was detrimental to Afghan and international security, as evidenced by their sanctioning of continuing narcotics production—despite its un-Islamic quality—and shelter for al-Qaeda and other terrorists. Acting Secretary Strobe Talbott described the danger of the Taliban in a February 1996 meeting with Pakistani Foreign Minister Assef Ali when he drew an analogy between Pakistani support for the Taliban in Afghanistan and the militants in Indian-controlled Kashmir. Talbott stated that while such support was undertaken to serve Pakistani interests, there were unintended consequences contrary to Pakistan's and the region's larger interests. These consequences became shockingly clear in coming years, on September 11, 2001.

These *September 11 Sourcebook Series* blame Pakistan more than US in the creation of Taliban. Pakistan saw an unstable Afghanistan as a boon for its internal security, allowing it a strategic depth against India. Initially, [See document 25] the Pakistanis supported the Pashtun-Islamicist Gulbuddin Hekmatyar, an incompetent commander from the Mujahideen days, in order to have influence over the Afghan political landscape. When Hekmatyar failed to deliver for Pakistan, the administration began to support a new movement of religious students known as the Taliban. The first document dates from November 1994, one month after the Taliban took the strategic post of Spin Boldak on the Afghan-Pakistan border, allegedly with cover fire provided by Pakistani Frontier Corps [see *Document 5 of September 11th Sourcebook Series*]. With that victory, the Taliban, who were being championed by a fellow Pashtun, Pakistani Interior Minister, began to make a name for themselves, and also gained a significant amount of military supplies. Pakistan supported the Taliban, not just to restore order to Afghan roads, which would open the way for a possible Trans-Afghan gas pipeline (TAP), but because they also saw the Taliban as a faction that they might have considerable influence over, and who might provide in Afghanistan, a strategic lever for Pakistan against India. But only quickly, the Pakistani authorities discovered they had made a blunder. The Taliban were not only uncontrollable, but unpredictable as well. In certain instances the Taliban would declare their desire for peace, willingness to work with the UN, and desire for a non-military solution for Afghanistan, then state that "anyone who gets in our way will be crushed."

Though Afghanistan itself has no known oil or gas reserves, but it is an attractive route for pipelines leading to Pakistan, India, and the Arabian Sea. In the mid-1990s, a consortium led by the California-based Unocal Corporation proposed a $4.5 billion oil and gas pipeline from Turkmenistan through Afghanistan to Pakistan. But this would require a stable central government in Afghanistan itself. Thus, began several years in which US policy in the region centered on "romancing the Taliban." In the post-1991 period, the US oil companies and their friends in the State Department were thinking at the prospect of gaining access to the huge oil and natural gas reserves in the former Soviet republics bordering the Caspian Sea and in Central Asia. These have been estimated as worth $4 trillion. The American Petroleum Institute calls the Caspian region "the area of greatest resource potential outside of the Middle-East." And while he was still CEO of Halliburton, the world's biggest oil services company, Vice-President during Bush presidency, Dick Cheney, told other industry executives, "I can't think of a time when we've had a region emerge as suddenly to become as strategically significant as the Caspian." The struggle to control these stupendous resources has given rise to what Rashid has dubbed the "new Great Game," pitting shifting alliances of governments and oil and gas consortia against one another. So, the reference to oil and gas pipelines explains everything related to US interest in the region and creation of Taliban as a potential supporter, which in the end backfired. The region also suited a major US strategic objective: isolating its nemesis: Iran

Thus, interest in oil in Persian Gulf region and other political considerations compelled continuation of tacit US support for the Taliban till 1998, when Washington blamed Osama bin-Laden for the bombing of the US embassies in Kenya and Tanzania and retaliated by launching cruise missiles at bin Laden's alleged training camps in Afghanistan. The Taliban's refusals to extradite bin-Laden—despite its atrocious human rights record—led to UN-imposed sanctions on the regime the following year. Whatever the US government's post-9/11 rhetoric about the repressive nature of the Taliban regime, its long history of intervention in the region has been motivated not by concern for democracy or human rights, but by the narrow economic and political interests of the US ruling class.

It seemed to be prepared to aid and support the most retrograde elements if it thought a temporary advantage would be the result. After 9/11 shocks, Washington launched a war against its former allies based on a strategic calculation that the Taliban can no longer be relied upon to provide a stable, US-friendly government that can serve its strategic interests.

September 11, 2001 Attacks on US and War on Terror: 2001-onwards

It was the period when Afghanistan was effectively ungoverned for seven years and was plagued by constant infighting between former allies and various *mujahideen* groups. Throughout the 1990s, Taliban emerged and grew unabatedly and brought in its loop the children of Afghanistan, many of them orphaned by the war, and many of whom had been educated in the rapidly expanding network of Islamic schools, *madarssas*, either in Kandahar or in the refugee camps on the Afghan-Pakistani border. Bin-Laden and his al-Qaeda enjoyed the Taliban's protection and a measure of legitimacy as part of their Ministry of Defense, although only Pakistan, Saudi Arabia and the United Arab Emirates recognized the Taliban as the legitimate government of Afghanistan.

However, there is another set of theory that denies any direct role of US in the creation of either al-Qaeda or Osama. For instance, in the BBC documentary *The Power of Nightmares*, writer and journalist Adam Curtis contends that the idea of al-Qaeda as a formal organization is primarily an American invention. Curtis contends the name "al-Qaeda" was first brought to the attention of the public in the 2001 trial of Osama bin-Laden and the four men accused of the 1998 United States embassy bombings in East Africa. The reality, according to Curtis, was that bin Laden and Ayman Zawahiri had become the focus of a loose association of disillusioned Islamist militants who were attracted by the new strategy. But there was no organization. These were militants who mostly planned their own operations and looked to bin Laden for funding and assistance. He was not their commander. There is also no evidence that bin Laden used the term "al-Qaeda" to refer to the name of a group until after September the 11th, when he realized that this was the term the Americans had given it [*The Power of Nightmares*, a BBC Documentary by Adam Curtis]. Similarly, CNN journalist Peter Bergen, known for conducting the first television interview with

Osama bin-Laden in 1997, calls the idea *"that the CIA funded bin Laden or trained bin Laden ... a folk myth. There's no evidence of this. ... Bin Laden had his own money, he was anti-American and he was operating secretly and independently. ... The real story here is the CIA didn't really have a clue about who this guy was until 1996 when they set-up a unit to really start tracking him.* [Bergen, Peter: 2006]. Bergen and others maintain the US aid was given out to the Pakistan government and then it went to Afghan, not foreign *mujahideen,* and that there was no contact between the Afghan Arabs (foreign *mujahideen*) and the CIA or other American officials, let alone, arming, training, coaching or indoctrination.

Whatever be the truth regarding CIA connections in creating Al-Qaeda, it is a fact that anti-American sentiments were brewing up in Islamist circles and American people and symbols were being targeted since the first World Trade Centre attacks in 1993, in which the justification for the attack was mentioned as US support for Israeli occupation of Palestinian territories. The chain of events since then, including the announcement of a *jihad* to expel foreign troops from Islamic lands in 1996, issuing of a *fatwa* (a binding religious edict) in 1998, which amounted to a public declaration of war against the United States and any of its allies, the bombing of the Khobar towers in Saudi Arabia on June 25th, 1996, the US Embassy bombings in East Africa in 1998 and the missile attack on USS Cole in 2000 ultimately led to the catastrophe of September 11, 2001. Inspired by the success of such brazen attacks, al-Qaeda's command core began to prepare for an attack on the United States itself.

The September 11 attacks were not only the most devastating terrorist acts in American history, killing nearly 3,000 people, destroying four commercial airliners, leveling the World Trade Centre towers, and damaging the Pentagon, the headquarters of American Department of defense, they were the first of its kind on American soil and a direct challenge to American economic and military prowess and exposing the vulnerability of the most powerful nation in the world. These attacks were conducted by al-Qaeda, acting in accord with the 1998 *fatwa,* issued against the United States and its allies by military forces under the command of bin Laden, al-Zawahiri, and their other associates.

National Security Strategy, Bush Doctrine

Here, a critical analysis and examination of the US policies is very much in order to find out if these policies were responsible for inciting Muslim hatred against US or was it the *vice versa* and to ascertain whether US policies thenceforth have gone in favor of world peace, democracy and human rights promotion, as United States widely claims, or whether its unilateralism has gone to complicate the situation further, leading to what some scholars, including Samuel P. Huntington, term as the beginning of a 'clash of civilizations', or whether it was a conspiracy of neoconservatives in President George W. Bush's administration to use the opportunity to America's economic and political advantage.

The heinous act of terrorism on September 11, 2001 shocked the entire world, including a large number of the followers of Islam. This act led to 'War on Terrorism', a redefinition of American foreign policy and its National Security Strategy, which jointly is referred to as *'Bush Doctrine'*, which claimed to combine the principles of 'just war' with a power-based national security strategy. Declaring these terrorist acts as the acts of war, the new *National Security Strategy of the United States* was devised which stated: "The United States of America is fighting a war against terrorists of global reach. The enemy is not a single political regime or person or religion or ideology. The enemy is terrorism—premeditated, politically motivated violence perpetrated against innocents." [*The National Security Strategy of the United States:* 2002]. It contains a strategy what neoconservatives have been insisting upon since the end of Cold War through their many documents and pronouncements and contained the principles such as: **preemption, regime change, unilateralism and benevolent hegemony**. After the tragedy of 9/11, it got implemented in the form of Bush Doctrine. As such, a new 'Department for Homeland Security' was created, the Congress passed the **'Patriot Act'** and a decision was taken to invade Afghanistan to depose Taliban regime that was sheltering al-Qaeda and its mastermind—Osama bin-Laden. Initially, such a decision had wide public support.

However, this point is contested by a number of scholars who believe that there cannot be a war against terrorism as it is merely a tactic, a methodology and not a person, institution or state against whom the war can be waged. In fact, the critics argue, the US is at war with all those individuals, groups and states who oppose its hegemony and unrivaled global leadership. Hence, WOT also has political origins. In this context, George W. Bush's remarks *at 2002 Graduation Exercise of the United States Military Academy, West Point,*

New York are important. [http://www.whitehouse.gov/news/releases/2002/06/20020601-3.html.]. In it, the war on terrorism was conceived as:

(1) an imminent threat, linked to the threat of the proliferation and potential use of weapons of mass destruction, perhaps raising the threat to the status of an emergency, wherein the very existence of the society is at risk, and
(2) in terms of a Manichean theory, it is a struggle between good and evil.

In short, Bush-Doctrine is consisted of:

- The Foreign Policy principles for United States pronounced by then President George W. Bush in the wake of the September 11, 2001 attacks on US;
- This policy gave United States the right to treat countries that harbor or give aid to terrorist groups as terrorists themselves;
- As such, it was used to justify the invasion of Afghanistan in 2001;
- Later, the Bush Doctrine came to include additional elements, including a policy of preemption, which held that the United States should depose foreign regimes that represented a threat to the security of the United States, even if that threat was not immediate—a justification for the invasion of Iraq in 2003; and
- The Doctrine also included a policy of supporting democracy around the world, especially in the Middle-East, as a strategy for combating the spread of terrorism, and a willingness to pursue US military interests in a universal way.

Some of these policies were codified in a *National Security Strategy of the United States* published on September 20, 2002.

As the September 11, 2001 attacks were thought to have been planned and executed by Osama bin-Laden and other members of al-Qaeda. Bush decided soon afterward that the proper response was not just military attacks against al-Qaeda bases, but deposing the Taliban altogether and installing in their place a US-friendly democratic regime. This presented a foreign-policy challenge, since it was not the Taliban that had initiated the attacks, and there was no evidence that they had any foreknowledge of the attacks. In an address to the nation on the evening of September 11, Bush stated his resolution of the issue by declaring that "we will make no distinction between the terrorists who committed these acts and those who harbor them." [President Bush's address to the nation, September 11, 2001]

Later, **two distinct schools of thought** arose in the Bush Administration regarding the critical policy question of how to handle dangerous countries such as Iran, Iraq and North Korea, which Bush called 'Axis of Evil' states. One School consisted of the then Secretary of State Colin Powell, who later resigned after US invasion of Iraq, and National Security Advisor Condoleezza Rice, as well as US Department of State specialists. It argued in favor of what was essentially the continuation of existing US foreign policy that was developed after the Cold War. It sought to establish a ***multilateral consensus*** for action, which would likely take the form of increasingly harsh sanctions against the problem states, summarized as the policy of containment. The other School, holding an opposing view, consisted of Vice-President Dick Cheney, former Secretary of Defense Donald Rumsfeld and a number of other influential Department of Defense policy makers such as Paul Wolfowitz and Richard Perle. They held that ***direct and unilateral action*** was both possible and justified and that America should embrace the opportunities for democracy and security offered by its position as sole remaining superpower. President Bush ultimately sided with the Department of Defense camp, and their recommendations form the basis for the Bush Doctrine. This camp is generally called the group of 'neoconservatives' or 'neocons', as they are popularly called. This group also consists of some very influential opinion-builder magazines and newspapers in media, such as, *Wall Street Journal, National Interest* and *Washington Post.*

The foreign policy pursued by US under President George W. Bush was largely based on the doctrine of **preemptive action**—a doctrine of preemptive war—that believed in taking the fight to the enemy and on **unilateralism**—meaning that if one is confident of its power, "it simply does not allow itself to be held hostage to the will of others" [Krauthammer: 2004]. Supporters of Bush Doctrine believed that in the bipolar world of Cold War, deterrence could work. But in a world of terrorists, terrorist states and WMD, the option of preemption is necessary, as the adversary working clandestinely is neither deterrable nor detectable—making preemption the only possible strategy. Rather, the doctrine of preemption, they said, was an improvement over the doctrine of deterrence or it is a road to deterrence itself as the examples of preemption would deter the potential acquirers or aspirants of WMD to abandon their dangerous ambitions. The argument was that a unipolar US could not wait for others to be offensive first and then would force America to be on defensive—as was the case after 9/11. America with all its political, economic and military might cannot allow the repetition of such a humiliating situation. Multilateralism is nice if possible; but it cannot be imposed on a great power, a unilateral power such as United States of today, to restrain it by a coalition of comparatively weak countries and render it inactive in the name of some higher, moral principles of internationalism and peaceful co-existence.

Proponents of this neocon ideology argued that in a world where international institutions are too weak to act, where there is no one to look to with trust and confidence of getting the justice done, the 'power' becomes the ultimate arbiter of what is just and right. And only if we agree with what the great theorist of international relations, Hans Morganthau, said about the necessity of preservation and extension of power for the sake of national interest, the neoconservative foreign policy under President Bush would stand some justification, at least in the name of 'national interest' if not for the sake of power itself. However, Bush's very often-expressed concern for promotion of democracy, freedoms, human rights and peace seemed not to be in harmony with what he actually did, particularly in Iraq and might raise doubts about his administration's interpretation of national interest as based on values. On its face value, this dichotomy between realpolitik and professed higher moral values made and still makes American foreign policy highly controversial and hence debatable.

This perception led the Bush administration to execute the strategic doctrine of preemptive action, central to the conduct of the war on terrorism. Given the nature of the terrorist threat, which is hidden and unpredictable, the ultimate and perhaps the only real means of security would be to eliminate the threat at its source prior to its execution. It is not possible to wait for the terrorists to strike in order to defend oneself, is the logic; United States must act first to preempt the threat. Preemption is given particular legitimacy when the linkage between terrorism and the proliferation of weapons of mass destruction is recognized—the Bush administration maintains under these conditions it would be irresponsible, a gross dereliction of duty, to wait for the enemy to attack before responding. The idea is that the threat must be eliminated before it is manifested. Though there is nothing objectionable in Bush's call to act before it is too late, what the critics look with suspicion are America's unilateralist and imperialist tendencies as the National Security Strategy (NSS) unambiguously mentions that while United States will always strive to enlist the support of international community, it will not hesitate to act alone if required, to exercise its right of self-defense by acting preemptively against the terrorists who would want to harm the US and its people anywhere in the world. Implicit in this strategy is also a radical doctrine of linkage between the terrorists who committed those acts and those who harbor them. The Doctrine clearly violated the principle of non-intervention in the affairs of other sovereign countries that is the basis of the nation-state system in the theory of international relations.

The policy of 'preemption' contained in Bush Doctrine indicated a major shift in US foreign policy: from 'containment' of Cold War period to 'preemption' in international terrorism era; from non-interventionist and isolationist foreign policy tradition to the quest for a new global strategy for global leadership. A large number of American foreign policy analysts saw clear imperialist tendencies in these set of doctrines that were pronounced under the name of 'War on Terrorism' [West : 2004].

Critics of the Bush Doctrine suspected the increasing willingness of the US to use military force unilaterally. Some published criticisms, such as those by Storer H. Rowley's June 2002 article in the *Chicago Tribune*, Anup Shah's in *Globalissues.org* and Nat Parry's April 2004 article in *ConsortiumNews.com*. These critical analyses and arguments reflect a turn away from international law, and mark the end of American legitimacy in foreign affairs [see: Tucker, Robert W. and David C. Hendrickson, "The Sources of American Legitimacy," *Foreign Affairs*, November/December 2004, pp. 18-32]. This doctrine is argued to be contrary to the 'Just War' theory and would constitute a war of aggression. Patrick J. Buchanan writes that the 2003 invasion of Iraq has significant similarities to the 1996 neoconservative policy paper: *A Clean Break: A New Strategy for Securing The Realm* [Buchanan, 2003].

Bush-Doctrine Implemented: Afghanistan, Iraq

The evidence collected after the September 11 attacks raised the suspicion towards Osama bin-Laden and his militant outfit, Al-Qaeda. Messages issued by bin-Laden after September 11, 2001 praised the attacks, and explained their motivation while denying any involvement. Bin-Laden legitimized the attacks by identifying grievances felt by both mainstream and Islamist Muslims, such as the general perception that the United States was actively oppressing Muslims. He asserted that America was massacring Muslims in 'Palestine, Chechnya, Kashmir and Iraq' and that Muslims should retain the 'right to attack in reprisal'. He also claimed the 9/11 attacks were not targeted at women and children, but 'America's icons of military and economic power'. In the immediate aftermath of the attacks, the United States government decided to respond militarily, and began to prepare its armed forces to overthrow the Taliban regime it believed was harboring al-Qaeda. Before the United States attacked, it offered Taliban leader Mullah Omar a chance to surrender bin-Laden and his top associates. The Taliban offered to turn over bin-Laden to a neutral country for trial if the United States would provide evidence of bins Laden's complicity in the attacks. US President George Bush responded by declaring his 'War on Terror' (WOT) and called upon the countries of the world to join WOT. Soon thereafter the United States and its allies invaded Afghanistan, and together with the Afghan Northern Alliance removed the Taliban government in the war in Afghanistan.

As a result of the United States using its special forces and providing air support for the Northern Alliance ground forces, both Taliban and al-Qaeda training camps were destroyed, and much of the operating structure of al-Qaeda disrupted. After being driven from their key positions in the Tora Bora area of Afghanistan, many al-Qaeda fighters tried to regroup in the rugged Gardez region of the nation. Again, under the cover of intense aerial bombardment, US infantry and local Afghan forces attacked, shattering the al-Qaeda position and killing or capturing many of the militants. By early 2002, al-Qaeda had been dealt a serious blow to its operational capacity, and the Afghan invasion appeared an initial success. Nevertheless, a significant Taliban insurgency remained in Afghanistan, and al-Qaeda's top two leaders, bin-Laden and al-Zawahiri, kept evading capture.

Here, a pertinent question arises: on what basis US pinpointed Osama bin-Laden and al-Qaeda as the forces behind 9/11? To this, US claim was that various sources confirm to the role of al-Qaeda in 9/11 catastrophe, such as US State Department's release of a videotape showing bin-Laden speaking to a small group of associates somewhere in Afghanistan shortly before the Taliban was removed from power, the 9/11 Commission's report investigating the September 11 attacks, which officially concluded that the attacks were conceived and implemented by al-Qaeda operatives and also the October 2004 videotape released through Al Jazeera in which bin-Laden appeared to claim responsibility for the attacks [see: 9/11 Commission Report: 2004].

Before opening hostilities in Afghanistan where the perpetrators of 9/11 catastrophes were believed to be hiding in safe havens, America persuaded many countries to break-off diplomatic and commercial relations with Taliban regime in Kabul. Pakistan, driven by considerations of *realpolitik* offered all help even at the cost of antagonizing Islamists within the country It was construed as joining forces against a close and friendly neighbor Afghanistan. As the third largest recipient of US aid, including counter-terrorism subsidies Pakistan has been critically dependent on such assistance and Washington fully played to that vulnerability. US needed Islamabad to employ Pakistan as a gateway to military operations in Afghanistan, a base for clandestine missions in Iran and a vehicle for other geopolitical objectives, including the one *vis-a-vis* India. Despite making democracy and freedom a rallying cry, Bush's democracy plank looked to merely target regimes that defy US. In the case of friendly but dictatorial regimes, like in Pakistan and most Arab nations, US foreign policy recognizes that such governments can further American interests only if they stay insulated from the popular pressures of a democracy. By accepting Pakistan's offer of alliance in WOT, Bush gave primacy to the narrow, short-term geopolitical interests and continued to rely on the very institution that is and has always been the part of the problem—Pakistan's military. Politically seen, such a strategy looks attractive because the US is distant and fallout of Pakistan's maneuvering and of US policies would remain largely confined to the region. In fact, Pakistani military rulers have always been politically expedient to advance American interests and hence America's neglect of India's concerns suffering world's highest rate of terrorism [Chellaney: 2007].

During the Cold War, and even until 9/11, the United States tolerated, applauded, or overlooked Pakistan's association with *jihadi* groups. In the context of Kashmir, Washington was as likely to criticize India for the heavy-handedness of its security forces as to condemn Pakistan's training and financing of "freedom fighters." In Afghanistan, the United States and Pakistan were partners in supporting the *mujahideen's* anti-Soviet struggle. And in the 1990s, nuclear proliferation concerns distracted Washington's attention from the counter-terrorism agenda. But after 9/11, the diplomatic costs of Pakistan's *jihadi* strategy started to mount. Overnight, terrorism became the White House's top priority, and Islamabad's semantic distinction between "freedom fighters" and 'terrorists' no longer held water. Overt official ties with Afghanistan's Taliban were the first casualty of the new "with us or against us" era.

Invasion of Iraq: The 'Operation Iraqi Freedom' or the 'Second Gulf War', popularly called the 'Iraq War' started in March, 2003 with the US-led invasion of Iraq, with the co-operation of a coalition of the troops of United States and United Kingdom mainly, along with smaller contingents from Australia, Poland and some other nations. The rationale for invasion, offered by President George W. Bush, and willingly or reluctantly supported by the coalition partners was the allegation, largely unsubstantiated, that Iraq possessed and was actively and aggressively trying to possess the Weapons of Mass Destruction (WMD) in violation of a 1991 agreement. The argument of US officials was that Iraq posed an imminent, urgent and immediate threat to United States, its people, allies and interests. The supporting intelligence, which later proved to be faulty as no evidence of Iraq possessing WMD was found, was widely criticized. Even the Iraq Survey Group concluded in 2005, after the invasion of Iraq, that Iraq had ended its WMD programs in 1991 and had none at the time of the invasion, but that they intended to resume production if and when the sanctions on Iraq were lifted. Although some earlier degraded remnants of misplaced or abandoned WMD were found, they were not the weapons for which the coalition invaded [Wikipedia search].

Another US suspicion that Saddam Hussein and al-Qaeda were co-operating could never be substantiated. Other reasons for the invasion stated by US officials included concerns about Iraq's financial support for the families of Palestinian suicide bombers, human rights abuses by Iraqi government, lack of democracy and Iraq's oil reserves. This last factor seems to be more realistic than most of the others for Iraq's invasion [see: *Washington Post*, September 15, 2002; *Guardian Unlimited*, June 4, 2003; *Stanford Daily*, October 15, 2007].

The invasion led to the quick defeat of the Iraqi army and flight of President Saddam Hussein, his capture in December 2003, and his execution in December 2006. The US-led coalition occupied Iraq and attempted to establish a new democratic government. But shortly after the initial invasion, violence against coalition forces and among various sectarian groups led to asymmetric warfare with the Iraqi insurgency, civil war between *shia* and *Sunni* Iraqis, and al-Qaeda operatives in Iraq. As the insurgency grew, there was a distinct change in targeting from the coalition forces towards the new Iraqi Security Forces, as hundreds of Iraqi civilians and police were killed over the next few months in a series of massive bombings. An organized Sunni insurgency, with deep roots and both nationalist and Islamist motivations, was becoming more powerful throughout Iraq. The Shia Mahdi Army, the military muscle of Iraq's urban Shia Muslims and perceived as a big threat to Iraq's security, also began launching attacks on coalition targets in an attempt to seize control from Iraqi security forces. The southern and central portions of Iraq were beginning to erupt in urban guerrilla combat as multinational forces attempted to keep control and prepared for a counteroffensive. In January 2005, Iraqis elected the Iraqi Transitional Government in order to draft a permanent constitution. Although some violence and widespread Sunni boycott marred the event, most of the eligible Kurd and Shia populace participated. Despite elections, the sectarian violence kept erupting every now and then resulting in heavy casualties of US soldiers, Iraqi civilians and security forces. The US military could do nothing much to improve the political and social situation there. Due to this grim scenario, Iraq was listed fourth on the 2006 Failed States Index compiled by the American *Foreign Policy* magazine and the *Fund for Peace* think-tank. Estimates of the number of people killed range from over 150,000 to more than 1.2 million [Burnham, Gilbert: 2006]. The financial costs of war were more than $474 billion to the US [see: National Priorities Project: Cost of War] and over £4.5 billion to the UK. Coalition nations began to withdraw troops as public opinion favoring troop withdrawal increased and Iraqi forces began to take responsibility for security. Yet, the war in Iraq remains one of the most controversial in modern history.

In fact, the US policy had begun to **shift away from containment to "regime change,"** since October 1998 as the US Congress passed and President Clinton signed the "Iraq Liberation Act." Signed in response to Iraq's termination of its cooperation with U.N. weapons inspectors the proceeding August, the act provided $97 million for Iraqi "democratic opposition organizations" to "establish a program to support a transition to democracy in Iraq" [See: *Iraq Liberation Act,* 1998: Library of Congress]. This legislation contrasted with the terms set out in United Nations Security Council Resolution 687, which focused on weapons and weapons programs and made no mention of regime change. One month after the passage of the "Iraq Liberation Act," the US and UK launched a bombardment campaign of Iraq called 'Operation Desert Fox'. The campaign's express rationale was to hamper the Saddam Hussein government's ability to produce chemical, biological, and nuclear weapons, but US national security personnel also hoped it would help weaken his grip on power.

There are theories supporting the thesis that Iraq invasion and a 'regime change' was on Bush's agenda much before his ascendancy to the US presidency [Mackay, Neil: 2002]. Part of the evidence for this claim is found in the document *Rebuilding America's Defenses: Strategy, Forces, and Resources for a New Century,* published in September 2000 by the Project for the New American Century (PNAC), a neo-conservative think tank that was formed by many people who went on to become insiders in the Bush administration, including Dick Cheney, Donald Rumsfeld, Paul Wolfowitz (Rumsfeld's deputy at the Defense Department), and Lewis "Scooter" Libby (Cheney's Chief of Staff). Some relevant passages of this document substantiate the thesis that the 2003 war against Iraq was really motivated by the perceived need to eliminate Saddam.

With the election of George W. Bush as US President in 2000, the US moved towards a more active policy of "regime change" in Iraq. The Republican Party's campaign platform in the 2000 elections called for "full implementation" of the Iraq Liberation Act and removal of Saddam Hussein by invading Iraq. The former Bush treasury secretary Paul O'Neill said that an attack on Iraq had been planned since the inauguration, and that the first National Security Council meeting involved discussion of an invasion, though he later backtracked [O'Neill: "Frenzy distorted war plans account", CNN.com, January 14, 2004]. Due to these undercurrent plans going on much before 9/11 attacks, the rationale for invading Iraq as a response to 9/11 has been widely questioned, as no direct cooperation between Iraq and al-Qaeda was known prior to 9/11 and subsequent intelligence has uncovered none [Smith, Jeffrey R. "Hussein's Pre-war Ties to al-Qaeda Discounted", *The Washington Post,* Friday, April 6, 2007; p. A01].

Shortly after September 11, 2001 some Bush advisors were insisting on an immediate invasion of Iraq, while others advocated building an international coalition and obtaining United Nations authorization. Bush eventually decided to seek U.N. authorization, while still holding out the possibility of invading unilaterally. Though the US Congress passed a 'Joint Resolution to Authorize the Use of United States' Armed Forces Against Iraq' in October 2002, which authorized the President to "use any means necessary" against Iraq, Americans polled in January 2003 widely favored further diplomacy over an invasion. Even some NATO allies of US, such as, Canada, France and Germany, together with Russia, strongly urged continued diplomacy instead of an invasion in February 2003 at United Nations. In face of such an opposition to invasion on Iraq, US feared a losing vote and a likely veto from France and Russia and therefore eventually withdrew its resolution asking for UN authorization for use of force against Iraq.

However, the US and UK abandoned the Security Council procedures and decided to pursue the invasion without U.N. authorization, a decision of questionable legality under international law. This decision was widely criticized worldwide, and opposition to the invasion coalesced on February 15 in a worldwide anti-war protest that attracted between six and ten million people in more than 800 cities, the largest such protest in human history according to the Guinness Book of Records [2004]. The **rationale for invasion** even without world body's authorization and much opposition at home and abroad was: (a) that Iraq possessed Weapons of Mass Destruction, including nuclear weapons of which it had to disarm; (b) regime change as Bush and Blair believed that the regime of Saddam was a very brutal and repressive regime; and (c) Saddam Hussein had ties to terrorists, specifically to al-Qaeda. Thus, removing Saddam Hussein from power in order to restore international peace and security was a major goal. Some commentators, on the contrary, have alleged that, while never making an explicit connection between Iraq and the September 11th attacks, the Bush Administration did repeatedly insinuate a connection, thereby creating a false impression on the American public.

Most of the mainstream Newspapers and magazines carried articles and write-ups, such as *Washington Post, New York Times, Boston Globe, Christian Science Monitor* and opinion polls represented the view that the Bush administration has succeeded in creating a sense that there is some connection between September 11 and Saddam Hussein, that he was "personally involved" in the September 11 atrocities though the sources knowledgeable about US intelligence say there is no evidence that Hussein played a role in the September 11 attacks, nor that he has been or is currently aiding al-Qaeda. The White House, however, was encouraging this false impression, as it was trying to get domestic support for a possible war against Iraq and demonstrate seriousness of purpose to Hussein's regime. For instance, the *BBC report* quotes Colin Powell, the then Secretary of State, in February 2003, stating that "We've learned that Iraq has trained al-Qaeda members in bomb-making and poisons and deadly gases. And we know that after 11 September, Saddam Hussein's regime gleefully celebrated the terrorist attacks on America." [*Transcript of US Secretary of State Colin Powell's Presentation to the U.N. Security Council* on "the US Case Against Iraq", cnn.com February 6, 2003]. The fact however is that the neoconservatives in Bush administration and his advisors were continuing to intentionally mislead the American public by drawing a link between Saddam Hussein and 9/11 in an attempt to make the invasion of Iraq part of the global war on terror.

Since the invasion, US and British claims concerning Iraqi weapons program and links to terrorist organizations have been discredited. While the debate of whether Iraq intended to develop chemical, biological, and nuclear weapons in the future remains open, no WMDs have been found in Iraq since the invasion despite comprehensive inspections. Similarly, the intelligence community has largely discredited assertions of significant operational links between Iraq and al-Qaeda, and Secretary Powell himself eventually admitted he had no incontrovertible proof [*New York Times*, January 9, 2004].

Basic Elements of Post-9/11 US Policies in Iraq

Preemption: As is argued by Kaplan and Kristol that "9/11 might not have changed the views of some of the President's diplomatic counselors. It may not even have changed the threat posed by Saddam Hussein. But it did change the President and the direction of his foreign policy". [Kaplan and Kristol: 2003: p. 71]. The catastrophic event shaped the entirely new concept of national security—intercepting threats before they could damage American citizens, and hence the logic of the neoconservatives' approval of preemption after 9/11, not only in the context of Taliban, al-Qaeda and Osama bin

Laden, but the crusade against evildoers everywhere. 'Operation Enduring Freedom' in Iraq completely overlooked the 'immorality' aspect of using force and immune US from the 'related obsession with avoiding casualties' reminiscent of Vietnam War and provided a compelling rationale for a sustained and proactive use of American power on a global scale justified as a necessary protective measure [Bacevich: 2003: pp. 229-31]. However, even after a decade of America's Iraqi project, the "irrefutable proof" that Iraqi regime possessed WMD has not been found and evidence of Iraqi collaboration with al-Qaeda is also not substantiated. [Suskind: 2004; Herbert: 2004; Risen and Miller: 2003]. Yet, to provide a justification for going to war in Iraq, a series of rationalizations have been offered ranging from alleged charge of possession of deadly weapons and plans to use them against West and America. Although the 'collateral damage' done in Iraq was tried to be erased from public memory by the 'embedded journalists' and mainstream media that hewed closely to the administration's language and position, but the historical records of the US-Iraqi relations provided evidence of pro-Iraqi American policies pursued by Reagan and first Bush administrations in the aftermath of Islamic revolution in Iran in 1979 and their role, thenceforth, in providing Iraq the required 'material' to produce chemical, biological and nuclear weapons (CBN).

Stopping rogue states: President George W. Bush might have justified US intervention in Iraq on the grounds of stopping rogue states and their terrorist clients from threatening US and its allies, "in practice it sent a clear signal to its allies that in exchange for their support they would be free to act as they chose" [Gendzier: "Who Rules the Middle-East Agenda?" in Crotty, William: 2005: p. 61]. In other words, it was an approval for friendly states like Israel and its policies in West Bank and Gaza to the utter disregard for the concerns either of the Palestinians or for their pressing issues for getting a state of their own and for the military dictators like President Musharraf of Pakistan to continue to sponsor cross-border terrorism in Kashmir as long as he could be relevant in containing Taliban and al-Qaeda terrorists in Afghanistan.

Rafiq Zakaria describes Saddam's execution as victor's vengeance, an unjust and controversial treatment to the Iraqi dictator—his capture, trial and execution—as a sad metaphor for America's occupation of Iraq. It needs to be mentioned here that Saddam Hussein was not a run-of-the-mill dictator. He created one of the most brutal, corrupt and violent regimes in modern history, something akin to Stalin's Soviet Union, Mao's China or Kim Jong II's North Korea. Whatever the strategic wisdom for the United States, deposing him might have been taken as something unquestionably good for Iraq [Zakaria: Newsweek: Jan. 8, 2007]. However, instead of trying him under international law or in a court with broader legitimacy, Bush administration breached canons of propriety by putting him in the hands of 'Iraqi government' dominated by Shiites and Kurdish politicians who had been victims of his reign. As a result, Saddam's trial, which should have been the judgement of civilized society against a tyrant, is now seen by Iraq's Sunnis and majority of Arabs as a farce, reflecting only the victor's vengeance. What was a 70 percent Iraqi support for American invasion in 2003 turned into a more than this percentage of opposition by the same Iraqi population in 2007, just after Saddam's execution. In effect, after the invasion, the United States dismantled the Iraqi state, leaving a deep security vacuum, administrative chaos and soaring unemployment. That state was dominated by Iraq's Sunni elite, who read this not as just a regime change but a revolution in which they had become the new underclass. For them, the new Iraq looked like a new dictatorship.

The sectarian nature of US policies in Iraq: US policy of building a national army and police force with blatantly obvious overtones making the forces overwhelmingly Shiite and Kurdish, mostly drawn from militias with stronger loyalties to political parties than to the state—backfired, as with a stronger Shiite army, the Sunni populace felt more insecure and more willing to support the insurgency. Washington's neoconservatives blamed it on the Iraqis, in reality the Bush administration was not quite so blameless. After thoughtlessly engineering a political and social revolution as intense as the French or Iranian one and then getting surprised that Iraq could not digest it happily, peaceably and quickly is really against all principles of 'just war'.

Washington's eagerness to go into Iraq rather than concentrate on hunting down al-Qaeda's leaders only helped the terrorist group to solidify its bases in other locations such as in Pakistan and in western Iraq. Osama bin-Laden mounted a successful propaganda campaign to make himself and his movement the primary symbols of Islamic resistance worldwide. By luring US into bleeding wars in the Muslim world, he wanted to bankrupt US as he did former USSR after its invasion of Afghanistan in 1980s. [Riedel: *Foreign Affairs*: 2007]. The US invasion almost confirmed al-Qaeda claims that the United States was an imperialist force and that the *jihad* against US forces was working. Due to its flawed policies in Iran, US almost fell into al-Qaeda'a trap that intended to force US to engage into wars one after another. Instead, US should actually have concentrated on routing out terrorism politically and diplomatically.

In fact, after failures of Iraqi occupation, Bush administration changed the rhetoric: it started concentrating on "freedom". Some commentators however maintain that the big oil companies, such as Halliburton, were behind it all as they thought that they could somehow make more money in Iraq under a US occupation than they did under a *Baathist* kleptocracy. The most solid reason for US occupation of Iraq seemed to be oil itself as nearly a ninth of the world's proven reserves lie there. However distasteful this explanation for war might seem, yet without that oil, all that remains is a terrifying landscape of sand and soldiers with nothing much to fight for.

War on Terror: A Critique

Following September 11 attacks on American territory, a two-pronged strategy, thought to be an antidote to future 9/11s was conceived: (1) ***retributive***, i.e., bringing the perpetrators to justice; and (2) ***deterrent***, i.e., to impose order and security coercively in order to prevent recurrence of such deadly incidents. Though this dual initiative was backed by a global community of states and also by international organization—United Nations and its Security Council, the fact remains that after nine years (in 2012) of 'pursuit of justice' under the banner of 'war on terror', justice and rule of law have actually been subverted under the heavy weight of blind search for security through the strong arm of law. "National and international laws and legal principles have been bypassed, transgressed or simply overruled by the new laws and executive decrees" [Mani: 2003]. The alarming consequences of this atavistic strategy are, that the state security has gained primacy over the emerging concept of 'human security'; 'law' has been decoupled from 'order' and the process of law works against the very justice it intends to deliver. For short-term political and security gains, the long-term risks and effects of the 'strikes' have been overlooked. None of the three dimensions of 'Justice' namely—*rectificatory* (righting a wrong), *protective* (protecting rights of individuals and institutions through rule of law) and *distributive* (providing social justice in terms of equal or equitable distribution of resources, opportunities and power)—have been achieved so far.

While the liberal stream of political and strategic analysts of 9/11 acknowledged the deficit of distributive justice amongst the desperate, aggrieved and marginalised followers of prosperous and educated perpetrators of terrorist acts and therefore pleaded, especially the European Union, for increasing development-aid flows and opening up trade for partially preventing the root causes of terrorism, US opted only a revengeful approach. In its retributive pursuit, the US and its partners abused the rule of law to track down the suspects, bypassing the rules and legality of principles, both national and international. Even the new laws were made to meet the objectives irrespective of their legality or morality. One such example is US attempt to transfer al-Qaeda suspects to Guantanamo Bay in Cuba where US laws prohibiting torture and requirement of fair trial do not apply. Even the *Amnesty International USA report* was quite apprehensive about the status of human rights of the captured suspects. Under the rubric of the "War on Terror," the United States has detained thousands of people in the US and around the world. Many were being held without charge or trial in violation of fundamental due process of rights. Others are being rendered to countries that engage in torture. Though some food packets were airdropped in Afghanistan after military action was initiated, it was scant. Geneva Convention of 1951 relating to the status of refugees was also violated as the coalition governments blocked the only escape route for refugees by directing the Iranian and Pakistani governments to seal their borders to check refugee flows to facilitate the coalition's hunt for Taliban and al-Qaeda men. Though security rationale justifies such a strategy, it is clearly devoid of all moral and humanitarian considerations.

In US psyche, international terrorism is portrayed solely as violence directed against US citizens, US commercial interests and white race of other nations, in complete disregard or recognition of the interests and rights of the non-white, non-US/European peoples and nations. The latter is being branded, for most of the times, not as victims but as perpetrators. In this context, the British writer George Manbiot's remarks are noteworthy: 'war on terror' has transformed into a veritable 'war on the Third World' [Manbiot: *The*

Guardian: 2002]. In the similar vein, Rama Mani believes that widespread racist, ethno-centrist and anti-Islamic attitudes in the West have risen apace with racist violence that has translated into dramatic electoral gains for ultra-nationalist political parties. Violation of fundamental human rights related to life and liberty are non-abrogable even during emergencies, and it is nothing short of a catastrophe that the international laws so painstakingly devised to protect civilians from war and unscrupulous governments after Second world war have been so blatantly contravened by some of their architects. The 'Patriot Act' in US which became law in October 2001 and the Anti-Terrorism, Crime and Security Bill in UK passed in November 2001 are examples of fabrication of new laws and decrees having potential to victimize innocent people in the name of security though at times its rightful usage cannot be denied and sometimes the perpetrators also take shield under protective international humanitarian laws and run scot-free, committing ever gruesome acts of terror and violence.

Despite all heavy-handed measures, the objectives of the post-9/11 US foreign policy remained unattained till 2011 in the context of capturing Osama when he could ultimately be captured and killed in a US military operation, Taliban in Afghanistan returned, al-Qaeda networks rebuilt with a wider network than before, peace and security in war-torn Afghanistan seems to be a distant dream. Iraq speaks of the same story.[1] US remained embroiled in these two countries at a very heavy price in terms of human and material losses with the menace of terrorist attack on America remaining intact as ever. Armed with a few contentions regarding weapons of mass destruction (WMD), the United States invaded Iraq and by doing so, US acts of war and the occupation have played, ideologically, tactically, and strategically, into the hands of the United States' enemies. For some Muslims, the invasion of Iraq confirmed Osama bin-Laden's argument that the United States and its allies seek to occupy Muslim countries, steal their wealth, and destroy Islam

The critics of US policies in the aftermath of 9/11, such as Rashid Ahmed and others working on conspiracy theories, would argue that though Bush had repeatedly proclaimed to capture Osama 'dead or alive', on several occasions when he could easily be captured, he was allowed to escape, at times with ISI support. Such a thesis is supported by the comment of an American official who is reported as saying that US could not risk "a premature collapse of the international effort if by some lucky chance Mr. bin-Laden was captured." [*Daily Mirror*, Nov. 16, 2001]. These critics argue that Osama's name has been used as a pretext to get the defense budget increased and also to get Congressional approval for President Bush's Missile Defense Program. As says Manbiot, Osama became the personification of evil required to launch a crusade for good: the face behind the faceless terror.... His usefulness to western governments lied in his power to terrify. When billions of pounds of military spending are at stake, rogue states and terrorist warlords become assets precisely because they are liabilities. With regard to Afghanistan, Rashid (Rashid: 2000] drawing on various sources, calls it a matter of public record that "corresponding with the growing shift in US policy against the Taliban, a military invasion of Afghanistan was planned long before 11th September". He suggests that at least one of the fundamental purposes behind this plan was to facilitate a huge project of a consortium of oil companies known as CentGas (Central Asia Gas Pipeline). This consortium, which includes Delta Oil of Saudi Arabia, was formed by Unocal, one of the oil giants of the United States, to build pipelines through Afghanistan and Pakistan for transporting oil and gas from Turkmenistan to the Indian Ocean. In September of 2000, a year before 9/11, an Energy Information Fact Sheet, published by the US government, said: "Afghanistan's significance from an energy standpoint stems from its geographic position as a potential transit route for oil and natural gas exports from Central Asia to the Arabian Sea. This potential includes proposed multibillion dollar oil and gas export pipelines through Afghanistan at one time, Unocal and Washington had hoped that the Taliban would provide sufficient stability for their project to move forward, but they had lost this hope."

Zbigniew Brzezinski, a former National Security Advisor in Carter administration and a Democrat 'Realist' who after 9/11 was criticized for his role in the formation of the Afghan *mujahideen* network, some of which would later form the Taliban and would shelter al-Qaeda camps, criticized War on Terror for creating a culture of fear in US, which has virtually undermined the America's ability to effectively confront the real challenges faced from fanatics who are hell-bent upon destabilizing US by their acts of terrorism. Though Brzezinski asserted that blame rightfully ought to be laid at the feet of the Soviet Union, whose invasion he claimed radicalized the relatively stable Muslim society, he however condemns the WOT, a vague term he believes was deliberately created by American politicians to intensify the emotions of the general public—an easier way to mobilize the public on behalf of the policies they want to pursue [Brzezinski, Z.: 2007]

The phrase 'war on terror' itself is meaningless. It defines neither a geographic context nor the presumed enemies. Terrorism is not an enemy but a technique of warfare—political intimidation through the killing of unarmed non-combatants. In fact, the war of choice in Iraq could never have gained the congressional support it got, says Brzezinski, without the psychological linkage between the shock of 9/11 and the postulated existence of Iraqi weapons of mass destruction. Support for President Bush in the 2004 elections was also mobilized in part by the notion that "a nation at war" does not change its commander-in-chief in midstream. The sense of a pervasive but otherwise imprecise danger was thus channeled in a politically expedient direction by the mobilizing appeal of being "at war."

It is also argued that to justify the "war on terror," the Bush administration crafted a false historical narrative that could even become a self-fulfilling prophecy. By claiming that its war is similar to earlier US struggles against Nazism and then Stalinism, while ignoring the fact that both Nazi Germany and Soviet Russia were first-rate military powers, a status al-Qaeda never had, the administration could be preparing the case for war with Iran. Such war might then have plunged America into a protracted conflict spanning Iraq, Iran, Afghanistan and perhaps also Pakistan. The term has made Americans potentially very susceptible to panic in the event of another terrorist act in the United States itself, which perhaps seems to be the justification for war in Iraq. In his 1997 book—*The Grand Chessboard: American Primacy and its Geostrategic Imperatives* Zbigniew Brzezinski argued that Central Asia, with its vast oil reserves, is portrayed as the key to the domination of Eurasia that can ensure continued "American primacy" if America gets control of this region. However, it will require a consensus on foreign policy issues within the American public which would be unlikely to approve an America that is too democratic at home but too autocratic abroad. This fact is likely to limit the use of America's power, especially its capacity for military intimidation. But, as Brzezinski explains, "the pursuit of power is not a goal that commands popular passion, except in conditions of a sudden threat or challenge to the public's sense of domestic well-being". Therefore, he counseled, the needed consensus on foreign policy issues will be difficult to obtain "except in the circumstance of a truly massive and widely perceived direct external threat" [Brzezinski: 1998: pp. 35-36]. The American public, which is ambivalent about the external projection of American power, had supported America's engagement in World War II largely because of the shock effect of the Japanese attack on Pearl Harbor. Clearly, Iraq did not represent any direct external threat to US or to its interests anywhere and therefore does not stand justification on rational analytical grounds.

Propaganda through media: A culture of fear mongering, reinforced by security entrepreneurs, the mass media and the entertainment industry, generated its own momentum. That America became insecure and more paranoid is hardly debatable as more and more studies report of the swell in ranks of terrorist outfits. But, there is need to grasp the ulterior motives of the politicians, security and entertainment industry. For instance, in more and more TV serials and films, the evil characters have recognizable Arab features, sometimes highlighted by religious gestures that exploit public anxiety and stimulate Islamophobia. Arab facial stereotypes, particularly in newspaper cartoons, have at times been rendered in a manner sadly reminiscent of the Nazi anti-Semitic campaigns. The atmosphere generated by the "war on terror" has encouraged legal and political harassment of Arab Americans, who, Brzezinski feels are generally loyal Americans, for conduct that has not been unique to them. Social discrimination, for example toward Muslim air travelers, has also been its unintended by-product. Not surprisingly, animus toward the United States even among Muslims otherwise not particularly concerned with the Middle-East has intensified, while America's reputation as a leader in fostering constructive inter-racial and inter-religious relations has suffered egregiously.

In the general area of civil rights too, WOT has spread a culture of fear that has bred intolerance, suspicion of foreigners and the adoption of legal procedures that undermine fundamental notions of justice. The rule of 'innocent until proven guilty' was diluted. People, including US citizens were held for lengthy periods of time without effective and prompt access to due process. The "war on terror" gravely damaged the United States internationally. For Muslims, the similarity between the rough treatment of Iraqi civilians by the US military and of the Palestinians by the Israelis has prompted a widespread sense of hostility toward the United States in general.

It's not the "war on terror" that angers Muslims watching the news on television; it's the victimization of Arab civilians. And such resentment is not limited to Muslims only.

Terrorism defined selectively: The events of 9/11 could have resulted in a truly global solidarity against extremism and terrorism. A global alliance of moderates, including Muslim ones, engaged in a deliberate campaign both to extirpate the specific terrorist networks and to terminate the political conflicts that spawn terrorism would have been more productive than a demagogically proclaimed and largely solitary US "war on terror" against "Islamo-fascism." Only a confidently determined and reasonable America can promote genuine international security, which then leaves no political space for terrorism. Though the likelihood of terrorist attacks in future cannot be denied or overlooked, but WOT has created hysteria of sorts. In fact, it is now increasingly recognized that insofar as the United States is waging a war on terrorism, "terrorism" is defined in a very selective, self-serving way, as any form of violent opposition to American leadership. It was waged against all those forces, state or non-state, those were perceived as hostile to American global interests. What is really going on, in other words, is "an empire-building project undertaken behind the smokescreen of the war on global terror" [Falk: 2002: p. 108]. Hence, the critics would say that the war [on terrorism] was never about bringing anyone to justice; it was about conquest and the mushrooming of US global power, all in the name of righteous vengeance. The notion that US foreign policy after 9/ 11 has simply been a reaction to those tragic events is quite wrong. Instead, it is product of certain conceptions, as propounded in neoconservative circles of Bush administration, of how the US should relate to the rest of the world. In fact, the invasion of Iraq was on the table from day one of President George W. Bush administration and hence, has little to do with Osama bin Laden, Al-Qaeda or the 'War on Terror' for that matter [Heine: 2007; Gordon and Trainer: 2006].

Colonel Dobrot, who served in Afghanistan for five years by participating in the global war on terrorism, suggests three root causes of the terror problem as manifested by al-Qaeda and other such groups: the lack of wealth-sharing in Islamic countries; resentment of Western exploitation of Islamic countries; and a US credibility gap within the Islamic community. Comparing the ends, ways and means of Washington's war on terror with those of al-Qaeda and other such groups, he concludes that the US is not achieving its long-term strategic objectives in its WOT, and recommends that US strategy must focus on the root causes of Islamic hostility. Dobrot also suggests that the US should combat radical Islam from within the Islamic community by consistently supporting the efforts of moderate Islamic nation to build democratic institutions that are acceptable in Islamic terms. In this ideological battle, where the enemy is a group of violent religious extremists, the United States cannot win it militarily [Dobort: 2006]. Two objectives identified in the National Security Strategy for combating terrorism are: defeating violent extremism and creating a global environment inhospitable to violent extremists. Though the first aim can be achieved by military means, it cannot create a future global environment that will be inhospitable to violent extremism. The United States' current policies and actions may, in fact, be creating more, not fewer extremists.

In order to build credibility for its policies amongst the Islamic *ummah* (community), the United States must communicate and promote democracy in terms that the Islamic world understands and respects. To achieve this long-term objective, it must consistently focus its reform efforts on those predominately Islamic nations with which it already has relationships. The United States must hold the Israelis, the Saudis, the Egyptians, and itself accountable to standards and policies perceived among mainstream Muslims as being consistent. Specifically, the United States must recognize democratically elected governments such as Hamas and actively engage them in public diplomacy, even if it disagrees with them. By taking up sincere steps to resolve Arab-Israel conflict, the US can reestablish much of its credibility in the Muslim world.

Neo-conservatism: A Realists' Approach to Foreign Policy

In post-cold war era, particularly under President George W. Bush, the American foreign policy remained largely influenced by a small clique or the core group in charge of policy-making institutions in US which consisted of neoconservative defense intellectuals. They were called "neoconservatives" because many of them started off as anti-Stalinist leftists or liberals before moving to the far right. These included: Paul Wolfowitz, the deputy secretary of defense, the defense mastermind of the Bush administration; Donald Rumsfeld, an elderly figurehead who held the position of defense secretary; Douglas Feith, No. 3 at the Pentagon; Lewis "Scooter" Libby, a Wolfowitz protégé who was Cheney's chief of staff; John R. Bolton, a right-winger in the State Department and Elliott Abrams—head of Middle-East policy at the National Security Council. On the outside were James Woolsey, the former CIA director, who has tried repeatedly to link both 9/11 and the anthrax letters in the US to Saddam Hussein, and Richard Perle. They call their revolutionary ideology "Wilsonianism", while in fact, the philosophical underpinnings of these 'Realists' were demonstrably Friedrich Nietzche and his advocacy of amoral power and Spinoza and his advocacy of a certain esoteric intellectual elitism [see, Kiracofe Jr., Clifford A.: 2002]. In fact, the genuine American Wilsonians would rather believe in self-determination for people such as the Palestinians.

In the debate between those who believe in the core values of traditional, multilateralist United States approach to foreign policy and those realists, called 'hawks', who advocate a radical break with American traditionalism/multilateralism and favor a revival of 19th century European imperialism as a permanent direction in US foreign policy, the neoconservatives belong to the latter category of realists. The intellectual roots of their thinking on foreign policy matters can be traced to Professor Leo Strauss and Professor Hans Morgantheau—both émigrés from Nazi Germany who advocated policies that came to be known as 'Realist School', whose leading practitioner in recent years has been Henry Kissinger. Realists believe that the international system is made up of states pursuing their own material self-interest in an anarchic environment. They generally discount the importance of ideology in motivating state action or the ability of international institutions to limit state power. Hence, to them, the US must take fullest advantage of its privileged position in world affairs. Accordingly, rather than co-ordinate its policies with other actors on the world stage, what Washington should do was to impose them on others unilaterally. They did not want to give any space to 'others' to disagree. According to their diagnosis, 9/11 was not the result of an excessive, but of an 'insufficient' involvement of United States in world affairs. The way forward to strengthen American imperialism was to strengthen even further the US military apparatus, promote more actively human rights and democracy, occupy, if necessary permanently, the countries like Afghanistan and Iraq. As Kaplan puts it, the view of neoconservative school of thought was that more successful the US foreign policy, more it will be looked at by historians as an Empire and as a Republic, however different US might be from Rome and from any other empire in the course of history [Kaplan, 2001].

These neocon defense intellectuals in George W. Bush's administration were in or around the Pentagon, are at the center of the Israel lobby and the religious right, plus conservative think tanks, foundations and media empires, such as the American Enterprise Institute (AEI) and belong to the class of intellectuals like Norman Podhoretz, Irving Kristol, Daniel Patrick Moynihan and Nathan Glazer. These neoconservatives had close association with pro-Israel lobby. It was this Zionist-neoconservative combine, called by Petras as 'Ziocons', that almost pushed America into wars on behalf of Israel in West Asia. James Petras argues that in the run-up to the Iraq war, Zionists in top strategic decision-making positions in the Pentagon, the Vice-President's office, the White House and the National Security Council designed and executed war policy, fabricated evidence, wrote presidential speeches, organized press conferences and presidential agendas, purged critics in the military and intelligence agencies and altered intelligence reports to suit their purposes [Petras: 2008]. Through such grip on US foreign policy, these Zio-cons though succeeded in destroying Iraq, but it came at a heavy cost in the form of military casualties, demoralization and an enormous expenditure of some trillions US$ of American taxpayers which invited widespread criticism, not only from the American public at large but also from senior active and retired military officers, ground troops and also from the entire range of top intelligence officials who were disgusted by the "Israel-first" lobby in Pentagon for distorting previous intelligence reports and 'fabricating' new intelligence based on disinformation over US intelligence. It is this Zionist-neoconservative combine which was also manipulating White House and Pentagon to target Iran despite the National Intelligence Estimate's (NIE) November 2007 report[2] confirming the non-existence of an Iranian nuclear weapons program. This is to establish Israel's hegemony in West Asia and to control natural resources in the region [Petras: 2008]. Under pressure from this Ziocon lobby, President Bush pronounced an apocalyptic message to the world in October 2007, a month before NIE report, proclaiming the advent of 'World War III' against Iran's nuclear weapons program and the threat of nuclear attack, somewhat similar to the 'holocaust', by Iran against the people of the US and Israel. Vice-President Dick Cheney frequently intervened to alter the content and conclusions of NIE findings and did all in his power to project Iran as an 'existential threat to Israel's survival'. In short, the critics of neoconservatives' influence on US foreign policy maintain that the neocon agenda is to protect the interests of Israel, rather than the security of United States. Many Bush opponents argue that Bush had *wily-nily* come under the control of a collective of Jewish advisors whose true purpose has always been furtherance of Israel's security interests.

Rejecting the traditional approach to respect international law and rule of law, the neoconservatives maintain that 'might makes right' and that the coercive force need no legal justification. Even the constraints on Executive power prescribed in the constitution are overlooked as is the place and status of Legislative branch, which under the separation of powers arrangements is co-equal of Executive branch. They advocate a 'preemptive', active warfare in their quest for global imperialism of US, as its military, economic and strategic prowess makes America overwhelmingly superior over others. Emboldened by their 'patronage' in the Bush administration, they even had plans to pervert United Nations and other international institutions into an engine of pursuing imperialist ambitions.

US Foreign Policy in the Middle-East

The **Middle-East** is a historical and political region of Afro-Eurasia with no clear boundaries and has been a major center of world affairs since ancient times. Being the geographic region of the origin of three of the world's great religions—Christianity, Islam, and Judaism—and being a region of the countries having large quantities of crude oil, a limited and crucial commodity, the Middle-East remains a region of strategic, economic, political, cultural, and religious importance. Mass production of oil began around 1945, with Saudi Arabia, Iran, Kuwait, Iraq, and the United Arab Emirates (UAE) having large quantities of oil. Estimated oil reserves, especially in Saudi Arabia and Iran, are some of the highest in the world, and Middle-Eastern countries dominate the international oil cartel OPEC. During the Cold War, the region was a theater of ideological struggle between the two superpowers: the United States and the Soviet Union, as they competed to influence regional allies and its resources. Within this contextual framework, the United States sought to divert the Arab world from Soviet influence. Throughout the 20th and into the 21st century, the region has experienced both periods of relative peace and tolerance and periods of conflict and war. Current issues include the rise of militant Islamic movements driven by Islamic fundamentalism, Iraq war, Israel-Palestine conflict, and the alleged Iranian nuclear program. The term "Middle-East" was popularized around 1900 in Britain; it has a loose definition traditionally encompassing countries or regions in Western Asia and parts of North Africa.

Oil: Nations' ever-increasing need for the Middle-Eastern oil and the oil-rich Arab states' eagerness for a rapid infusion of modern technology make not only for a blending of interests that neither side can afford to ignore, it also creates high stakes of emerging as well as well-established economies in the Middle-Eastern region so as to meet their energy demands. It is not only the superpowers that want their dominance in the region. Equally important is the striking role of some of the region's political figures that have left an indelible mark on the chain of events in the international politics. The figures such as Col. Nasser (Egypt's President, 1954-70), Yassir Arafat (Palestinian leader of PLO and President of Palestine National Authority), Anwar Sadat (Egyptian President, 1970-81), King Hussein (ruler of Jordan), King Faisal (Saudi Arabia), Saddam Hussein (President of Iraq, 1979-2003) and many more have made their impact felt by the outside world. A combination of these factors demonstrates the logic of an outside interest in the Middle-East that goes beyond the great powers to embrace the remoter corners of the Third World. For the world's powerful states, UK and France till World War II and United States thereafter, control of oil abroad has been and still remains the main foreign policy objective in the Gulf region in order to procure natural resources and trading privileges, and to seek economic rent and excess profits through political dominance which the corrupt and despotic leaders of the region's states, lacking legitimacy, are ready to offer in order to buy foreign support for their positions through oil.

The United States has had critical interests in the Middle-East for as long as it has been a global power. Securing the flow of the region's oil to the world economy has always been a central priority. During the Cold War, competition with the Soviet Union for Middle-Eastern allies was another. And helping to protect Israel while keeping the Arab-Israeli conflict from escalating to Armageddon has long been a third. Washington has dealt with crises ranging from Iranian hostage-taking to Iraqi aggression to Arab-Israeli fighting in the region. The principal US concerns still emerge from the Middle-East. The war on terrorism may be global, but its roots are there. Iraq remained in a mess even after United States' withdrawal in 2011. Iran is important on US foreign policy agenda due to its radical Islamist postures and the alleged nuclear program. "The United States has for decades sought to play a more permanent role in Gulf regional security. While the unresolved conflict with Iraq provides the immediate justification, the need for a substantial American force presence in the Gulf transcends the issue of the regime of Saddam Hussein" [PNAC: 2000: p. 17]. The main thing, in other words, was getting a substantial American force presence in the Gulf, with Saddam providing the "immediate justification."

There is one another important perspective pertaining to Middle-Eastern region. Throughout much of the twentieth century, this desperately poor region needed foreign investment and protection against hostile neighbors and therefore, many countries in this region kept competing with each other. Simultaneously, the energy requirements of the advanced countries made them dependent on the oil-rich Middle-Eastern countries, which enhanced their bargaining position *vis-a-vis* western countries that needed their oil. Due to this oil commodity, a number of foreign countries have become politically active in the region, using not only the Middle-Eastern governments but also various non-governmental groups, such as Islamic fundamentalists and terrorist groups, for their own purposes. The combination of these two factors: outsiders' interest in the oil of the region and intra-regional disputes has caused extraordinary levels of tensions in the region with recurrent open conflicts as well as a multitude of other latent national, ethnic and territorial disputes. Indeed, since the end of Cold War, Middle-East remains the world's most conflict-ridden and militarized region. In other words, the domestic factors on the one hand, such as, the absence of democracy, prevalence of autocratic rulers, excessive centralization of power and the excessive interest of foreign players in the region on the other, remains at the roots of much of the international politics played here in this region or elsewhere in the world. Since almost all the major powers have been trying to get a foothold in the Middle-East with varying degrees of success, they got allies and partners in the despots of these countries and their military, which could ensure order to ensure constant flow of oil from Gulf and military equipment from the arms-producing countries. Promotion of democracy, therefore, was nowhere on the agenda of these westerners. Till 9/11, the most stable link was between Saudi Arabia and United States, the other being French quest for Iraqi oil.

But, Iraq's relations with its regional rival Iran and with Saudi Arabia have caused much of instability in the region and have affected the oil market also. Iran also causes regional instability due to its internal tensions and shifting of relations with outside powers.

However, this strong desire to control the region's oil wealth unfortunately led to the creation of some artificial states, such as, Kuwait, and states with mixed Kurdish and Arab populations, such as, Syria and Iraq. The arbitrary creation of borders and the installation of unpopular pro-colonial leaders served the purpose of the outside great powers who embarked upon a plan of dividing the local populations and ensuring the establishment of impotent client-regimes whose patterns of administration were subservient to colonial interests. In 1945, when Britain was still a major colonial power, US and British coordination and cooperation were highlighted in the following memo: "Our petroleum policy towards the United Kingdom is predicated on a mutual recognition of a very extensive joint interest and upon control, at least for the moment, of the great bulk of the free petroleum resources of the world... US-UK agreement upon the broad, forward-looking pattern for the development and utilization of petroleum resources under the control of nationals of the two countries is of the highest strategic and commercial importance." (See: *Memorandum by the Acting Chief of the Petroleum Division*, 1 June 1945, FRUS, 1945, Vol. VIII, p. 54).

Democracy: After the Second World War, it became impossible to prevent the wave of democratization that swept the colonies, and one by one, the old puppet governments in the region collapsed. Britain and France lost their colonies, but the US stepped in as the new and dominant neo-colonial power in the region (Heikal: 1987, p. 38). US imperial goals were expressed without mincing any words in a 1953 internal US document: *"United States policy is to keep the sources of oil in the Middle-East in American hands."* In 1958, a secret British document described the principal objectives of Western policy in the Middle-East: "The major British and other Western interests in the Persian Gulf [are] (a) to ensure free access for Britain and other Western countries to oil produced in States bordering the Gulf; (b) to ensure the continued availability of that oil on favorable terms and for surplus revenues of Kuwait; (c) to bar the spread of Communism and pseudo-Communism in the area and subsequently to defend the area against the brand of Arab nationalism." (See: File FO 371/132 779. *Future Policy in the Persian Gulf*, 15 January 1958, FO 371/132 778).

US foreign policy shapes events in every corner of the globe. Nowhere is this truer than in the Middle-East, a region of recurring instability and enormous strategic importance. In the first decade of the 21st century, the Bush Administration's attempt to transform the region into a community of democracies helped produce a resilient insurgency in Iraq, a sharp rise in world oil prices, and terrorist bombings in Madrid, London, and Amman. With so much at stake for so many, there is a need to understand the forces that drive US Middle-East policy [Mearsheimer and Walt: 2007]. The politics of US policy in the Middle-East is a subject that is far too complex, emotional, and important a topic to be sidelined. Despite the physical distance between the United States and the Middle-East, US influence has been felt in every country within the region. Throughout the 20th century, strategic interests, including a longstanding competition with the Soviet Union, have provoked a variety of US interventions ranging from diplomatic overtures of friendship to full-blown war. American economic interests—particularly in assuring access to Middle-Eastern oil—have long motivated presidents and lawmakers to intervene in the region. In addition, strong cultural ties bind American Jews, Arab Americans, Iranian Americans, and Turkish Americans, among others, to the area, and these interest groups seek to make their voices heard in the US foreign policy arena. As in 20th century, in 21st century too, the US has had global interests and a global reach to match. In the Middle-East, the US has made itself a key player by using its diplomatic, economic, and military power in support of its national interests. Though US has enjoyed a generally positive reputation in the region at the end of World War I, it began to involve itself more deeply in regional politics in the late 1940s. This was not only to fight communism in the cold war period, but also to ensure the steady supply of oil, and making sure that no single power dominated the region. After 9/11, it added fighting terrorism. The US has supported leaders and governments it considered being stable allies, like the Saudi royal family, Israel, and Egyptian governments since Anwar Sadat.

Since 9/11, the great impact of the neoconservative ideology on the foreign policy of US in general and of the Republican Party in particular can be clearly noticed who aligned with the elements of the fundamentalist 'Christian Right'. Their main piece of advice to President Bush was to put Iraq on the front burner, and the Palestine question on the back-burner, if not in the freezer [Kiracofe, 2002].

Though the traditional US approach in the Middle-East has been of peaceful and friendly relations with the nations of the region,[3] these neoconservatives advocated a policy of destabilization and Balkanization of the Middle-East, regime-change at some places, redrawing of borders and the reallocation of the control of hydrocarbon resources of the region in such a way that serves American strategic and commercial interests. They were thinking to use Turkey, a NATO ally of US, as a counterweight to the Arab and Muslim world in the Middle-East.

Now, let us look at the US relations with some strategically important countries of the Middle-East

US-Saudi Arabia: Since the 1930s, when oil was discovered in the Arab peninsula, this became one of the key pillars of US foreign policy in the region, and the US government did everything in it's power to maintain the feudal regime of the House of Al Sa'ud. Since Saudi Arabia possesses both the world's largest known oil reserves and produces the largest amount of the world's oil, it is perhaps the best example of a contemporary energy superpower, in terms of having power and influence on the global stage due to its energy reserves and production of not just oil, but natural gas as well. It is often referred to, as the world's only "oil superpower". Considered being the leading state of OPEC, its decisions to raise or cut production almost immediately impact world oil prices. However, despite the fact that the royal clan of Al Sa'ud, which has been propped up since 1943 and has been called the largest family business in the world, has been operating without any regard for democratic norms or civil rights for it's citizens or millions of migrant workers that produce most of the nation's wealth and despite lacking any popular mandate, the fact remains that it has survived largely on the basis of US military strength.[4] According to some analysts, the US government has pumped in over $33 billion in weapons, military supplies and equipment so as to preserve the authority of this decadent and oppressive monarchy. The uncritical US support for autocratic Gulf monarchies and their human rights abuses exposes the hypocrisy in American rhetoric about democracy and human rights and creates the perception among Gulf subjects that their countries are being ruled in the interests of an outside power. [See: Mamoun Fandy 'US Policy in the Middle-East', *Foreign Policy In Focus*, Vol. 2, No. 4, January 1997].

As of 2005, Saudi Arabia has 25 percent of the worlds proven oil reserves which are the world's largest and has the worlds fourth largest natural gas reserves. Saudi Oil accounts for about 13 percent of US oil imports. Former US ambassador to Saudi Arabia, Chas Freeman, said of Saudi Arabia that one of the major things the Saudis have historically done, in part out of friendship with the United States, is to insist that oil continues to be priced in dollars. America's energy needs in lieu of Saudi Arabia's needs for capital defines the relationship between the two. The Saudis wish to modernize and beautify their country into a western-style paradise, as well as to create long-term investments throughout the world for use once their oil reserves become depleted. Successive American presidents, before 9/11, have provided 'red carpet' treatment to the Saudis. The American posture toward Saudi Arabia and many other OPEC countries has been touted as a special relationship in the media, though this has been shaken by the rise of Islamic militancy, and most acutely by the events September 11, 2001. For the first time in close to a century, the leadership of the United States as well as many of the American people, began to weigh the benefits versus costs of those relationships, and reliance upon an energy source that was costly, easily interruptible, polluting, and which would eventually run out.

Yet, despite America's inevitable reliance on the oil giant—Saudi Arabia, an excessive dependence on it is growingly becoming unpopular due to Saudi Arabia's religious and social orthodoxy and also for its projected enmity towards Israel, a sensitive matter for many neo-conservatives in particular. And although a long-established ally, it has poured money into the support of the *mullahs* and *madrassas* which acted as the recruiters and the academies for terrorists, including many of the 9/11 bombers, most of whom were Saudi citizens.

To date, the Saudis alone have invested approximately 70 billion dollars around the globe, 60 percent of which was invested in the United States. These investments in the United States have traditionally been a welcome counterweight to the systemic US trade deficit with the Kingdom. As American demand for Saudi oil continues at 1.5 million barrels per day, US service and merchandise exports revenues to the Kingdom cover nowhere near the level of expenditures for petroleum. One enabler of US consumption has been the historic Saudi Arabian willingness to finance this trade deficit by investing in the United States. This relationship, while symbiotic, and necessary to a US economy addicted to consumption, is viewed by many as 'golden hand-cuffs' voluntarily worn by the United States [see: Energy Policy of United States: *Wikipedia Search*]. In economic terms, the balance of payments is one reflection of a nation's financial economic stability. The current US account balance is a 'negative value.' As of 2004, the account balance in the US was minus (-) 665.5 billion dollars. This borrowing on the part of the United States has, predictably, led to an enormous foreign debt. In contrast, Saudi affluence is soaring, with a record 70 billion-dollar budget surplus for 2006 [see: *Financial Times*: 26.03.07]. The IMF has forecast a Saudi fiscal surplus of 9.4 percent of gross domestic product in 2011 and 8.0 percent in 2012 [Al Arabia News, 21 December, 2011]

United States policies in Iraq: Though the US-Iraq relations reached at their lowest ebb after 9/11 in 2001 and the subsequent American preemptive war in Iraq and a widely clear role of US in the execution of the Iraqi dictator, Saddam Hussein, America supported Iraq's same Saddam Hussein during the Iran-Iraq War (1980-1988), when Iran's new post-revolutionary Islamic regime appeared to be the region's biggest threat. Hussein, however, had since become a significant focus of American anger because of his invasion of Kuwait in 1990—which led to the Gulf War—in an effort to control more of the region's oil. Even the international response to this threat and the underlying goal of the U.N. force, which included 500,000 American troops, was to ensure continued and unfettered access to petroleum. Saddam's known desire to develop weapons of mass destruction became a serious matter of concern for US, which began bombing Iraqi targets during the second Gulf War and continued to enforce a no-fly zone. The US-led economic embargo of Iraq, intended to force Hussein from power and keep Iraq from rearming and further developing weapons of mass destruction had a devastating impact on the health and living conditions of the Iraqi people, and sympathetic Arabs hold this grievance against the United States. As far as the US military action in the Middle-East is concerned, before the 1990 Gulf War, it was severely limited. As peacekeepers in Lebanon after Israel's 1982 invasion, US forces fared poorly. Two hundred forty-one Marines were killed when their barracks were targeted by a suicide truck-bomb in October 1983, prompting a US withdrawal from Beirut to offshore warships. After a 1986 discotheque bombing in West Berlin was traced to Libya, the US bombed that country, killing three-dozen civilians.

The most significant direct US military intervention came in response to the Iraqi invasion of oil-rich Kuwait in August of 1990, which led to the Gulf War. Although the invasion didn't directly threaten American territory, a vital US economic interest—oil—was at stake. Along with this, the principles of international law that protects the sovereignty and territorial integrity of nations was also at stake. The Gulf War won the US the gratitude of the oil-rich states of the Persian Gulf for eliminating the Iraqi military threat, but these regimes have had to deal with increased internal criticism for allowing US troops to remain in Saudi Arabia. The Gulf War also left charges that the US had abandoned some of its most vulnerable allies. The Kurds and *Shiis* of Iraq were encouraged to revolt against Saddam Hussein by the US, with assurances of US support. But little support materialized when the uprising actually got under way, and Iraqi retaliation against both rebelling groups was harsh. Limited US intervention allowed the creation of Kurdish safe havens in the north and assisted *Shii* refugees fleeing into Iran in the south, but charges that the US abandoned its regional allies linger to this day, leading to skepticism that America's (George W. Bush's) call for a new government in Iraq would be accompanied by full American support.

It is argued that the tragic war in Iraq is in many ways the first oil-depletion and oil-currency war of the 21st century [Klark: 2005]. Clark argues that he ill-fated unilateral invasion of Iraq that was designed to maintain US dominance of the global oil supply and enforce petrodollar supremacy has had the ironic effect of encouraging momentum towards petroeuros and other petrocurrencies, along with new geopolitical and energy alignments unfavorable to the US. Iraq, possessing the world's second-largest oil reserves, was therefore already a target of US geostrategic interests. Together with the fact that Iraq had switched to paying for oil in Euro—rather than US dollars—the Bush administration's unreported aim was to prevent further OPEC momentum in favor of the Euro as an alternative oil transaction currency standard. In order to maintain dollar hegemony and the unsustainable macroeconomics of 'petrodollar recycling', America's Iraq policies were devised. The Iraq war applies to geostrategic tensions between the United States and other countries, including the member states of the European Union, Iran, Venezuela and Russia. Clark warns that without changing the current course of US unilateralism, the American experiment will end the way all empires end—with military overextension and subsequent economic decline. Kupchan in his *End of American Era* argues that war in Iraq was a dispute about first-order principles, i.e., war and peace, and not a dispute about a pipeline, a weapons system, trade [Kupachan: 2003]. It goes to the heart of the international order that the Americans, Europeans, Japanese, and others worked so hard to put into place, starting in 1941 and into the Cold War. In the meantime, the American military remained trapped in a tragic quagmire in Iraq, even as the Bush-Blair administrations severely obfuscated their real geostrategic and macroeconomic agenda regarding Iran by engaging in a propaganda campaign about an Iranian "nuclear weapons program" that according to both the IAEA and CIA simply does not exist.

The five reasons, or as the critics would term it, 'the five pretexts' of waging war on Iraq are mostly based on unsubstantiated grounds. The ***first reason***: the possession of weapons of mass destruction has long been discredited. The many years of United Nation's weapon inspections and United State's own intelligence agencies have gone on record stating Saddam Hussein was not in possession of such weapons. The ***second reason***: Iraq was a danger to its neighbors and such a charge was based on Iraq's invasion of Kuwait in 1990, but after a crushing defeat in the 1991 Gulf War and twelve years of suffocating international trade sanctions, that included a total arms embargo, Iraq has not even a plausible military threat to any of its neighbors. The ***third reason*** that Saddam Hussein had alleged links with al-Qaeda, has never been substantiated and is now widely rejected, as a secular dictator was not likely to fit very well in avowedly Sunni al-Qaeda driven by religious frenzy and extremism. The ***fourth reason*** that Iraq opposes US policy pertaining to Palestine-Israel conflict is based on Iraq's incessant support for Palestinian cause; this however does not hold a ground for a full-fledged US invasion as many countries, including the non-Muslim, have sympathetic attitude towards Palestinian cause for statehood. Rather, it has been US that has abandoned the peace process every now and then whenever a peace deal was just round the corner. Only the ***fifth reason*** that Saddam Hussein was a cruel oppressor of a section of Iraqi people, Kurds in particular, had some substance, though a flawed one and this argument has its own problems. White House under Bush issued several statements on Saddam's use of chemical weapons in Halabja in March 1988, "with a clear goal to systematically terrorize and exterminate the Kurdish population in Northern Iraq, to silence his critics and to test the effects of his chemical and biological weapons. Hussein launched chemical attacks against 40 Kurdish villages and thousands of innocent civilians in 1987-88, using them as testing grounds" [see: US Department of State, Washington, March 13, 2003 Communiqué at www.state.gov/documents/organization/18817.pdf]. However, a close scrutiny reveals that these accusations made by US that the incident was predetermined military experiment on the Kurds are fabricated, deliberately made to meet US policy objectives [Lin, Sharat G: 2004, pp. 3625-32]. Though March 1988 Halabja chemical attacks are well-documented, but why US punished Saddam for these crimes after fifteen years and not when the incident took place is quite intriguing. Also, whereas the other members of UN Security Council condemned the use of chemical weapons in 1988, US ambassador to UN did not even mention on the official record about the charges of Iraqi chemical weapons use, let alone condemning it [*Security Council Official Records,* for the years 1983-87, UN, New York, 1985-88]. In fact, after the Iranian Revolution in 1979, US policy had a tilt towards Iraq which was maintained throughout Iran-Iraq war, despite the official position of neutrality. Thus forgotten for fifteen years, Halabja was dug out when US government accelerated its search for pretexts for war against Iraq after 9/11. Both CIA and DIA (Defense Intelligence Agency) reports about facts in Halabja were overridden by the White House to justify an invasion on Iraq in 2003 [Lin, *ibid.*].

US policies towards Iran: US-Iran relations have seen many ups and downs since the days of Cold War. According to R. Nicholas Burns, Under Secretary of State for Political Affairs under George W. Bush administration, the United States had no relationship as unique, complex and difficult as it had with Iran. Iran is the only country in the world today with which the United States has no sustained direct contact. Ayatollah Khomeini, the hostage crisis, and terrorism dominate Americans' popular image of Iran. The official Administration line is that Iran is developing weapons of mass destruction. Ignoring the fact that US allies such as Pakistan and Israel are doing exactly the same, the understanding is that Iran has built, under Russian supervision, a nuclear power plant to ease its power crisis. To date about 60 outside inspections have been conducted including the National Intelligence Estimates Report, published in November 2007 and all have deemed Iran clean. Iran is also a party to the Nuclear Non-Proliferation Treaty (NPT), which India, Israel, and Pakistan all refuse to sign. Owing to these reasons, Iran remains a crucial country in a region of incomparable geostrategic [Milani: 2005, pp. 41-56].

Historically, the growing Soviet influence in Iran during Cold War era raised US enxieties and suspicion. The US toppled the regime of Iran's elected Prime Minister Mohammed Mossadeq in 1953, who intended to nationalize the Iranian oil industry. The US-backed coup reinforced the power of the young Mohammed Raza Pehalavi who was pro-Western and was viewed by many in Iran as increasingly autocratic and oppressive [Kinzer: 2003]. He tried to institute many Western social reforms by decree, and his secret police viciously silenced opposition voices. However, a 1979 Islamist revolution against the Shah's regime swept a new kind of Islamic state into power, the Islamic Republic of Iran, governed by Islamic jurists and scholars. The popular hatred of the Shah also tarred his American supporters, and the revolution's anti-American passion led to the storming of the US Embassy in Tehran, where 53 hostages were held for more than a year.

Iran is the world's fourth largest producer, and fifth largest exporter of oil. It has the third largest proven reserves of oil in the world. It is also the sixth largest producer of natural gas, with the second largest proven reserves. These features give Iran a hybrid status between Russia and Saudi Arabia as a potential Energy superpower. If its current relations with United States become less tense, increased Foreign Direct Investment (FDI) in the energy sector may hold the potential for it becoming an energy superpower in the combination of gas and oil energy. British journalist Julian Evans has written that the sooner Iran lets go of its outdated ambitions to be a nuclear superpower, the quicker it can become a modern energy superpower.

American fears that Iran is developing nuclear weapons have been a major militating factor in relations since shortly after the revolution. Today there are no formal diplomatic relations between Iran and the United States. Since 1995, the United States has had an embargo on trade with Iran. On January 29, 2002, American President George W. Bush gave his "Axis of evil" speech, describing Iran, along with North Korea and Iraq, as an axis of evil and warning that the proliferation of long-range missiles developed by these countries constituted terrorism and threatened the United States. The speech caused outrage in Iran and was condemned by reformists and conservatives. Since 2003, the United States has been flying unmanned aerial vehicles, launched from Iraq, over Iran to obtain intelligence on Iran's nuclear program, reportedly providing little new information. The Iranian government has described the surveillance as illegal. In January 2006, James Risen, a *New York Times* reporter, stated in his book *State of War* that the CIA carried out a Clinton-approved operation in 2000 (Operation Merlin) intended to delay Iran's nuclear energy program. [Risen: 2006].

Iran's President Ahmadinejad has staunchly defended Iran's right to a nuclear program on the basis that its nuclear program is only for peaceful purposes, a right protected under international law. Amidst much debate internationally about how to respond to this issue, United Nations Security Council has imposed sanctions and worked to resolve the conflict through multilateral talks. Even the intelligence reports that Iran has frozen its nuclear program since 2003 does not convince US and the relations are tense between the two countries.

After the Bush administration vowed to transform the Middle-East in the aftermath of 9/11, the region is indeed profoundly different. Washington's misadventures in Iraq, the humbling of Israeli power in Lebanon, the rise of the once-marginalized Shiites, and the ascendance of Islamist parties have pushed the Middle-East to the brink of chaos. In the midst of this mess stands the Islamic Republic of Iran. Its regime has not only survived the US onslaught but also managed to enhance Iran's influence in the region. Iran now lies at the center of the Middle-East's major problems—from the civil wars unfolding in Iraq and Lebanon to the security challenge of the Persian Gulf. Tehran's cooperation is crucial in the resolution of these problems. Ever since the revolution that toppled the Shah in 1979, the United States has pursued a series of incoherent policies toward Tehran. At various points, it has tried to topple the regime—even, on occasion, threatening military action. At others, it has sought to hold talks on a limited set of issues. Throughout, it has tried to limit its influence in the region. But none of these approaches has worked, especially containment, which is still the strategy of choice in the Iran policy debate.

As for US strategy to contain Iran's nuclear program, some strategic analysts would argue that Iran has dispersed its nuclear facilities throughout the country to protect them from a possible US strike and placed them deep underground, though all intelligence reports including the November 2007 National Intelligence Estimates'

(NIE) report has ruled out possession of any nuclear weapons by Iran as of now.

US policies in Israel/Palestine: A product of a Zionist effort to find a national home for Jews and a place for them to return to their roots, Israel became home for many, including nearly 75,000 European Jews escaping persecution from Nazi Germany during World War II, but at the same time created a permanent problem of sorts with the Arabs who were either displaced by Jewish settlers from areas where they had been living or became unwilling citizens of Israel. US support for Israel began when President Harry S. Truman extended US recognition to the Jewish State immediately after its 1948 declaration of independence. Continued US support for Israel has varied in form and intensity over time, but this support has remained a pillar of US foreign policy in the Middle-East. US support for Israel is based on several factors: a commitment to one of the few democratic states in the region, a need for stable allies, a sense of a shared Judeo-Christian religious tradition, and as a market for the products of the American defense industry. The US-made aircrafts were critical to the Israeli victory in the 1967 Six-Day War that pitted Israel against an alliance of Arab powers. And when the Yom Kippur War of 1973 again threatened the Jewish state, a massive US airlift of war material was crucial to Israel's survival in the conflict.

Under pressure from international community and more specifically from the Arab states, the US has been active for many decades in its attempts to broker peace between Israel and its Arab neighbors. Notable achievements include the 1978 Camp David meeting that negotiated peace between Egypt and Israel and the 1993 Oslo interim peace agreement that established a framework for negotiating peace between the Israelis and Palestinians and set in motion the process for achieving a Palestinian state. Supporters of the Palestinians, however, believe that the US has not done all that it could to bring about peace. After all, because much of the support to Israel is in the form of American military equipment, the American economy and American jobs are tied to a continually upgrading Israeli army. Some Palestinians argue that the United States is too committed in its support for Israel to make unbiased decisions and is unwilling to pressure the Israelis to negotiate a fair peace.

In recent past, the US backed Ariel Sharon and his Likud government in Israel, even as Sharon authorized military strikes against the Palestine Authority (PA) and militant groups in the

Occupied Territories of the West Bank and Gaza Strip. At a time when Israeli soldiers are regarded by many Arabs as agents of an oppressive army of occupation, unconditional US support for the Jewish state in its struggle with the Palestinians has challenged American relationships with nations long considered allies, like Egypt and Saudi Arabia. These Arab allies argue that American principles like human rights and freedom of the press are not promoted in Israel in the same way that Americans push for reform elsewhere.

In both the US and the Middle-East, the Israel-Palestine question is unfortunately filtered through the lens of politics, which is too often framed by zealots and violence. The deafening power of extremism suppresses all other voices of reasoning. In November 2007, when the voices were raised for a 'last call for a two-state solution' to this problem, it was accepted neither by the Israeli Prime Minister, Ehud Olmert, nor by the then US President—George W. Bush—despite some positive hope of its acceptance by the Arab side. Such mistrust and state of affairs on both sides leads to an unending violence that could go on with peace prospects remaining quite uncertain in near future [Wallerstein: 2007; p. 4].

The American foreign and domestic policy establishment, pro-Israel organizations, political action committees (PACs), and individuals do play significant roles in the US political process. United States needs to find ways to bridge the gap between its current policies and the national aspirations of Palestinians and other Arabs. John Mearsheimer and Stephen Walt in their 2007 book *Israel Lobby* claim that they want the Israel Lobby and US foreign policy "to foster a more clear-eyed and candid discussion of this subject" [Mearsheimer and Walt: 2007]. But Walter Russell Mead doubted such an honest effort and argued that unfortunately, that is not going to happen as "The Israel Lobby" always hardens and freezes positions rather than open them up. This goes to delay rather than hasten the development of new US policies in the Middle-East. These scholars claim that the United States has been willing to set aside its own security in order to advance the interests of another state. The Israel-Lobby's perspective on widely reflected in the mainstream media and in debate in American academia. When the Lobby succeeds in shaping US policy in the Middle-East, the Israel's enemies get weakened or overthrown, Israel gets a free hand with the Palestinians, and the United States does most of the fighting, dying, rebuilding, and paying. The "core of the Lobby" is "American Jews who make a significant effort in their daily lives to bend U.S foreign policy to advance Israel's interests.

The level of American support to Israel as it has been the largest in terms of direct US economic and military assistance since 1976 which amounts to well over $140 billion US dollars as in 2003 dollars. Israel receives about $3 billion in direct foreign assistance each year, which is roughly one-fifth of America's foreign aid budget. It is suspected that Israel uses roughly twenty-five percent of its aid allotment to subsidize its own defense industry and is the only recipient that does not have to account for how the aid is spent, an exemption that makes it virtually impossible to prevent the money from being used for purposes the United States opposes. The United States gives Israel access to intelligence that it denies its NATO allies and has turned a blind eye towards Israel's acquisition of nuclear weapons. Washington also provides Israel with consistent diplomatic support. A number of US vetoes in Security Council have been to protect Israel. Even the invasion of Iraq is at least partly viewed as an effort to improve Israel's geo-strategic situation. But, it should also be noted that very often, Israel has been a liability and a strategic burden. For example, the US decision to give Israel $2.2 billion in emergency military aid during the October War triggered an OPEC oil embargo that inflicted considerable damage on Western economies. Nevertheless, Israel is seen as a crucial ally in the war on terror, because its enemies are America's enemies. This new rationale seems persuasive, but Israel is in fact a liability in the war on terror and the in broader effort to deal with rogue states.

As far as US war on terrorism is concerned, it can be argued that terrorism is a tactic employed by a wide array of political groups: it is not a single unified adversary. The terrorist organizations that threaten Israel (e.g., Hamas or Hezbollah) do not threaten the United States, except when it intervenes against them (as in Lebanon in 1982). Moreover, Palestinian terrorism is not random violence directed against Israel or "the West"; it is largely a response to Israel's prolonged campaign to colonize the West Bank and Gaza Strip. More importantly, saying that Israel and the United States are united by a shared terrorist threat has the causal relationship backwards: rather, the United States has a terrorism problem in good part because it is so closely allied with Israel, not the other way around. US support for Israel is not the only source of anti-American terrorism, but it is an important one, and it makes winning the war on terror more difficult.

The unconditional US support for Israel has always made it easier for extremists like bin Laden to rally popular support and to attract recruits. The citizens in the countries of Middle-East are genuinely distressed at the plight of the Palestinians and at the role they perceive the United States to be playing. As for so-called rogue states in the Middle-East, they are not a dire threat to vital US interests, apart from the US commitment to Israel itself. Israel's nuclear arsenal is one reason why some of its neighbors want nuclear weapons, and threatening these states with regime change merely increases that desire. In short, treating Israel as America's most important ally in the campaign against terrorism and assorted Middle-East dictatorships both make US efforts at WOT more difficult.

The logic provided for unqualified US support to Israel is based on various assumptions, such as: (1) it is weak and surrounded by enemies, (2) it is a democracy, which is a morally preferable form of government; (3) the Jewish people have suffered from past crimes and therefore deserve special treatment, and (4) Israel's conduct has been morally superior to its adversaries' behavior. However, these claims do not stand scrutiny. Today, Israel is the strongest military power in the Middle-East. Its conventional forces are far superior to its neighbors and it is the only state in the region with nuclear weapons. Egypt and Jordan signed peace treaties with Israel and Saudi Arabia has offered to do so as well. Syria has lost its Soviet patron, Iraq has been decimated by three disastrous wars, and Iran is hundreds of miles away. The Palestinians barely have effective police, let alone a military that could threaten Israel. These facts of the strategic balance in favor of Israel, widens the qualitative gap between its own military capability and deterrence powers and those of its neighbors.

The claim that Israel is a fellow-democracy surrounded by hostile dictatorships and therefore qualifies for a favorable US support is flawed. There are many democracies around the world, but none receives the lavish support that Israel does. The United States has overthrown democratic governments in the past and supported dictators when this was thought to advance US interests, and it has good relations with a number of dictatorships today. Thus, being democratic neither justifies nor explains America's support for Israel. Whereas the United States' liberal democracy gives equal rights to people of any race, religion, or ethnicity Israel was explicitly founded as a Jewish state and citizenship is based on the principle of blood kinship. There is no denying that Jews suffered greatly from the despicable legacy of anti-Semitism, but the creation of Israel involved additional crimes against a largely innocent third party: the Palestinians who continue to suffer till date due to a biased and unjustified US favoritism and patronage to Israel.

Pressure from extremist violence and the growing Palestinian population has forced subsequent Israeli leaders to disengage from some of the occupied territories and to explore territorial compromise, but no Israeli government has been serious to offer the Palestinians a viable state of their own. It is nevertheless true that Palestinians have used terrorism against their Israeli occupiers, and their willingness to attack innocent civilians is wrong. This behavior cannot be approved despite Palestinians' belief that they have no other way to force Israeli concessions.

The US and Egypt: Though Egypt's foreign policy operates along the non-aligned lines, it exercises an extensive political influence in the Middle-East politics, has been a key-partner in search for a peaceful resolution of Palestine-Israel conflict and being a prominent player in the Arab world, United States looks at Egypt as an important US ally. The relationship between the two has gone through a sea-change from a policy of distrust during Suez Crisis in 1956 to a policy of friendship with the Camp David Accords in 1978 which resulted in a treaty between Egypt and its neighbor Israel making the US reward President Sadat's peace initiative with a substantial, long-term aid package. However, after 9/11, the Egyptian unwillingness to send troops to Afghanistan and Iraq in peace stabilization missions caused some bitterness in relations between the two countries. Despite this, US policy makers see Egypt as a leader and as having a moderating influence among many Arab, African, Islamic and Third World states.

In view of the popular uprising in Egypt since early 2011, the American strategic analysts are engaged in reframing a coherent policy initiative *vis-a-vis* Egypt. After the resignation of President Hosni Mubarak from the Egyptian Presidency of 29 years, the question at present is how Egypt's transition to a more democratic system will take place and what implications will it have for American foreign policy in Middle-East. In fact, the Arab Spring that has engulfed most of the countries of MENA region is bound to have implications and may set in motion some positive trends in the context of weakening terrorist movements by Islamists in the region.

US and politics of Oil: Dictating Foreign Policy of Oil-Rich Sovereign States

The history of oil in Middle-East began with the British Navy's plans for what became known as the Great War of 1914-18. That Empire intended to use petroleum extracted from this region, to provide its navy the crucial strategic advantage, of a change to oil-burning, from coal-burning warships. Since that time, this region has been dominated by the great powers' struggles over control of the special, strategically significant economic advantages of oil extracted from this region. But, it was never oil alone which shaped the fate of the Middle-East; for as far back as known history of civilization reaches, long before the discovery of oil, the Middle-East has been the strategic crossroads of Eurasia and Africa combined, as it is today. With or without petroleum, the historic strategic significance of the Middle-East would remain. In any ordering of the world's strategic economic affairs, Middle-East oil will continue to be an outstanding factor in the petroleum supplies of the world economy for at least a generation or more yet to come. This would be so for the obvious economic reasons. As such, in all matters of current world affairs, the states powerful as well as not as powerful will have relevant strategic interests in the region.

As some countries are blessed with abundant natural resources, such as, oil, natural gas, phosphates, and other minerals, this factor, however, make Middle-Eastern economies face some significant obstacles to successful development. One major obstacle has been Western efforts to control the region's resources. The discovery of enormous oil deposits in the Middle-East coincided with increasing dependence upon oil in the West in the early 20th century. Money from oil has created enormous opportunities for development in those countries where it is concentrated, such as Saudi Arabia, Kuwait, Bahrain, the United Arab Emirates, Qatar, Iraq, Iran, and Algeria. It, however, has simultaneously created enormous economic stakes in the region by the foreign powers, especially the United States, resulting in a politics of oil that has had a devastating effect on economies and polities in the Middle-East. Maintaining a secure supply of oil remains a problem for the United States, often a political one.

Hence, oil policy, or more precisely energy policy, has been an important driving force behind American foreign policy under various US Presidents, to a far greater extent than is fully understood. Many non-Americans have held the idea that American power derives[5] from unlimited reserves of cheap oil, coal and natural gas; the picture of an America which could afford to leave the lights on all night in the skyscrapers and drive SUVs to its hearts' content dies hard. In fact, the United States is now dangerously dependent on imported energy.

Taking a historical line, we find that in 1940, the United States produced over two-thirds of all the world's oil and natural gas and it was in the late 1940s, for the first time, that the US began to import oil. However, whereas in the 1950s only 10 percent of US oil consumption was imported, by 2004, the end of George W. Bush's first term, imports were 57 percent of consumption. The US policy exoerts warn that by around 2025, the United States will have to import more than two-thirds of its oil and gas; current trends suggest that point may be reached even earlier. To understand the oil politics of United States, one must also take into consideration certain other statistical facts, such as, that about 40 percent of the energy consumed by the United States comes from oil. The United States, with about 5 percent of the world's population, is responsible for 25 percent of the world's oil consumption while only having 3 percent of the world's proven oil reserves The United States, as of 2006, imports more than 65 percent of the oil it consumes, a steep growth since a 10 percent in 1970. At the current rate of unchecked import growth, Americans will be 70 percent to 75 percent reliant on foreign oil by the middle of the next decade. In short, the US is now more dependent on oil, and less secure in its supply, than at any time in the 145-year history of oil consumption. The top four suppliers of crude oil to the US are: Canada, Venezuela, Mexico and Saudi Arabia. This economics of petroleum shapes and defines US politics of oil and its relationship and role in the oil-rich countries everywhere, particularly in the oil-rich Middle-East.

Taking cognizance of this grave scenario, the American administrations under various Presidents have been making their energy policy giving it a highest priority. The need to build up American military power to protect American, and in theory other western powers', access to middle-eastern oil was implicit in *the Project for a New American Century*, in June 1997. Within weeks of George W. Bush's election in November 2000, the National Security Council (NSC) staff in the White House were directed to cooperate with the National Energy Policy Development Group in producing a report, duly published on May 17, 2001 [Mayer: 2004]. A significant chapter in it talked about "strengthening global alliances" that literally meant increasing imports. In effect American military power was to be "projected" so as to induce middle-eastern states to double their production from 22.4 to 45.2 barrels a day by 2025. In other words, it meant United States dictating the fundamental economic policy of supposedly sovereign states in the Middle-East.

The NSC acknowledged the need to put pressure on "rogue states", including Iraq and Iran, to increase American access to their oil production, and also on the possibility that it might be necessary to use military force to seize oilfields or other installations. The sensitivity of American policy-makers to the fact that something like two-thirds of all American oil imports passes through the narrow Straits of Hormuz at the entrance to the Persian Gulf cannot be exaggerated. "In the Middle-East and Southwest Asia", the document said, "our overall objective is to remain the predominant outside power in the region and preserve US and western access to the region's oil". Michael T. Klare adds that it was hard to distinguish US military operations designed to fight terrorism from those designed to protect energy assets [Klare: 2004]. Alan Greenspan, former chair of the Federal Reserve, declared that "the Iraq war is largely about oil" in his recently released memoirs [Greenspan: 2007]. Though some Americans hold the view that America was not fighting for oil, it is far from truth.

Since America's economy depends largely upon its use of energy to run the day-to-day lives of its citizens, a feeling is growing that if US doesn't lock up Middle-Eastern oil for itself now, it won't have it for the use by Americans in the very near future. However, cost of ensuring this oil supply is a hefty one as the war launched by US in Iraq in the name of "The War on Terror" has seen not only as many as a million Iraqi deaths and many more wounded, and displaced 4.4 million Iraqis, the Americans are also losing lives there and the war has created so many US enemies around the world. As a result of this war, many Americans have come to believe that US policy is devoid of moral and ethical considerations of war. The feeling is that when we have not been attacked by the people we invade, there is no justification for invading a weakened nation for the resource we desire.

The elimination of Saddam Hussein was part of this oil policy who had allowed France to make inroads into Iraqi oil. Even in the case of Iran, how American policy establishment is more concerned about Iranian oil than about Iran acquiring nuclear weapons is clear from Henry Kissinger's statement that "An Iran that practices subversion and seeks regional hegemony—which appears to be the current trend—must be faced with lines it will not be permitted to cross. The industrial nations cannot accept radical forces dominating a region on which their economies depend, and the acquisition of nuclear weapons by Iran is incompatible with international security." [Wiessman: 2007]. US political leaders however denied such an addiction to Iraqi oil. But in American scheme of things, the Middle-East has always been the key to preventing the world running out of oil.

Under a system known as "production-sharing agreements", or PSAs, oil majors such as BP and Shell in Britain, and Exxon and Chevron in the US, would be able to sign deals of up to 30 years to extract Iraq's oil. Critics fear that given Iraq's weak bargaining position at the moment, it could get locked in to deals on bad terms now for decades to come. However, both Britain and US have repeatedly denied that Iraq war was fought for oiland expressed that oil revenues be put in a trust fund for Iraqi people to be administered through the United Nations, that they have no intention to seize Iraqi oil and that their oil will be used for Iraq's reconstruction But the optimism apart, the reality is that since the invasion, Iraqi oil production has dropped off dramatically. In a country where economy is dependent on oil—as oil accounts for 70 per cent of the economy—control of the assets has proved a recipe for endless wrangling. Most of the oil reserves are in areas controlled by the Kurds and Shias, heightening the fears of the Sunnis that their loss of power with the fall of Saddam would be compounded by economic deprivation. Not only is Iraq's whole oil infrastructure creaking under the effects of years of sanctions, insurgents have constantly attacked pipelines, so that the only steady flow of exports is through the Shia-dominated south of the country. Whatever little operations exist, these are rife with corruption and smuggling. The Iraqi people refuse to allow the future of their oil to be decided behind closed doors. A just negotiation cannot be expected at a time when Iraqis are still under conditions of occupation.

There are reports that the rich nations are trying to use even the World Trade Organization (WTO) to create a new global policy framework for "energy security" that would fundamentally redefine, under the logic of "free trade in services," who will access energy resources, which ones are used and how, and who will benefit most from their exploitation. Some influential countries at WTO, such as, United States, the European Union, and the Kingdom of Saudi Arabia are pressurizing a group of developing countries including many OPEC nations to "offer" specific commitments to liberalize their energy sectors under expanded world trade rules [Menotti: 2006]. Though energy sector is relatively new to the arena of multilateral trade negotiations, the WTO is becoming the site of what might be called "the other oil war" to ultimately decide who controls our energy future. The world is entering a period of historic transition, where energy security has become the new lens through which governments view international relations. European officials are panicking over Russia's recent restrictions on gas exports, Brazil feels vulnerable after Bolivia's decision to nationalize its energy industry, China's hunt for energy resources has many world leaders on edge, while the US continues to mobilize even more diplomatic and military might to secure supplies. If current proposals are approved, foreign energy services companies could win WTO rights to unilaterally decide which energy resources, energy workers, and energy technologies to use, extending even to the protection of their rights to perform "monitoring services" that enable remote control over pumping. Indeed, a WTO takeover of the energy sector could be best understood as the corporate response to mounting concerns about peak oil and the end of cheap energy because it aims to allow energy companies to switch to whatever type of energy is most profitable without government control. In short, for energy-exporting nations, WTO's expansion into energy policy could reduce governments' abilities to set conditions on foreign investors and foreign service providers and also to develop domestic policies related to employment, conservation of natural resources, engaging local service providers and in matters of acquiring technological expertise. Subjecting energy to WTO rules could result in the breakup of state-owned companies, which as of now control seventy percent of the world's one trillion barrels of proven petroleum reserves. WTO terms would remove or reduce the "barriers" that foreign companies dislike most, such as limits on foreign investment, requirements to partner with domestic firms, hire local labor, or transfer new technology. And all this is going to favor the rich, oil-importing nations such as US.

As such, US energy security which so much depended on the Middle-East makes it harder to fight the war on terror. Today, despite the fact that the newly discovered oil fields in Brazil, the Gulf of Mexico, the Caspian Sea, and off the coast of West Africa are changing the nature of world market and the global awareness towards energy conversation and new technologies have slowed the growth of energy consumption, keeping prices relatively low, the energy use has also multiplied simultaneously in the last quarter century. And procuring the oil of these new oil-hubs will not be easy as most of these potential sources of supply export only a small part of their production. Moreover, potential supplier-countries like Kazakhstan, Nigeria, Venezuela and Angola are hardly models of political stability; and the US faces stringent competition for oil supplies from energy-hungry China. Therefore, maintaining a secure supply of oil from Middle-East remains a problem, often a political one.

Some strategic analysts think that the American era in the Middle-East has ended and a new era in the modern history of the region has begun. It will be shaped by new actors and new forces competing for influence, in which Washington will have to rely more on diplomacy than on military might. In this, the local actors will enjoy an upper hand who might be committed to changing the *status quo*. Shaping the new Middle-East from the outside will be exceedingly difficult, and will be the primary challenge of US foreign policy for decades to come. The June 1967 war forever changed the balance of power in the Middle-East. The use of oil as an economic and political weapon in 1973 highlighted US and international vulnerability to supply shortages and price hikes. And the Cold War's balancing act created a context in which local forces in the Middle-East had significant autonomy to pursue their own agendas. The 1979 revolution in Iran, which brought down one of the pillars of US policy in the region, showed that outsiders could not control local events [Haass: 2006].

The end of the Cold War and the demise of the Soviet Union brought about an era in the region's history, during which the United States enjoyed unprecedented influence and freedom to act. But this American primacy has almost come to a close primarily owing to Bush's 2003 Iraq project. It is one of history's ironies that the first war in Iraq, a war of necessity, marked the beginning of the American era in the Middle-East and the second Iraq war, a war of choice, has precipitated its end.

Moreover, the failure of traditional Arab regimes to counter the appeal of radical Islamism and very often, a hasty push for elections in the name of promotion of democracy by US, regardless of the local political context—has provided terrorists and their supporters with more opportunities for advancement than they had before. With advancing globalization, the region of Middle-East was also impacted, as has been the case with all other regions. Now, it is less difficult for radicals to acquire funding, arms, ideas, and recruits. The rise of new media, and above all of satellite television, has turned the Arab world into a "regional village" and politicized it. Much of the content shown—scenes of violence and destruction in Iraq; images of mistreated Iraqi and Muslim prisoners; suffering in Gaza, the West Bank, and Lebanon—has further alienated many people in the Middle-East from the United States. As a result, governments in the Middle-East now have a more difficult time working openly with the United States, and US influence in the region has waned. In this scenario, though the US would enjoy a better credibility than any other outside power, it would certainly be less than what it earlier enjoyed. Since American agenda of pushing democracy in the region seems to be backfiring at the moment, authoritarianism, religious intolerance and anti-Americanism are likely to stay in near future. Islam will increasingly fill the political and intellectual vacuum in the Arab world and provide a foundation for the politics of a majority of the region's inhabitants.

The present status of Middle-East as a region fraught with terrorism, religious extremism, civil wars, authoritarianism and so on can be changed for better if the meddling by foreign powers is contained at least in military terms. Even democracy cannot be artificially and immediately generated in a social and political milieu that has long tradition of autocracies based on religion. Applying diplomacy and not military option in Iran, and reviving peace-talks on Israel-Palestine genuinely and honestly can revive the trust of the people of the Middle-East in US which only serving one's vested interests cannot achieve.

(Mis)managing the Iraqi Oil: Immediately after Iraq's occupation, the Bush White House made no secret of its plans to quickly dismantle that country's strong public sector though it could not be put into action due to its realization of the fact that denationalizing the oil industry would be a blatant violation of the Geneva Conventions which bar an occupying power from altering the fundamental structure of the occupied territory's economy.

Equally important was the stand taken by Ayatollah Ali Sistani, the senior Shiite cleric in the country whom the occupying Americans could not afford to alienate, who disapproved of the wholesale privatization of Iraq's major companies as, according to him, they belonged to the "community," meaning: the state. As a religious decree issued by a Ayatollah, his statement carried immense weight. Even more effective was the violent reaction of the industry's employees to the rumors of privatization. An increase in the bombing of oil facilities and pipelines was noticed which was built on the premise that privatization is coming.

US Quest for Oil in Central Asia: This kind of US interest in oil prompted it to have a special interest in the Central Asia as well, where the newly emerged Central Asian Republics (CAR) after the disintegration of former USSR in 1991 are rated as rich in some essential energy resources, such as, gas. In the Soviet era, Moscow's dominant influence over Central Asia was beyond dispute. But the emergence of the five independent Central Asian Republics (CAR) in 1991 has brought competition, not only from other great powers which hope to exert their influence in the region, including China and the United States, but also from energy companies around the world that want to compete for the region's vast oil and gas resources. Due to the geographical location of the region, Russia and China remain the two natural partners in most of the trading in the region. However, the growing US military presence in the region as part of the US-led war on terror became a cause of concern both for China and Russia and also for the governments of this volatile region, though to counterbalance US influence, China and Russia have developed Shanghai Cooperation Organization (SCO). The ethnic, religious and linguistic affinity of the CAR region with Turkey and Iran give them opportunity to explore new political and economic roles in the region. Whereas Turkey shares ethnic and linguistic ties with the Turkic states of Central Asia, the Islamic Republic of Iran shares a common language with Tajikistan and a border with Turkmenistan. It also shares the same religion, Islam, with all the states though, unlike them, it practices Shi'ite rather than Sunni Islam.

The United States, as the world's leading oil importer harboring the major part of the world's oil industry is Russia's chief contender in the region and has a persistent interest and a stake in the world's major oil provinces, wherever they are. Having plans to see Central

Asian oil reaching the world market and to exploit the investment opportunities for US oil companies, the United States' evident interest in getting an economic and political foothold in Azerbaijan and Central Asia can be understood. Since the region is landlocked, transit routes are required to export oil and gas to other countries. In the US government's scheme of things, Iran which can serve as a potential transit point is nevertheless seen as a major adversary and therefore, US policy has been to avoid any understanding between the region's oil and gas exporters and Iran. Instead, it wants Turkey to serve as a transit route so as to help it economically with transit revenues and also to ensure a regular supply of oil to Israel from the nearby Turkish ports. United States seems to have an overriding concern to avoid the oil from Azerbaijan and Central Asia reaching the Gulf. US interest in finding an alternative in the Central Asian oil seems to be partly due to put a downward pressure on Atlantic crude prices and partly to reduce the overall supply and price risk in the world oil market due to the dependence on the Gulf, which is politically a fragile region because of the protracted breakdown in the peace process between Israel and Palestine and the role of US seen in it.

Interestingly, Afghanistan also has an oil connection as a primary future route to oil from Caspian Sea region and Central Asia to reach other locations. The United States has made numerous attempts to help those US oil companies that want to build up a pipeline from Central Asia to Pakistan, so that they could avoid transporting oil through both Russia and Iran—both politically uncomfortable countries for US. Unocal, the California-based global energy-resource corporation, had plans in 1990s to have an oil pipeline from Turkmenistan through Afghanistan to Pakistan and possibly to India for the purpose of avoiding transportation through Iran and Russia. However, the terrorist attacks on US facilities overseas interrupted the pipeline project and Unocal halted its plans and withdrew staff from Afghanistan. When George W. Bush occupied Presidency, his administration tried to revive relations with Taliban on the condition of giving up Osama bin-Laden and in exchange US was ready to cut billions of dollars of revenue due on Taliban that would be generated through the pipeline project. Even a neighboring Iran, with its oil resources, nuclear ambitions and religious fundamentalism nowadays maintains influential positions within the regions of the Caspian Sea, the Persian Gulf and Central Asia. Some political analysts see a connection between the recent US military intervention in

Afghanistan, though in the name of war on terror, and unexploited energy reserves in the region [see: Burleigh: South Asia Voice: October, 2001; Said Aburish: 1998; Cohen: 2007].

Democracy Promotion or Regime Change?: Another Counter-Productive Attempt by US

The promotion of democracy in the Arab world, an area to date resistant to effective political liberalization, has become a central pillar in American Middle-East foreign policy as well as an integral element in the Global War on Terrorism (GWOT). The underlying assumption of US policy is that democracy will moderate some of the anti-American sentiments from the region as well as undermine terrorist activities and support. The idea was that greater political freedom can undercut the forces of Islamic radicalism and indoctrination. The available evidence so far however suggests that American attempts at promotion of democracy are proving to be counterproductive because it has polarized the political discourse and brought Islamist groups to the forefront as a powerful and popular political force in the Arab world.

The issue can be understood from two standpoints: (1) the reservations over US sincerity, depth of commitment, and consistency in behavior on the democratization issue, and (2) the strengths and weaknesses of the approach in the context of democracy as an antidote to Islamist terrorism.

The critics assert that the United States focuses too heavily on holding elections rather than on the structural underpinnings commonly found in liberal democracies across the world. Francis Fukuyama, a neoconservative turned a critic of US policy of preemption, sees a connection between US bid for oil and the policy of promotion of democracy in the Middle-Eastern autocracies. The important commodity 'oil' and the constant flow of cheap labor, both skilled and semi-skilled, from the historical to post-historical countries creates stakes in the democratization of historical countries, because if the oil weapon is used for political reasons by the underdeveloped countries of the historical world to disrupt its production, it can have grave consequences for the industrialized countries of the post-historical world. Due to the constant threats posed by these historical countries, still mired in religion, ethnicity and narrow questions of nationality, to their neighbors and to others in developed countries, it is in the interest of the post-colonial countries to prevent the spread of certain deadly technologies to the historical world that are highly prone to conflict and violence. And hence, the interest of post-colonial countries to promote the cause of democracy in the underdeveloped, oil-rich countries of the Middle-East and of other regions as well may be justified [Fukuyama: 1992: pp. 278-81].

In fact, America's obsession with spread of democracy is also an extension of national interest, indispensable for securing America's interests globally. The reasons are that: democratic regimes by their very nature are friendlier to US; they are better suited to work in harmony with other democratic nations and are consistent with development, human rights and liberties. But the critics would question if democracy were altruistically good, then why US selectively ignores or overlooks its absence in Pakistan, Saudi Arabia or for that matter in Russia. Rather, US forge alliances with these countries every now and then revealing its hypocrisy through duality of behaviour. But the hardcore neocons' answer to this question is that US should be very clear in pinpointing where it will support democracy in the framework of national interest, which Krauthammer, a neoconservative supporter of Bush policies would call: *where it counts.* This kind of pure 'national interest'-driven selective support for democracy invites criticism from various quarters, particularly from the religious fundamentalists in the Islamic countries who have a vested interest in continuation of confused situation within and outside their countries as such a situation help breed terrorism and implement their violent projects in the name of God and religion. America's widely justified military action in Afghanistan to overthrow the barbaric Taliban regime and bringing in a new, democratic government were taken as important steps in fighting Islamic radicalism and fundamentalism. However, the repetition of such a preemptive military action against the secular, though dictatorial too, regime of Saddam Hussein on flimsy grounds and installing a pro-US government there through the so-called 'democratic elections' raises doubts about the seriousness and validity of America's democracy projects in other non-democratic countries, particularly in the 'oil-rich' Middle-Eastern region.

The arguments are that there exists little evidence to demonstrate that Bush was able to export either freedom or democracy. On the contrary, there are signs of authoritarian governments gaining strength or they are in chaotic state. Afghanistan, Iraq and Pakistan are the most recent and living examples of this thesis [Cole: 2007]. On the contrary, as a backlash to democracy agenda, the authoritarian regimes and dictators are attacking civil society and adopt various strategies to resist US menace. These strategies include: increasing their military capabilities, blackmailing USA with acts of terrorism or using international organizations to weaken American sanctions against them [see: Walt S.: 2005]. They perceive US idea of democracy promotion as a code word for regime change.

One more institution that comes under the scanner of the critics of US efforts at democracy promotion is 'National Endowment for Democracy' (NED), which was created in 1984 by the government, but not as an arm of government. NED's main goal is to assist the supporters of democracy in undemocratic countries where the democracy-supporters are in conflict with their governments. Since the US government otherwise has good diplomatic relations with the governments in these countries, no material support can be given through 'proper' channels to the opponents of the government. The funds that originate in US Treasury but are distributed by an independent private agency, such as NED, not tied up with any particular US administrative unit are easier to receive as well as to distribute. However, it is exactly this mechanism that raises doubts on the US intentions of its efforts at promotion of democracy in non-democratic countries. Due to such a free status and with enormous funds, NED has a history of corruption and financial mismanagement; its functioning is superfluous and often destructive. Through the endowment, the American taxpayer has paid for special-interest groups to harass the duly elected governments of friendly countries, interfere in foreign elections, and foster the corruption of democratic movements [Corney B.: 1993]. The critics also question the wisdom of giving a quasi-private organization the fiat to pursue what is effectively an independent foreign policy under the guise of "promoting democracy."

Still another aspect of the debate is whether NED is simply a relic of the Cold War that should be eliminated for that reason. During the Cold War, the Soviet Union led a powerful ideological campaign against democracy, but there is no longer any such pervasive, systemic threat to freedom. Critics contend, therefore, that even if there was once a national security rationale for funding NED, that rationale no longer exists.

Another argument is that democracy agenda is grounded in petroleum-based politics and that is the reason Bush's efforts in Arab-

Muslim world since 9/11 have only helped to bring hard-line fundamentalists to power as is the case with Iran, Palestine and Iraq. Unless America breaks its addiction to oil, Middle-East's addiction to authoritarianism will continue [Thomas L Friedman, "Addicted to Oil", in *New York Times,* Op-Ed column on February 1, 2006]. In the Arab world, oil and authoritarianism are inextricably linked. Even the secular autocratic regimes, like those in Egypt, Libya, Syria and Iraq, never allowed the emergence of any truly independent judiciary, media, progressive secular parties or civil society groups—from women's organizations to trade associations. The mosque became an alternative power center because it was the only place the government's iron fist could not fully penetrate. That is why the outcome of the free and fair elections in any Arab country, be it Egypt or Palestine, the Islamists burst ahead. Both the mullah and secular dictators in Arab-Muslim world could sustain themselves, without empowering the people and without any progressive measures, because they had oil and massive foreign aid. In contrast, the East Asian military regimes quickly turned into civilian democracies because there already existed vibrant free markets with independent economic centres of power and thankfully, no oil. US is criticized, and rightly so, for selectively choosing the countries in the region for promotion of democracy, Saudi Arabia being generally exempted from such attempts given the benefits accrued from Saudi's trillions of dollars being invested in US stock markets that prohibits US from taking any real action in terms of democracy and human rights promotion.

Another important question is whether democracy is an antidote to terrorism? Though the academic literature on the relationship between terrorism and other sociopolitical indicators, such as democracy, is surprisingly scant and there are very few case studies that try to determine whether more democracy leads to less terrorism, the preliminary conclusions from the academic literature seem to discredit the supposedly close link between terrorism and authoritarianism that formed the basis of the Bush administration's logic. Although Bush's call for promotion of democracy in the Arab world was intended not only to end terrorism, but also to strengthen American security, the evidence suggests that such a thesis does not stand scrutiny and is not based on any sound empirical premise. Terrorism appears to stem from factors much more specific than regime type. Nor is it likely that democratization would end the current campaign against the United States. Al-Qaeda and like-minded groups are not fighting for democracy in the Muslim world; they are fighting to impose their vision of an Islamic state. Nor is there any evidence that democracy in the Arab world would "drain the swamp," eliminating soft support for terrorist organizations among the Arab public and reducing the number of potential recruits for them. Based on public opinion surveys and elections in some countries of the Arab world, the advent of democracy there seems likely to produce new Islamist governments that would be much less willing to cooperate with the United States than are the current authoritarian rulers.

A new survey of US public opinion on foreign policy conducted in January 2006 by a non-profit forum—*Public Agenda*—shows that the war in Iraq and terrorism are not the only problems on Americans' minds. Public concern over the United States' dependence on foreign oil may soon force policy-makers to change course. Surprisingly, the Republicans were the strongest critics of President Bush's democracy promotion agenda, though they were also the surest supporters of his policies. Though the war in Iraq was the primary foreign policy issue on which public pressure continued to mount, the issue of energy dependence on foreign countries also remains high on the mind of American public.

Despite the current spate of Arab Spring, the transition to democracy in the Arab world is a daunting task as is clear from the fact that there exists a familiar litany of Muslim-world elections—such as in Afghanistan, Iraq, Lebanon, the Palestinian territories, and Saudi Arabia—all pointing to the trend that for non-Islamist political forces, it will have to be patient and long wait to compete for power in these elections. A hushed-up democratization policy is bound to result in Islamist domination of Arab politics. America's initiative at elections in the Arab world is bound to deliver very different results *vis-a-vis* democratic transitions in Eastern Europe, Latin America, and East Asia. In those regimes, liberalism prevailed and its great ideological competitor, communism, was thoroughly discredited, whereas the Arab world offers a real ideological alternative to liberal democracy: the movement that claims as its motto "Islam is the solution." Washington's hubris should have been crushed in Iraq, where even the presence of 140,000 American troops has not allowed politics to proceed according to the US plan. If the United States really does see the democracy-promotion initiative in the Arab world as a "generational challenge," the entire nation has to learn these traits [Gause III: 2005].

US Militarism: Another factor in Unilateralism

The US military reach and extension determines its influence, control and dominance in the new century. With 700 to 800 military bases of United States in other countries, as on in 2005, and with America's increasing deployments in Iraq, as was repeatedly requested by the President (George W. Bush) to the Congress in January 2007, the ever all-encompassing imperial footprints and militarism are too clear, particularly in the context of pursuit of President Bush's strategy of preemptive wars in a unipolar world, who was being profusely supported by his neocon (neoconservative) colleagues in administration and in mainstream media.

"Interestingly enough, the thirty-eight large and medium-sized American facilities spread around the globe in 2005—mostly air and naval bases for our bombers and fleets—almost exactly equals Britain's thirty-six naval bases and army garrisons at its imperial zenith in 1898. The Roman Empire at its height in 117 AD required thirty-seven major bases to police its realm from Britannia to Egypt, from Hispania to Armenia. Perhaps the optimum number of major citadels and fortresses for an imperialist aspiring to dominate the world is somewhere between thirty-five and forty" [Johnson: 2006]. The worldwide total of US military personnel in 2005, including those based domestically, was 1,840,062 supported by an additional 473,306 Defense Department civil service employees and 203,328 local hires. Johnson's study believes that these staggering numbers do not include American garrisons in Kosovo, Afghanistan, Iraq, Israel, Kyrgyzstan, Qatar and Uzbekistan and colossal base structures in Persian Gulf and Central Asian areas since 9/11. The US government tries not to divulge any information about the bases US uses to eavesdrop on global communications, or nuclear deployments, which, as William Arkin, an authority on the subject, writes, "[have] violated its treaty obligations. The US was lying to many of its closest allies, even in NATO, about its nuclear designs. Tens of thousands of nuclear weapons, hundreds of bases, and dozens of ships and submarines existed in a special secret world of their own with no rational military or even 'deterrence' justification." [Johnson: 2006]. In some cases, foreign countries themselves have tried to keep their US bases secret and the Pentagon itself seems to want to play down the building of facilities aimed at dominating energy sources, or, in a related situation, retaining a network of bases that would keep Iraq under US hegemony regardless of the wishes of any future Iraqi government.

After the collapse of the Soviet Union in 1991, the huge concentrations of American military might in Germany, Italy, Japan, and South Korea were no longer needed to meet possible military threats from the Soviet Union or any of its allies close to these countries. Though some minor efforts were made by the Clinton administration to close some bases in Germany, such as those protecting the Fulda Gap, once envisioned as the likeliest route for a Soviet invasion of Western Europe, but nothing substantial was really done in those years to plan for the strategic repositioning of the American military outside the United States. Since US emerged as a sole superpower after collapse of Soviet Union, efforts were made to enhance military strength to promote, what critics term, overt imperialism that included preventive and preemptive unilateral military action, spreading democracy abroad at the point of a gun, obstructing the rise of any "near-peer" country or bloc of countries that might challenge US military supremacy, and a vision of a "democratic" Middle-East that would supply US with all the oil it wanted. A component of military enhancement was redeployment and streamlining of the military. The initial rationale was for a program of transformation that would turn the armed forces into a lighter, more agile, more high-tech military suitable for 21st century challenges.

After 9/11 terror attacks in 2001, this military strategy became urgent. After its 2001 campaign against the Taliban and al-Qaeda—the military/defense planners in Bush administration started working on strategies to defend the United States by "deterring aggression and coercion" in four "critical regions": Europe, Northeast Asia (South Korea and Japan), East Asia (the Taiwan Strait), and the Middle-East so as to be able to defeat aggression or to "win decisively" in the sense of "regime change" and occupation, as the case may be. In other words, as the military analyst William M. Arkin commented, such a strategy goes beyond preparing for reactive contingencies and reads more like a plan for picking fights in new parts of the world. After the collapse of the Baathist regime in Baghdad the Department of Defense came up with an Integrated Global Presence and Basing Strategy, known informally as the "Global Posture Review."[5]

The Defense Department made plan of re-structuring America's global forces as in present times the military has to operate with an entirely different concept and therefore needs to be prepared for the whole range of military operations, from combat to peace-keeping, anywhere in the world. pretty quickly [Johnson: 2006]. But criticisms apart, there is a different viewpoint supporting US enlarging its defense infrastructure. The argument is that the United States, the world's leading military power throughout the twentieth century, requires a defense budget sufficient enough for the country to confront the threats of the twenty-first century. The supporters of an enhanced defense budget though admit the excesses in some parts, but believe that those are inherent features of any government activity. And unlike some areas of government spending where it is possible to consider privatization or a shift of responsibility to states and localities, defense must remain a federal responsibility, unchecked by the disciplining forces of competition. Moreover, deterring other great powers, such as Russia and China, will require Washington to maintain its dominance in conventional warfare and therefore at least to maintain its current level of military spending. But in addition, the United States now faces three new types of threats for which its existing military capacity is either ill suited or insufficient. First, there are relatively small regional powers, such as North Korea, Iran, and Pakistan that can or might be able to strike the United States and its allies with weapons of mass destruction (WMD). Second, there are global non-state terrorist networks, such as al-Qaeda, with visions of re-creating the world order. And third, there are independent terrorists and groups motivated less by a long-term vision of global conquest than by hatred, anti-Americanism, and opposition to their own governments. Each of these threats is exacerbated by the relative ease with which crude WMDs can be developed due to the diffusion of modern technology and the potential emergence of a black market in fissile material.

Furthermore, there seems to be general agreement that the United States has committed so much of its war-fighting capacity to Iraq and Afghanistan that it could not fight in Iran or North Korea or elsewhere if that were deemed necessary. That limit on capacity encourages US adversaries to behave in ways that are contrary to US interests. Dealing appropriately with diverse enemies requires an increase in intelligence capabilities, both human and electronic. And the new threats mean that US will have to have today's IT-based

Revolution in Military Affairs (RMA).[6] According to Boot, this is because the US military has, so far at least, "gone the furthest fastest" in exploiting the new information technologies. It has outstripped all competitors by combining new and sophisticated hardware, volunteer professionals capable of using it, superb training under realistic conditions, and the right tactical doctrines. The result has been an enormous yield in conventional military power. Although the United States' economic power and "soft" power have also contributed to its strong global position, it is its military power—based partly on the mastery of information technologies—that truly distinguishes the country today [Boot: 2002].

However, US difficulties in capturing bin-Laden (though it succeeded in this task in 2011) and problems in the US occupation of Iraq have reason to question RMA's build-up as a military nirvana. When reviewing the gamut of theories, **three fundamental versions of Revolution in Military Affairs** come to the forefront. The **first** perspective focuses primarily upon changes in the nation-state and the role of an organized military in using force. This approach highlights the political, social, and economic factors worldwide, which might require a completely different type of military and organizational structure to apply force in the future. The **second** version deals with the nature of the emerging international order that requires the evolution of weapon technology, information technology, military organization and military doctrine among advanced powers. This tends to include advanced intelligence, surveillance, command, control, communication and reconnaissance (Network-centric warfare that aims to connect all troops on the battlefield.) and precision force. An advanced version of RMA incorporates other sophisticated technologies, including unmanned aerial vehicles or *UAVs*, nanotechnology, robotics, and biotechnology. Finally, the **third concept** is that a "true" revolution in military affairs has not yet occurred or is unlikely to. Several critics expressing their concern point out that a "revolution" within the military ranks might carry detrimental consequences, produce severe economic strain, and ultimately prove counterproductive. Such authors tend to profess a much more gradual "evolution" in military affairs, as opposed a rapid revolution. To the suspicion of the critics of RMA, if a catastrophic and catalyzing event like new Pearl Harbor were to occur, RMA could be justified to get the massive funding needed. It was in response to this prediction that John Pilger in *New Statesman,* December 12,

2002 made the assertion that "[t]he attacks of 11 September 2001 provided the 'new Pearl Harbor'."

In fact, RMA targets to weaponize and hence dominate space. A document called "Vision for 2020," which begins with this mission statement: "US Space Command—dominating the space dimension of military operations to protect US interests and investment" [see: www.spacecom.af.mil/usspace]. The real purpose is to dominate space to protect the commercial interests of America's elite class and according to current projections, over $1 trillion will be required from American taxpayers. Discussing this American project of global domination associated with the weaponization of space, Richard Falk says: "The empire-building quest for such awesome power is an unprecedented exhibition of geopolitical greed at its worst, and needs to be exposed and abandoned before it is too late." Falk continues: "If this project aiming at global domination is consummated, or nearly so, it threatens the entire world with a kind of subjugation, and risks encouraging frightening new cycles of megaterrorism as the only available and credible strategy of resistance" [Falk: 2002]. It reflects Reagan administration's Strategic Defensive Initiative, which today is called the Missile Defense Shield. Though the system is designed to shield America from attacks, its real purpose is to prevent other nations from deterring America from attacking *them.* This statement further suggests that Iran, Iraq, and North Korea were later determined by President Bush to deserve the title "axis of evil" because of their perverse wish to develop the capacity to deter the United States from projecting military force against them.

Intelligence: A Foreign Policy Tool

Since America's entry into the World War and realization of its new responsibilities as a leading super power ever after, a robust peacetime intelligence infrastructure was put in place through the National Security Act on July 26, 1947 which served as an important foreign policy tool, particularly during the cold war period. As a result, the Central Intelligence Agency (CIA) was created in 1947 with the signing of the National Security Act by President Harry S. Truman. The act also created a Director of Central Intelligence (DCI) to serve as head of the United States intelligence community; to act as the principal adviser to the President for intelligence matters related to the national security; and serve as head of the Central Intelligence Agency. It enabled the US administration to put spy planes in the sky, satellites into the space and listening posts in strategic locations around the world. A professional cadre of analysts, case officers, linguists, technicians and program managers was built up that served Washington's interests very well throughout the cold war era, but which, however, failed to meet the newly emerging challenges of the post-cold war world infested with the brutal acts of terrorism and violence generated largely by religious fundamentalism. The state-sponsored terrorist groups as well as the non-state actors that committed catastrophic acts such as 9/11 have necessitated an overhauling of intelligence apparatus based to a collaborative approach of information gathering and sharing that enables the government and policy-makers to take timely and tactical decisions. Intelligence Reform and Terrorism Prevention Act (IRTPA) 2004 was one such step in this direction but not fully sufficient to meet the current and forthcoming challenges. As the 9/11 Commission report also highlights the dismal failure of intelligence to prevent the tragic events of September 11, 2001, the country was unprepared even after enough warnings from terrorist attacks on World Trade Centre in 1993, on Khobar Towers in Saudi Arabia in 1996, on US Embassies in East Africa in 1998 and on USS Cole on 2000.

In addition to the sixteen agencies which formally comprise the US Intelligence Community (IC), it also includes selected tactical military intelligence and security organizations, as well as those responsible for security responses to transnational threats, to include terrorism, cyber warfare and computer security, covert employment of weapons of mass destruction, narcotics trafficking, and international organized crime. It conducts intelligence activities considered necessary for the conduct of foreign relations and the protection of the national security of the United States.

Disregard for International Law and Institutions: A close comparative analysis of the central argument of Hobbes' Leviathan and the designing and implementation of US foreign policy during Bush Presidency indicates that the way the current international political maneuvering of international law and institutions by the so-called hegemon of the world has taken place, the dilemma of the 'state of nature' does not seem to be resolved [Duffy: 2005]. The blatant disregard for international laws and agencies, national and international court decisions, and broadly accepted customary law have been used, manipulated, and flouted before and after 9/11. This shows that there is no sovereign power and no order other than a balance of fear, a balance exemplified by the Cold War. The way the international law confronted the crimes and criminals in Yugoslavia and Rwanda is nowhere seen in the current conflicts in Afghanistan and Iraq.

In spite of the UN Security Council's Resolution 1373 of September 28, 2001, which "established the obligation of all states to, among other things, afford other states the greatest measure of assistance in connection with criminal investigations or proceedings in relation to terrorism", the US and its allies have displayed no scruples in pressuring detainees to provide information. Undoubtedly, the issue of international responsibility in reaction to terrorism, especially when non-state actors are involved, is a thorny issue. Nevertheless, there are proven instruments for the peaceful resolution of conflict, the use of international humanitarian law, and the extent of international human rights law. Yet, a case study of the US detention of what the Bush Administration described "enemy combatants" at the US naval base in Guantanamo, Cuba demonstrates the need for a proper understanding of the international legal regime. The way in which the Guantanamo detainees have been kept outside the framework of international law has done a great damage to the status of the United States and the United Kingdom in the international community.

Overarching Themes/Major Schools of US Foreign Policy Tradition

Now, having discussed the major theoretical and practical aspects of US foreign policy that intentionally or unintentionally has gone to exacerbate terrorism in the post-cold war period, let us find out what have been the major rallying points in the formation of American foreign policy, particularly in the context of 'realist' vs. an 'idealist' approach.

Henry Kissinger, a proponent of *realpolitik*, who played a dominant role in United States foreign policy making in 1970s, wrote in his book *Diplomacy* [Kissinger: 1994] that at the turn of the last century the United States faced a choice between two fundamentally different approaches to international relations, one represented by Theodore Roosevelt and the other by Woodrow Wilson. Theodore Roosevelt, President from 1901 to 1909, urged the nation to establish its relations with the rest of the world solidly on the concept of national interest, based on military might and balance of power diplomacy. Woodrow Wilson, President from 1913 to 1921, pressed the nation to support a foreign policy grounded in law and deriving strength from cooperation with others. Throughout twentieth century, one can notice the swing between these two approaches to international relations in United States foreign policy, i.e. (1) Traditional/Idealist/*Isolationist*[7] and (2) *Realist*[8]/Pragmatist/ Radical, both of which drove America to successfully prevail over its cold war rival, the Communist Russia and the world ushered into a new era of the post-Cold War, posing new challenges for the world as well as for the only superpower—the United States of America. In addition to these two foreign policy schools, Idealist and Realist, there are other foreign policy doctrines also that have guided various US Presidents from time to time and in accordance with the requirements of that particular time. These schools can be identified as "Liberal Internationalist"[9] and "Neo-Realist or Neoconservatist". The overlapping over certain issues amongst all these schools cannot be ruled out.

The US foreign policy between post-World War II up to the end of Cold War swung between idealism and pragmatism. Cliff Staten in his book *US Foreign Policy since World War II: an Essay on Reality's Corrective Qualities* traces the post-World War II course of US foreign policy through what he interprets as alternating periods of activist intervention and less active pragmatism. Examples of the former he sees as the policies of Presidents Truman and George W. Bush; examples he cites of the latter, a less ambitious approach to foreign policy, are those of Presidents Eisenhower and George H.W. Bush. The book argues that America has a split personality when it comes to foreign policy. On one hand, it is rooted in pragmatism: If one method does not work, try another until the problem is solved. The logic behind such an approach is that in order to be successful, the US must, at a minimum, balance its resources and capabilities with its objectives and liabilities. The other side of American personality grows out of the mythology of American exceptionalism—that shining beacon for the rest of the world to see the New World. It is this Wilsonian idealism that leads US to believe that it can and should remake the world in its image. This split personality of realism and idealism has historically manifested itself in US foreign policy. But whenever a critical choice had to be made between limits of resources and missionary/idealistic goals, America has always switched on to the pragmatic side of its foreign policy character to redeem itself. In other words, the redemptive qualities of reality save US from the sins of missionary instinct.

The noted Cold War historian John Lewis Gaddis has argued that the history of America's containment policy toward the Soviet Union also reflected the swing of a pendulum between idealist and realist sides. Whereas in the early years of the Cold War, US foreign policy focused on containing communism in Europe where its military strength was greatest, the idealist side of US foreign policy motivated it to revert to its traditional isolationist character elsewhere, although it is important to note that this isolationism during this time resulted in the fall of China to the communists, the red scare in the United States, and the development of the atomic bomb by the Soviet Union. However, the invasion of South Korea by the communist North led President Truman without Congressional approval to mobilize and place US troops on the Korean peninsula. Military aid was also extended both to the French in Indochina and to the Philippine government that was facing an internal rebellion by the Huks. The Seventh Fleet was ordered to prevent any Chinese attack on Taiwan. Defense spending nearly tripled that year. Containment was no longer limited to Western Europe. It now included East Asia. Thus, from 1950, pragmatism was the guiding principle of US foreign policy.

By December of 1952 Eisenhower had come to the conclusion that the United States should not be engaged in a conventional war on the Asian mainland. As a Realist, he recognized the limits of the country's capabilities. He ushered in a pragmatic foreign policy by redefining and limiting the goals of containment and redirected support for the French efforts in Indo-China and initially promised only economic aid to South Vietnam. He came to rely more upon the resources of other countries through alliances, such as the Baghdad Pact and the Southeast Asian Treaty Organization, to counter the communist threat. Eisenhower failed to offer support for the Hungarian uprising clearly recognizing that the United States did not have the military resources to challenge the Soviet Union in its sphere of influence. He used low-cost CIA-engineered coups to achieve goals in Guatemala and Iran. He created the United States Information Agency in 1953 with the idea of using American culture, film, music and theatre to counter the appeal of communism. Given his administration's fiscal conservatism, Eisenhower actually reduced the size of the armed forces and developed the strategic doctrine of massive retaliation.

However, certain events at the end of Eisenhower's presidency paved the way for idealist, missionary foreign policies of Presidents Kennedy and Johnson. They committed US men and money in Vietnam, all in the name of survival of liberty. Embarrassed by the Bay of Pigs fiasco and helpless to stop the construction of the Berlin Wall, Kennedy expanded US commitments in Indo-China, pledged to defend West Berlin from attack, and launched the Alliance for Progress in Latin America. By the end of the year 1965, President Johnson had dramatically expanded US resources, commitments, and objectives in Southeast Asia. By the end of his term in office he had committed more than a half a million American soldiers to Vietnam. And sending 22,000 troops to the Dominican Republic in 1965 reflected his goal to preempt any future Cuban style revolution in American backyard.

Under the influence of the realist Henry Kissinger as Nixon's closest foreign policy adviser, US was led to the adoption of a more limited foreign policy referred to as détente. A gradual withdrawal from the Southeast Asian quagmire, while playing the "China card" and linkage politics as new and less costly resources against the Soviet Union, allowed for the successful arms limitations and trade agreements between the two super-powers. An implied agreement over conduct and spheres of influence (the Helsinki Accords) permitted the United States to refocus, set new priorities, and limit its foreign policy objectives to the maintenance of a balance among the major powers of the world.

Watergate, a Congress reasserting itself into the foreign policy processes, the first oil crisis, and the 'secretive and amoral' realist foreign policy practices of the Nixon and Ford years among other factors led to the election of Jimmy Carter as president under whom the US foreign policy began to make the shift back to its idealist side. Initially this was evident in Carter's emphasis on human rights. The rise of the right wing of the Republican Party led by Ronald Reagan, the second oil crisis brought on by the fall of the Shah of Iran coupled with the hostage crisis, the Sandinista revolution in Nicaragua, and the Soviet invasion of Afghanistan influenced President Carter to pull the SALT II Treaty from Senate consideration and to ask for a dramatic increase in defense spending, including strategic forces. The so-called *Carter Doctrine of 1980* committed the United States to the defense of open waterways in the Persian Gulf.

An expansionist empire of the Soviet Union prompted the

'Reagan Doctrine' that promised not only containment but also the rollback communism worldwide, leading to a dramatic increase in defense spending and a renewed arms race with the Soviet Union. Central America and the Caribbean became the testing ground for his crusade. Yet, by Reagan's second term, it was becoming evident that this policy could not be sustained. Reagan's second term was marked by arms agreements with the Soviet Union and Congressional efforts to rein in his prolific spending habits.

President George H.W. Bush, who presided over the end of the Cold War, the demise of the Soviet Union, and a concerted effort by the Congress to balance the budget, directed a foreign policy that witnessed success in the Gulf by masterfully putting together the largest and most successful war coalition since the Second World War. Financially speaking, the Gulf War cost the United States very little and Bush refused to expand the war beyond the limits set by United Nations resolutions and the US Congress. President Bill Clinton, who was elected primarily on a domestic policy platform, was under continuous Congressional and public pressure to balance the budget and to reap the major benefit of the end of the Cold War—reduced spending on defense or the so-called peace dividend. This policy was followed also by his successor, George W. Bush till 9/11, who initially promised an even more limited role for the United States in world affairs. But, the post-Cold War presidents before 9/11: George H. W. Bush, Bill Clinton, and George W. Bush, at least until 9-11, lacked the unifying theme equivalent to the Cold War anti-communist crusade that had served Truman from 1950 to 1952, Kennedy and Johnson from 1961 to 1969 and Reagan from 1981 to 1985. Anti-communism allowed those presidents to gain public support for their idealistic foreign policies, which the post-Cold War presidents lacked.

But, the horrific events of 9/11 and the unifying theme of a war against global terrorism again provided the initial public support, almost of the cold war era, for President George W. Bush to embark on a renewed, more realistic foreign policy. Several other factors also explain the dramatic change in foreign policy under Bush. One was the failure of the Clinton administration to define a strategic vision for the United States in the post-Cold War era. The strategic vision of neoconservatives supported the use of American forces in a pre-emptive and, if necessary, unilateral approach to achieve a "new American century."

The neoconservatives (neocons) wanted to exercise a more assertive, pre-emptive, and unilateral foreign policy that drew upon the hegemonic power and military capabilities of the United States to reshape the global political system. Since the end of the first Gulf War many in this group had already made clear their desire to topple Saddam Hussein in Iraq. The events of 9/11 led to Bush's declaration of a global war against terrorism and the struggle against al-Qaeda, bin-Laden, and the Taliban led to broad and open-ended nation-building goals in Afghanistan. In Iraq, failure to find weapons of mass destruction, coupled with the failure to plan adequately for the post-Saddam era, led to the redefinition and dramatic expansion of US objectives. These objectives now fall under the amorphous goal of bringing democracy to the entire Middle-East and much of the developing world.

Undoubtedly, the United States faces very real dangers today and potentially bigger ones might come in the future, but these are not threats that can be tamed by increased spending on the most expensive components of military power. A more modest and sensible national security strategy can and should be put in place.

In fact, the war in Iraq that pushed parts of the US armed forces close to the breaking point, demonstrated how the soldiers and volunteers pay a high price for the politicians' miscalculations. The efforts to combine the war against Saddam Hussein with the "war on terror," were another mistake as the two were not the same. Moreover, the main challenge for US foreign policy makers is not killing the terrorists but finding them, and the capabilities most applicable to this task are intelligence and special operations forces. It requires recruiting, training, and effectively deploying a limited number of talented and bold people with the relevant skills. As regards the other major security threat to the United States, such as, the proliferation of nuclear and biological weapons of mass destruction (WMD), this problem too cannot be solved through the deployment of large and expensive conventional forces. At best, bombing can set back a program temporarily; at worst, it can energize it.

The notion that the United States has the right and the responsibility to regulate regional peace, discipline violators of civilized norms, and promote democracy and world order is the product of the emergence of a unipolar world. This myth is going to be pretty costly for the US as well as for the entire world. The global-policing ambition also runs the risk of generating resistance in local situations as the local actors rarely see the dominant power's actions as benign or disinterested; external interventions often provoke resentment and nationalist reactions. As a result, muscular military activism tends to multiply enemies, whereas sound strategy should reduce and divide them. Spending a great deal of national blood and treasure on remote problems elsewhere in the name of humanitarian intervention has generally proved to be quite expensive. Iraq experience is a case in point.

In addition to this, it needs to be noted that the two main engines of change are at work in the global system namely, *first:* the rise of Europe as a counterweight to the United States; and *second*, the erosion of liberal internationalism, a moderate, centrist international behaviour that manages the international system through compromise, consensus, and international institutions. A Europe that made way for US to play the central role in world affairs after World War II is gradually stitching together its separate nation states into a collective whole, and might soon usurp what it lost to US six decades ago. The evidence of this rising counterweight Europe is its GDP which is about $8.5 trillion while that of US is 10 trillion—a close competition. The Euro is not only steadily gaining ground against the dollar, it has replaced the dollar as the main currency in Central Europe. The Euro is emerging as a rival to the dollar as a global reserve currency.

At political level, Europe is projecting the voice of Europeans on the global stage and is trying to carve a niche for European geopolitical map of the world. The EU is gradually moving into the realm of geopolitics, diplomacy, and military ambition. It replaced the US as the main diplomatic arbiter in the Balkans. It was the EU that brokered the dissolution of Yugoslavia into Serbia and Montenegro and is gradually assuming the responsibility for most peace-keeping in that part of the world. The EU might not be a military rival to US, at least in near future, but it will have an impact that might lessen the power of US to call all the shots. Apart from EU, the other groupings such as BRIC with emerging China, India and Brazil are already recognized as potential power centers.

Europeans' growing belief, especially in the wake of 9/11, that the US is in the process of decamping from the European continent, and that they therefore need to assume responsibility for their own defense is going to expedite the process of more intense unification of Europe and EU.

Also, there is a growing resentment in Europe against American power and policy, which makes them want to get rid of their American pacifier and protector: they no longer want to hitch their wagon to a superpower that they see as having run-off the rails. As for the US foreign policy under President Bush and the crisis over Iraq is concerned, the first casualty was "Liberal Internationalism" of US.

Conclusion

Orientalism versus Occidentalism Debate

Having discussed the elements and factors in US behavior and policy prescriptions that are widely held responsible for a violent terrorist reaction, let us now see what the 'Orientalist' *versus* 'Occidental' debate has to say about western views on East's religion, language, culture, peoples and society. Edward Said in his controversial book *Orientalism* (Said: 1978) brings forth the negative connotations of whatever is Eastern, particularly Islam largely due to the fact that western views were shaped during the era of European imperialism in the 18th and 19th centuries in which kind of a high self-perception emerged at the cost of 'inferior' Eastern way of thinking that was mired in destruction of a democratic order in the West. This puts Islam on the defensive pleading apologetically for its humanism, its contribution to civilization and development and perhaps even to democratic niceness. The fact is that since Islam represents not only a formidable competitor but also as a late-coming challenge to Christianity, it is seen with hostility and fear.

Said found it surprising why Europe's advance into modern scientific age that freed it from superstitions and ignorance did not include Orientals in their modernization project. In the contemporary global political system, feels Said, Islamic Orient is important for its resources and for its geographical location and for this reason US after World War II being in hegemon dominant position started investing heavily in sponsoring the burgeoning academic and journalistic, one-sided studies in Islam and Middle-East, which historically has not been a realm of interest for the western scholars and academics. Though US does not merely see Muslims and Arabs as oil suppliers or potential terrorists, they are projected in a way to make them vulnerable to military aggression. The resurgence of Islam in some parts of Islamic world, such as Iran, Afghanistan and

Palestine are seen as encroachments upon traditional Western hegemony and hence the Western/American endeavors to project this image: Islam being anti-human, anti-democratic, anti-Semitic and anti-rational associated with the kinds of Ayatollah Khomeini, Col. Muammar e-Qaddafi, Palestinian terrorists and the frightening notions of '*jihad*', slavery, subordination of women, without any mention of Arabic writers, intellectuals, musicians and other positive contributions made by a host of Islamic intellectuals from the Muslim world.

In other words, the whole swatches of Islamic history, culture and society simply do not exist except in the truncated, tightly packaged forms made current by the media. In the similar vein, Samuel P. Huntington had also cautioned against the dangers of Western universalism. Democracy is promoted but not if it brings Islamic fundamentalists to power; non-proliferation is preached for Iran and Iraq but not for Israel; free trade is the elixir of economic growth but not for agriculture; human rights are an issue for China but not with Saudi Arabia; aggression against oil-owning Kuwaitis is massively repulsed but not against non-oil-owning Bosnians. Double standards in practice are the unavoidable price of universal standards of principle. He further argues that in the emerging world of ethnic conflict and civilizational clash, Western belief in the universality of Western culture suffers three problems: it is false; it is immoral; and it is dangerous...Imperialism is the necessary logical consequence of universalism.

The problem with western scholars, writers and journalists in all these decades have been that they are hardly familiar with non-western culture and civilization and have formed a habit to take for granted the universality of western values and norms and hence construct a particular view of the world that does not want to look beyond secular democracy, individual human rights, secularity and rule of law as the only valid values. This west-centrist approach creates a sort of myopia for the norms and values non-western. Therefore, those scholars who were obsessed with communism as the greatest hindrance to democracy and justice, etc. were shocked to see that even the end of communism did not bring the much thought-out change in favor of democracy. Even the much-hyped and certainly good in many ways project of structural globalization brought only institutional dimension of modernity related with science and technology and the economic benefits resulting from them—though that too has its critics. Though the science and technological aspect of modernity was accepted by other non-western civilizations, some of these had their strong objections and opposition to be receptive to cultural globalization. Hence, what Tibi would say, more structurally the globalized world, the more culturally fractured it would be [Tibi: 2002: p. 66]. The failure of this project of modernity to move beyond its present abode—US and Western Europe, whose thought-processes it dominated in the post-colonial period—is the main cause for the ongoing process of cultural and religious revival in non-western societies, particularly the Islamic societies who have come to understand that modernization is not necessarily the westernization.

A sort of anti-Americanism has increased in recent years and America's popularity in the rest of the world has declined precisely due to these universalistic pretensions. The United Statesis strong enough to do as it wishes with or without the world's approval and should simply accept that others will envy and resent it. The world's only superpower does not need permanent allies; the issues should determine the coalitions, not *vice-versa*. Though US has recovered from unpopular policies in the past, such as those regarding the Vietnam War, but that was often during the Cold War, when other countries still feared the Soviet Union as the greater evil. By using its soft power, it was to draw others into a system of alliances and institutions that has lasted for 60 years. The Cold War was won with a strategy of containment that used soft power along with hard power. But winning terrorism without the cooperation of other countries, despite their considerations of self-interest, is very difficult [Nye: 2004].

The unrealistic neoconservative goal of global domination has resulted in the current and future geopolitical tensions. The concept of the United States openly violating international law with unilateral "preventive wars" in the oil-rich regions of the world may simply not be tolerated forever by other industrialized nations. Unfortunately, recent US foreign policies are exacerbating global tensions and could lead to even worse warfare than the Bush years have already wrought. There must be no misunderstanding—global resource warfare *will* ultimately leave even the so-called "winner" in a ruined state of energy deprivation along with economic and moral bankruptcy. As Robert Freeman observed, the destiny of the US will be decided by how willing the American people are to adjust to the realities of hydrocarbon depletion. The reduction of fossil-fuel consumption and the transition to a more sustainable energy paradigm will be decades long and very difficult. However, the Americans are largely blinded by government propaganda through pervasive corporate media renders the people unaware that terrorism and other international frictions are too often the result of 50 years of US meddling overseas, including the overt or covert replacement of governments with puppet regimes and dictators [Johnson, Charmer http://www.thenation.com/doc/20011015/johnson, 27 September, 2001].

The three factors that led to the erosion of American power are: overuse of its military, a strained economy and loss of legitimacy and the solidification of anti-Americanism as a political idea. *Jihadist* Islamism is the most violent expression of this [Power: 2002; Cohen: 2007]. This downward spiral owes much to America's failure to grasp the New World as it is. The attempt to blackball the changing realities is what some commentators would term as neo-McCarthyism. Due to the burgeoning union between oil politics, grassroots fundamentalism, and the financialization of the national economy, some strategic analysts predict a "perfect storm" on the America's horizon [Phillips, Kevin: 2006]. There certainly is a connection between American global hegemony and oil, a single energy source on which the economic power of nation-states has relied historically. Various American governments instead of confronting the inevitable oil problem through long-term planning try to resort to 'unholy' alliance with powerbrokers in oil industry.

Though there is nothing unusual for a country such as US to take robust and effective action to address the immense strategic vulnerability that results from the widening gap between American energy production and American energy consumption, what invites criticism is its arrogant, unilateral way which is more concerned about the profits of American corporations than anyone else. As Hodgson rightly argues, "It would have been one thing to put the United States at the head of a coordinated, international effort to reduce dependence on carbon fuels; it is quite another to send American forces around the world—from Colombia to Uzbekistan, from Angola to the Persian Gulf—to oblige the world to meet an American demand for almost half the world's oil supplies" [see: Hodgson: 2005].

Middle-East's apprehensions: In the Middle-East, the United States has led some of the greatest modern military operational victories: Operation Desert Storm, Operation Enduring Freedom (OEF), and Operation Iraqi Freedom (OIF) and has successfully exercised the ability to wage war and achieve decisive military operational victories. However, the US successes of removing militarily oppressive regimes are quickly overshadowed by skeptical Islamic populations and regimes that are quick to oppose US efforts to save or free other people from the same type of oppression and predicament. This should raise **a few questions**:

- Why do Muslims, saved from genocide in Bosnia by US-supported forces, condemn the US trying to liberate Iraqis from an oppressive regime?
- Also why do Afghans, freed from the oppression of the Taliban, one year later condemn and protest US efforts to remove Saddam Hussein's oppressive government, wave pictures of Saddam Hussein in the streets and refer to Saddam Hussein as their "Muslim brother"?
- How does a US strategy with nearly ten years of engagement in the Middle-East result in the strong collective public opposition to US?

The answers to some of these questions could be that

- US strategy has inadequately applied its instruments of power to effectively conduct engagement operations and promote policies in the Middle-East;
- While oil remains the main focus, democracy and human rights in the Middle-East are not the priority in the region;
- Concerns about its impact on their natural resources as well as on their politics;
- Omission in addressing specific problems such as rooting out terrorism and promoting human rights and democracy;
- US policies and actions in the Israel-Palestine conflict shape the perceptions of the population in the region; due to such actions;
- Islamic Jihad and radical fundamentalists continue to threaten US interests and increase the requirements for force protection.

So, it can be concluded that the neoconservative experiment for radical transformation through militarism and unilateralism has ended in failure, having proved itself poorly adapted to the realities of the 21st century. As all wars come to an end, terrorism is also bound to end at some point of time. The Cold War ended not with US forces but when the realization of the uselessness of such a war dawned upon the leaders of both the rivals. The major challenge for American foreign policy in the twenty-first century lies in "the management of relations among contending centers of power and the consequent rivalries which will ensue. The War on Terror and post-9/11 orientations of US foreign policy have displayed a kind of political fundamentalism that has *willy nilly* gone to exacerbate Islamic fundamentalism in the post-cold war period. In brief, the state of affairs between the West, especially the US, and the Muslim World is dire. There is a need for a radical reconsideration of the most basic propositions of the use of American power towards the other countries, particularly the Muslim countries, if a cataclysmic confrontation with the Muslim World—in which the Americans seem to be increasingly standing alone—is to be avoided.

NOTES AND REFERENCES

1. The United States has concluded its military withdrawal from Iraq after nearly nine years of occupation. At the peak of the invasion, around 170,000 US troops were manning the tight-fisted occupation that was being conducted from more than 500 bases around Iraq. The occupation killed anywhere between a 100,000 to a million Iraqis. Nearly 4,500 US troops also died. In terms of treasure, the war cost Washington around one trillion dollars. After this withdrawal, only 157 troops will stay behind, mainly to train Iraqi forces. A handful would also remain to guard the monstrous US embassy, the largest American diplomatic facility anywhere in the world [*The Hindu*, 18 December, 2011].
2. November 2007 National Intelligence Estimate (NIE) report, titled – *Iran: Nuclear Intentions and Capabilities*, judged that Tehran halted an alleged nuclear weapons program in fall 2003, and that it remained halted as of mid-2007. The estimate further judged that US intelligence did not know whether Iran intended "to develop nuclear weapons," but that "Iran probably would be technically capable of producing enough HEU [highly enriched uranium] for a weapon sometime during the 2010-15 time frame" if it chose to do so. Iran continues to maintain that its nuclear program is peaceful [see: NIE Report, 2007].
3. Refer to 1787 US treaty with Morocco for peaceful cultural and commercial relations; with Sultanate of Oman during 1840s and the efforts of Christian missionaries to promote education and progress there, leading to the establishment of the American University of Beirut, the American University of Cairo; President Franklin Roosevelt's efforts in establishing firm relations with the Kingdom of Saudi Arabia and so on.
4. By some accounts, as much of 40 percent of the country's oil revenues goes

straight into the pockets of the ruling family. Lacking a free media and freedom of press, civil society institutions like rights' advocacy groups, trade unions and political parties, bar associations and any other democratic institutions, secrecy and fear permeate every aspect of the state structure in Saudi Arabia, and most Gulf Kingdoms. In the absence of an independent judicial process, the political and religious opponents of these governments are detained indefinitely without trial or are imprisoned after grossly unfair trials. Torture is endemic, and foreign workers, particularly non-Muslims are most at risk. No criticism of Islam, the ruling family or the government is tolerated. Forms of punishment are often quite barbaric and include public executions and amputations. Also frequent are private acts of vendetta and rape against the docile immigrant workers, particularly women. As the citizenry feels a growing sense of alienation *vis-a-vis* these highly unpopular and repressive regimes, the US government is now being held responsible for this state of affairs.

5. Top ten countries with maximum oil reserves in order of their highest reserves: Saudi Arabia, Iraq, Kuwait, U.A.E., Iran, Venezuela, Russia, Mexico, Libya and US.

6. RMA designed to establish American military supremacy over all other nations and groups, is the vision of defense and military related analysts and think-tanks. They suggest that an information age technology combined with appropriate doctrine and training would enable a small but very advanced US military to protect national interests worldwide with unprecedented efficiency. It is a theory about the future of warfare, often connected to technological and organizational recommendations for change in the military/defense apparatus tied to modern information, communications, and space technology, taking into account the total system integration in the United States military. Renewed interest was placed on RMA theory and practice after what many saw as a stunning, one-sided victory by the United States in the 1991 Gulf War against Iraq through use of superior satellite, weapons-guiding, and communications technology. After the Kosovo War where the United States did not lose a single life, RMA is visualized as the future face of warfare.

7. **Isolationism** is a foreign policy, which combines a ***non-interventionist military policy*** and a political policy of economic nationalism. In other words, it asserts to avoid wars not directly related to territorial self-defense. But though economic nationalism is often interpreted as 'economic protectionism', in US the Libertarian Isolationism can best be defined as a policy of non-participation in foreign political relations, ***but free trade and affability to all in economic matters.*** The United States' policy of non-intervention in political affairs was maintained throughout most of the 19th century. The first significant foreign intervention by the US was the Spanish-American War, which saw the US occupy and control the Philippines. Since this was the first take-over of non-contiguous territory where people speak a different language, this is generally considered the first colonial act of the US.

   Today, non-interventionists argue that the United States is far removed from its earlier history of non-intervention. They point to both Republican and Democratic presidents who, since the 1950s, have often used intervention as a tactic of foreign policy, including: Harry S. Truman's intervention in Korea,

1950; Dwight Eisenhower, John F. Kennedy, Lyndon B. Johnson and Richard Nixon's in Vietnam in 1960-70s; John F. Kennedy's in Cuba; Nixon's in Chile; Jimmy Carter's in Soviet invasion of Afghanistan in 1979; Ronald Reagan's in Grenada in 1983; George H.W. Bush's in Panama, Kuwait and Somalia; Bill Clinton's intervention in Haiti in 1994 and in Bosnia in 1995, in Kosovo in 1999; George W. Bush's intervention in Afghanistan in 2001, in Iraq in 2003. Some scholars might argue that through America's decades of membership in the United Nations, multilateral interventionism has become the dominant policy of the United States government. However, unilateral interventionism was articulated as the preferred policy of the George W. Bush administration for the invasions of Afghanistan and Iraq.

The isolationist view of America considered US to be spiritually superior to the Old World. It came from the Founding Fathers as they felt that US had big oceans on both side and nice small neighbors to the north and south, so why imperil our natural security by roaming around the world and getting involved in great-power intrigues? The unilateralism came from two things: one, *American exceptionalism*, the sense that US was a new, unique nation, and therefore need not engage itself in the world, or if it does, it has to be on its terms; and two, *American populism*: the Jeffersonian and Jacksonian tradition of fierce jealousy over sovereignty, liberty, and autonomy. Isolationism and unilateralism triumphed over internationalism even in the period between two great wars of 20th century when the internationalism was most pressingly required. That is precisely the reason why the idealist Woodrow Wilson essentially failed to sell liberal internationalism to the American people which they just refused to buy and hence Senate's disapproval of Treaty of Versailles of which League of Nations—Wilson's brainchild—was a part. However, in today's world, which is driven by export-driven economies, massive population flows and globalization reducing the world into a 'global village, Isolationism is outdated, intellectually obsolete and politically bankrupt. [Krauthammer: 2004]. But there are other foreign policy analysts who believe that in the wake of the Cold War and with the ending of the threat from the 'evil; empire', one can see a return to a more traditional American foreign policy that is characterized by unilateralism and neo-isolationism more than by liberal internationalism [Kupachan: 2003]

8. **Realism**, in the context of international relations, encompasses a variety of theories and approaches, all of which share a belief that states are primarily motivated by the desire for military and economic power or security, rather than ideals or ethics and that the mankind is not inherently benevolent but rather self-centered and competitive. This term is often synonymous with power politics. The theory believes that the international system is anarchic with no authority above states capable of regulating their interactions; states must arrive at relations with other states on their own, rather than it being dictated to them by some higher controlling entity. Sovereign states are the principal actors in which each state is motivated by their own national interest and security in pursuit of which the states strive to amass resources. Their comparative level of power derived primarily from their military and economic capabilities determines relations between states. There are no universal principles, which

all states can use to guide their actions. The classical realists can be cited as Chankaya, the Indian philosopher of statecraft and the author of *Arthasastra,* Thomas Hobbes who in his work *Leviathan* highlights the selfish nature of man in the pre-political state of nature, Niccolo Machiavelli, a Florentine political philosopher, who wrote *The Prince,* in which he held that the sole aim of a prince (politician) was to seek power, regardless of religious or ethical considerations.

Modern realism, whose precursor can be taken as Hans Morgantheau, began as a serious field of research in the United States during and after World War II. George F. Kennan and Walter Lippmann belong to the class of modern Realists. *Neorealism*, deriving from classical realism, also advocates state security in which the states must be on constant preparation for conflict through economic and military build-up. The prominent neo-Realists are Robert Jervis, Kenneth Waltz, Stephan Walt and John Mearshiemer. *Modern Realist* Statesmen are Henry Kissinger, Zbigniew Brzezinski and Brent Scowcroft [Wikipedia search]. The *neoconservatives* in the current George W. Bush administration belong to this school of Realists, who following the tradition of Henry Kissinger, respect power, believe that democracy is important and tend to downplay the internal nature of other regimes and their human right concerns.

9. By "liberal internationalism," is meant a moderate, centrist internationalism that manages the international system through compromise, consensus, and international institutions. It is a foreign policy doctrine that argues that liberal states can intervene in other sovereign states in order to pursue liberal objectives of maintaining international peace and security. Such intervention includes military intervention and humanitarian aid. This view is contrasted to isolationist foreign policy doctrines, which oppose such intervention. The goal of liberal internationalism is to achieve global structures within the international system that are inclined towards promoting a liberal world order. To that extent, global free trade, liberal economics and liberal political systems are all encouraged. In addition, liberal internationalists are dedicated towards encouraging democracy to emerge globally. Once realized, it will result in a 'peace dividend', as liberal states have relations that are characterized by non-violence, and that relations between democracies is characterized by the democratic peace thesis. In the US, it is often associated with the American Democratic Party. This School believes that, through multilateral organizations such as the United Nations, it is possible to avoid the worst excesses of "power politics" in relations between nations.

Examples of liberal internationalists include Woodrow Wilson, Franklin D. Roosevelt, Bill Clinton and George W. Bush before 9/11. Proponents of the 'Realist' tradition in international affairs, on the other hand, are skeptical of liberal internationalism. They argue that it is power—diplomatic clout and military force (or the threat of it)—that ultimately prevails.

Though it failed in preventing World War II in face of a stronger 'primitive imperialist realism', but we must remember this world would have been an immensely dangerous place had realism prevailed at the time. Since 1945, Realism emerged as a dominant stream of foreign policy processes, particularly in US and was followed every now and then by various US Presidents.

In the years following World War II, isolationism gave way to internationalism once again, a dominant philosophy undergirding American foreign policy, implying both conflict and co-operation with other nations including intervention if necessary though after Vietnam debacle, both mass public and leadership segments were divided on the issue of America's interventionist role. The other fundamental principals of this school having almost a consensus were themes like globalism, anti-communism and containment (till 1991) and military might along with belief in multilateralism, international institutions and law and disarmament. Depending on the requirements of time and situation, internationalism has had two faces: co-operative and militant. Post-World War II history of international politics by and large confirms to this. But the 21st century's realists and neoconservatives ridicule some aspects of liberal internationalism calling it an ideology of passivity at times when very urgent American intervention and preemption are opposed by 'liberal internationalists' who, Realists blame, see the pursuance of national interest as merely self-interest writ large. But such a neoconservatives' criticism of liberal internationalists, such as, that of President Clinton's policies in 1990s, cannot be substantiated as he involved US in four military actions: deepening intervention in Somalia, invading Haiti, bombing Bosnia and going to war over Kosovo.

REFERENCES

Ansari, Ali M., *Confronting Iran: The Failure of American Foreign Policy and the Next Great Crisis*, London, Oxford University Press, 2005.

Armstrong, David and Joseph Trento, *America and the Islamic Bomb: The Deadly Compromise*, Hanover: Steerforth Press, 2007.

Baker, Anni P., *American Soldiers Overseas: The Global Military Presence*, Praeger, 2004.

Basevich, Andrew J., *American Empire: The Realities and Consequences of US Diplomacy*, Cambridge, Harvard University Press, 2003.

____, *The New American Militarism: How Americans Are Seduced By War*, New York & London, Oxford University Press, 2005.

Bergen, Peter, *Bin-Laden, CIA links hogwash*, CNN, 15.08.2006.

Betts, Richard K., "A Disciplined Defense: How to Regain Solvency", *Foreign Affairs*, Nov./Dec. 2007.

Biddle, Stephan D., *American Grand Strategy after 9/11: An Assessment*, A Monograph, Strategic Studies Institute, 2005: http://www.carlisle.army.mil/ssi www.carlisle.army.mil/ssi

Blum, William, *Killing Hope: US Military and CIA Interventions Since World War II.*

Bricker, Bonnie and Adil E. Shamoo, "The costs of War for Oil," *Foreign Policy in Focus (FPIF)*, October 19, 2007.

Boot, Max, *The Savage Wars of Peace: Small Wars and the Rise of American Power*, Basic Books, 2002.

Brands, H.W., *What America Owes to the World?: The Struggle for the Soul of Foreign Policy*, Cambridge University Press, 1998.

Briody, Dan, *The Halliburton Agenda: The Politics of Oil and Money*, May 2004.

Bruce, Steve, *God is Dead: Secularism in the West*, Blackwell: Oxford, 2002.

Brzezinski, Zbigniew, *The Grand Chessboard: American Primacy and Its Geostrategic Imperatives*, New York: Basic Books, 1997.

____, *Second Chance: Three Presidents and the Crisis of American Superpower*, New York, Basic Books.

Brzezinski, Z., "Terrorized by 'War on Terror': How a Three-Word Mantra Has Undermined America", *Washingtonpost.com*, March 25, 2007; B01.

Buchanan, Patrick J., "Whose War?" *The American Conservative*, March 24, 2003.

Byrne, Malcolm and Jeffrey Richelson (eds.), "Terrorism and US Policy, 1968-2002: From the Dawn of Modern Terrorism to the Hunt for Bin-Laden", *Chadwyck-Healey/Proquest*, 2002. In the Middle-East, 2006.

Cameron, Fraser, *US Foreign Policy After the Cold-War: An Introduction*, UK: Routledge, 2002.

Cheney, Dick, Donald Rumsfeld, Paul Wolfowitz, Jeb Bush, and Lewis Libby, "Rebuilding America's Defenses: Strategy, Forces and Resources For a New Century", *Project for New American Century Report*, September 2000.

Chellaney, Brahma, 'Out of Sync, Out of Mind', *Hindustan Times*, Editorial, 6.03.07.

Chomsky, Noam, *Rogue States: The Role of Force in World Affairs*, New Delhi: Indian Research Press, 2000.

Chossudovsky, Michel, "War and Globalisation: The Truth Behind September 11", *Global Outlook and Center for Globalisation (CRG)*, 2002.

Clarke, William R., *Petrodollar Warfare: Oil, Iraq and the Future of the Dollar*, New Society Publishers, 2005.

Cogan, Charles, "Partners in Time: The CIA and Afghanistan since 1979," *World Policy Journal*, Summer 1993.

Cohen, Roger, "Let's face the New Core Facts", *International Herald Tribune*, 12 November 2007.

Cole, N. Scott, "Hugo Chavez and President Bush's Credibility Gap: The Struggle Against US Democracy Promotion" in *International Political Science Review*, 28:4, September, 2007.

Conry, Barabara, "Loose Cannon: The National Endowment for Democracy", in *Cato Foreign Policy Briefing No. 27*, November 8, 1993.

Cooley, John K., *Unholy Wars: Afghanistan, America and International Terrorism*, ND: Penguin, 2001.

Cronin, Patrick M., *From Globalism to Regionalism: New Perspectives on US Foreign and Defense Policies*, Diane Publishers, 1994.

Coll, Steve, "Anatomy of a Victory: CIA's Covert Afghan War," *Washington Post*, July 19, 1992.

Coll, Steve, Ghost Wars: The Secret History of the CIA, Afghanistan, and Bin-Laden, From Soviet Invasion to September 10, 2001, NY: The Penguin Press, 2004.

Congressional Research Service (CRS) Report, US Democracy Promotion Policy in the Middle-East: The Islamist Dilemma, The Library of Congress, June 15, 2006.

Cook, Robin, "The Struggle Against Terrorism cannot be Won by Military Means", *Guardian*, 8 July, 2005.

Cooley, John, *Unholy Wars: Afghanistan, America and International Terrorism*, New York: Pluto Press, 2002.

Cordovez, Diego and Selig S. Harrison, *Out of Afghanistan: The Inside Story of the Soviet Withdrawal*, NY, Oxford University Press, 1995.

Crenshaw, Martha, *Terrorism in Context, Pennsylvania*, PSU Press, 1995.

Cronin, A.K., "Behind the Curve: Globalization and International Terrorism", *International Security*, 27 (3): 30-58, 2002.

Crotty, William, *Democratic Development and Political Terrorism: The Global Perspective* (ed), Boston : Northeastern University Press, c2005.

Curtis, Adam, *The Power of Nightmares*, a BBC Documentary, 2004.

Daalders, Ivo H., James M. Lindsay, *America Unbound: The Bush Revolution in Foreign Policy*, Brookings Institution Press, 2003.

Dobrot, Colonel Laurence Andrew, "The Global War on Terrorism: A Religious War", *Strategic Studies Institute of the US Army War College at Carlisle*, Pennsylvania: November 2006.

Dobson, Alan P. and Marsh, Steve, *US Foreign Policy Since 1945*. 1. Auflage, Routledge: 2000, 2006.

Dolan, Chris J., *In War We Trust: The Bush Doctrine And The Pursuit Of Just War*, Burlington, VA, Ashgate, 2005.

Dolan, Chris J. and Betty Glad (eds.), *Striking First: The Preventive War Doctrine and the Reshaping of US Foreign Policy*, New York & London, Palgrave Macmillan, 2004.

Domke, David, God Willing? Political Fundamentalism in the White House, "The War on Terror" and the Echoing Press, London and Ann Arbor MI, Pluto Press, 2004.

Dreyfus, Robert, *Devil's Game: How the United States Helped Unleash Fundamentalist Islam*, New York, NY: Metropolitan Books, 2005.

Duffy, Helen, *The "War on Terror" and the Framework of International Law*, Cambridge, UK: Cambridge University Press, 2005.

Elan, Evan, *The Empire Has No Clothes: US Foreign Policy Exposed*, The Independent Institute, 2004.

Etzioni, Amitai, *Security First: for a Muscular, Moral Foreign Policy*, Yale University Press, 2007.

Falk, Richard, *The Great Terror War*, Northampton, Mass.: Olive Branch Press, 2002.

Fandy, Mamoun, 'US Policy in the Middle-East', *Foreign Policy In Focus*, Vol. 2, No. 4, January 1997.

Feldman, Noah, *After Jihad: America and the Struggle for Islamic Democracy*, NY: Strauss, 2003.

Freedman, Lawrence, "A Choice of Enemies: America Confronts the Middle-East," *Public Affairs*, 2008.

Fukuyama, Francis, *End of History and the Last Man*, New York; Free Press, 1992.

Fukuyama, Francis, *America at the Cross Roads: Democracy, Power and Neo-Conservative Legacy*, 2006.

Gaddis, John Lewis, *The US and the Origins of the Cold War* 1941-47, 1972.

____, *Surprise, Security and the American Experience*, 2004.

Gannon, Kathy, *I is for Infidel: From Holy War to Holy Terror*, New York: Public Affairs, 2005.

Gasper, Phil, "Afghanistan, the CIA, bin-Laden, and the Taliban", *International Socialist Review*, November-December 2001.

Gause III, F. Gregory, "Can Democracy Stop Terrorism?", *Foreign Affairs*, September/October 2005.

Gilbert, Burnham *et al.*, Mortality After the 2003 Invasion of Iraq: A cross-sectional Cluster Sample Survey, *Lancet*, 2006, 368(9545):1421.

Gordon, Michael and General Bernard E. Trainer, *The Generals' War: the Inside Story of the Conflict*, Sage, 2003.

Gordon Philip H., *Winning the Right War: The Path to Security for America and the World* : Brookings Institution Press, 2007.

Greenspan, Alan, *The Age of Turbulence: Adventures in a New World*, Penguin, 2007.

Griest, P. and S. Mahan, *Terrorism in Perspective*, London: Sage, 2003.

Griffin, David Ray, *The New Pearl Harbor: Disturbing Questions About Bush Administration and 9/11*, Massachusetts: OliveBranch Press, 2004.

Gunaratana, Rohan, *Inside Al-Qaeda: Global Network of Terror*, New Delhi: Roli Books, 2002.

Hahn, Peter L., *Crisis and Crossfire: The United States and the Middle-East Since 1945*, 2005.

Halliday, Fred, *Two Hours That Shook the World: September 11, 2002: Causes and Consequences, End of the Republic*, Metropolitan Books, 2004.

Haass, Richard N., "The New Middle-East" *Foreign Affairs*, November/December 2006.

Heikal, Mohamed H., *Cutting the Lion's Tail: Suez Through Egyptian Eyes*, Harper Collins, 1987.

Heine, Jorge, "Empire Defanged? Non-US Perspectives on US Foreign Policy", *International Political Science Review*, Vol. 28, No. 5, November, 2007.

Herring, George, *America's Longest War*, (4th edition), New York, 2001.

Hiro, Dilip, "How the Bush Administration's Iraqi Oil Grab Went Awry: Greenspan's Oil Claim in Context" in *Global Policy Forum*, September 25, 2007.

____, Blood of the Earth: The Global Battle for Vanishing Oil Resources, Politico's Publishers, 2008.

Hodgson, Godfrey, "Oil and American Politics", openDemocracy.net, 2 October, 2005; also produced in *Energy Bulletin*, 8 Oct. 2005.

Huntington, Samuel P., *The Clash of Civilizations and the Remaking of World Order*, 1996.

Ikenberry, John (ed.), *America Unrivaled: The Future Balance of Power*, NY: Cornell Univ. Press, 2002.

Jaco, Charles D. *et al. The Complete Idiot's Guide to the Politics of Oil*, Alpha Books, 2003.

Jane's Intelligence Review, *Why Was Russia's Intelligence on Al-Qaeda Ignored*? 5 October, 2001.

Jackson, John S. III and Chris Barr, "Who Rules the Middle-East Agenda?" in Crotty William: 2005.

Jervis, R., "Why The Bush Doctrine Cannot be Sustained", *Political Science Quarterly*, 120 (3), 2005.

Johnson, Charmer, *Blowback: The Costs and Consequences of American Empire*, New York: Metropolitan Books, 2002, 2005.

Johnson, Charmer, *The Sorrows of Empire: Militarism, Secrecy and the End of the Republic*, Metropolitan Books, 2004.

Johnson, Charmer, *Nemesis: The Last Days of American Republic* (reprint): Metropolitan, 2006, 2007.

Joffe, Josef, Uberpower: The Imperial Temptation of America, New York and London: Norton, 2006.

Juergensmeyer, Mark, *The New Cold-War?: Religious Nationalism Confronts Secular State*, University of California Press, 1994.

Kabbani, Muhammad Hisham, "Islamic Extremism: A Viable Threat to US National Security", at Open Forum at the US Department of State, January 7, 1999.

Kabhe Unipolar Moment: Realist Theories and US Grand Strategy after the Cold War", *International Security*, Vol. 21, No. 4 (Spring, 1997), pp. 49-88.

Kagan, Robert, *The Return of History and the End of Dreams*, Knopf, 2008.

Kaplan, Lawrence F. and William Kristol, *The War over Iraq: Saddam's Tyranny and America's Mission*, Encounter Books, 2003.

Kaplan, Robert D., *Warrior Politics: Why Leadership Demands a Pagan Ethos*, New York: Random House, 2001.

Kepel, Gilles, *Jihad: The Trail of Political Islam*, London: I.B.Tauris, 2000.

Kepel, Gilles, *Jihad*, Harvard University Press, 2002.

Kinzer, Stephen, *All the Shah's Men: An American Coup and the Roots of Middle-East Terror*, John Wiley & Sons, 2003.

Kiracofe Jr., Clifford A., "Challenges to US Foreign Policy", a Lecture at *National Council of Arab-US Relations*, Washington D.C., September, 2002.

Kissinger, Henry A., *Diplomacy*, New York: Simon and Schuster, 1994.

Kissinger, Henry A., *Does America Need a Foreign Policy?: Toward A Diplomacy for 21st Century*, Simon & Schuster, 2002.

Klare, Michael T., *Blood and Oil: The Dangers and Consequences of America's Growing Dependency on Imported Petroleum*, Henry Holt, 2004.

Krauthammer, Charles, "Democratic Realism: An American Foreign Policy for a Unipolar World", 2004 Irving Kristol Lecture, *American Enterprise Institute (AEI)*, Feb. 10, 2004.

Kupchan, Charles and Joanne J. Myers, *The End of the American Era: US Foreign Policy and the Geo-politics of 21st Century*, Carnegie Council, 2003.

Kux, Dennis, *The United States and Pakistan: Disenchanted Allies*, Woodrow Wilson Press, Washington D.C., 2000.

Lewis, Bernard, *The Crisis of Islam: The Holy War and Unholy Terror*, NY; Modern Library, 2003.

Levering, Ralph B., *The Cold War: A Post-Cold War History*, 2nd edition, 2005.

Libschutz, Ronnie D., *After Authority: War Peace and Global Politics in 21st Century*, New York, Suny Press.

Lin, Sharat G., "US Lying About Halabja: Justifying the Invasion of Iraq", *Economic and Political Weekly*, September 8, 2004, pp. 3625-32.

Lippman, Walter, *The Cold War: A Study in US Foreign Policy* (1947).

Lippman, Walter, "The Cold War," *Foreign Affairs*, 1987, 65:4:869-84.

Mackay, Neil, "Bush Planned Iraq 'Regime Change' before becoming President", *Sunday Herald*, 15 September, 2002.

Mani, Rama, "In pursuit of an Antidote: The Response to September 11 and the Rule of Law", *an Opinion, Conflict, Security and Development*, 3:1, April, 2003.

Maley, William (ed.), *Fundamentalism Reborn?: Afghanistan and the Taliban*, New York University Press: New York,1998.

Mamdani, Mahmood, *Good Muslim, Bad Muslim: America, The Cold-War and the Roots of Terror*, New York: Pantheon, 2004.

Manbiot, George, "War on the Third World", *The Guardian*, March 5, 2002.

Mastanduno, Michael, "Preserving the Unipolar Moment: Realist Theories and US Grand Strategy after the Cold War", *International Security*, Vol. 21, No. 4 (Spring, 1997), pp. 49-88.

Mayer, Jane, "Contract Sport: What did the Vice-President do for Halliburton", *New Yorker*, 16-23 February 2004.

Mc Coy, Alfred W. and Cathleen B. Read, Leonard P. Adams, *Politics of Heroin in Southeast Asia: CIA Complicity in the Global Drug Trade*, Lawrence Hill Books, 1972, 1991 (revised).

Mead, Walter Russell, *Mortal Splendor: The American Empire in Transition*, Houghton Miffin, 1988.

_____, *God and Gold: Britain, America and the Making of the Modern World*, Random House, 2007.

Mearsheimer, John J., *The Tragedy of Great Power Politics*, NY: W.W. Norton & Co., 2001.

Mearsheimer John J., Stephan M. Walt, *The Israel Lobby and US Foreign Policy*, Farrar, Straus.

_____, "Soft Power: The Means to Success in World Politics," *Public Affairs*, 2004.

Menotti, Victor, *The Other Oil War: Halliburton's Agenda at the WTO*, A Policy Brief on the Energy Services negotiations in the World Trade Organization (WTO), *International Forum on Globalization*, 2006.

Milani, Abbas, "US Foreign Policy and the Future of Democracy in Iran", *The Washington Quarterly*, Summer, 2005, pp. 41-56.

Munthe, P, "Terrorism: Not Who But Why?" *Royal United Service Institute Journal*, August, 2005: 8-12.

Neack, Laura, *The New Foreign Policy: US and Comparative Foreign Policy in 21st Century*, NY: Rowman & Littlefield, 2002.

Nye, Joseph Jr., *The Paradox of American Power: Why the World's Superpower Can't Go It Alone*, Oxford University Press, 2002.

_____, "Soft Power: The Means to Success in World Politics," *Public Affairs*, 2004.

The National Security Strategy of the United States, National Security Council, The White House, September 2000 [http://www.whitehouse.gov/nsc/nssall.html]

National Intelligence Estimate: Iran—Nuclear Intentions and Capabilities, November 2007.

9/11 Commission Report, The National Commission on Terrorist Attacks Upon the United States (also known as the 9/11 Commission), 2002.

O'Donnell, Tom, *Global Political Economy of Oil: US Policy in the Persian Gulf.*

Petras, James, "Zioniosts and the US Join Intelligence Report", *Economic and Political Weekly*, Vol. XLIII, No. 1, January 5, 2008.

Phillips, Kevin, *American Theocracy: The Peril and Politics of Radical Religion, Oil and Borrowed Money in 21st Century*, New York: Viking, 2006.

Power, Samantha, *A Problem from Hell: America and the Age of Genocide*, New York: Basic Books, 2002.

Prados, John, "Notes on the CIA's Secret War in Afghanistan", *The Journal of American History*, Vol. 89, No. 2, Sept. 2002.

Rashid, Ahmed, *Jihad: The Rise of Militant Islam in Central Asia*, Yale University Press, 2002.

____, *Taliban: Militant Islam, Oil and Fundamentalism in Central Asia*, Yale University Press, 2000.

____, *Al-Qaeda in 2007: Striving to Regain the Initiative*, 2006.

____, *Taliban: The Story of Afghan Warlords*, London: Pan Books, 2001.

____, *Descent into Chaos: The United States and the failure of Nation Building in Pakistan, and Central Asia*, New York: Viking Adult, 2008.

Riedel, Bruce, "Al-Qaeda Strikes Back", *Foreign Affairs*, May-June 2007.

Risen, J. and Judith Miller, "No illicit arms found in Iraq, US inspector tells Congress", *The New York Times*, 2003.

Said, Edward E., *Orientalism*, New York: Random House, 1978.

____, "Islam Through Western Eyes", *The Nation*, 28th April, 1980 issue or visit: http://www.thenation.com/19800426/19800426said

____, *Covering Islam: How the Media and the Experts Determine How We See the Rest of the World*, New York: Pantheon; London: Routledge & Kegan Paul, 1981.

____, *End of Peace Process*, NY: Pantheon Books, 2000.

Said, Aburish, *A Brutal Friendship: The West and the Arab Elite*, London: Indigo, 1998.

Schindler, John R., *Unholy Terror: Bosnia, Al-Qaeda and the Rise of Global Jihad*, New York: Zenith Press, 2007.

Scott, Peter Dale, *Drugs, Oil, and War: The United States in Afghanistan, Columbia and Indochina*, Rowman & Littlefield, 2003.

Scott, Peter Dale, *The Road to 9/11: Wealth, Empire and the Future of America*, Berkeley, California: University of California Press, 2007.

Shapiro, Ian, *Containment: Rebuilding a Strategy Against Global Terror*, Princeton: Princeton University Press, 2007.

Shuster, Mike, "Middle-East and the West: The Clash with Islam", NPR, August 24, 2004.

Smith, Jeffrey R., "Hussein's Pre-war Ties To Al-Qaeda Discounted", *The Washington Post*, April 6, 2007.

Staten, Cliff, "US Foreign Policy Since World War II: An Essay on Reality's Corrective Qualities", http://www.americandiplomacy.org"

Stiglitz, Joseph E. and Linda J. Bilmes, *The Three Trillion Dollars War: The True Cost of the Iraq Conflict*, W.W. Norton, 2008.

Suskind, Ron, *The Price of Loyalty*, NY: Simon & Schuster, 2004.

Tanzer, Michael, *The Political Economy of International Oil and the Underdeveloped Countries*, Boston: Beacon, 1969.

Telhami, Shibley, *The Stakes: America and the Middle-East*, Cambridge: Westview Press, 2002.

Tellis, Ashley J., "Pakistan: Conflicted Ally in the War on Terror", *Carnegie Endowment Policy* Brief No. 56, November 2007.

Tenet, George, *At the Center of the Storm: My Years at the CIA*, New York: Harper Collins, 2007.

Tibi, Bassam, *The Challenge of Fundamentalism: Political Islam and the New World Disorder*, Berkeley, University of California Press, 1998, updated edition, 2002.

____, *Islam Between Cultures and Politics*, Palgrave, 2005.

Tucker, Robert W. and David C. Hendrickson, "The Sources of American Legitimacy," *Foreign Affairs*, November/December 2004, pp. 18-32.

United Nations Press Release, 8 March, 2001.

Vanaik, Achin, "Introduction to "Selling US Wars", *Transnational Institute*, a TNI Publication, March 2007.

Wallerstein, Immanuel, "Last Call for Two-State Solution", *International Herald Tribune*, November 2, 2007, p. 4.

Walt, S., *Taming American Power: The Global Response to US Primacy*, New York: Norton, 2005.

West, Cornel, *Democracy Matters: Winning the Fight Against Imperialism*, New York: Penguin, 2004.

Wiessman, Robert, "Greenspan, Kissinger: Oil Drives US in Iraq, Iran" a Blog from *Huffington Post*, Posted September 17, 2007.

Woodward, Bob, *Pan of Attack*, Simon & Schuster, 2004.

Wright, Steven, *The United States and Persian Gulf Security: The Foundations of the War on Terror*, Ithaca Press, 2007.

Yankelovich, Daniel, "The Tipping Point", *Foreign Affairs*, May/June, 2006.

Yousaf, Mohammed, *Afghanistan: the Bear Trap: The Defeat of a Superpower*, Casemate, 2001.

Zakaria, Farid, "Vengeance of Victors", *Newsweek*, January 8, 2007.

____, *The Post-American World*, W.W. Norton and Company, Inc., 2008.

Chapter 4

Expanding Terror Networks: Asian Context

Those who call us terrorists wish to prevent world public opinion from discovering the truth about us and from seeing the justice on our faces. They seek to bide the terrorism and tyranny of their acts, and our own posture of self-defense.

— *Yasser Arafat*

There exists a global consensus on the 'rise of Asia' as the most dynamic region in world politics and economics. This region accounts for 60 percent of world's more than 6.1 billion population constituted by a variety of ethnic, linguistic, religious and cultural compositions and nearly $ 30 of GDP outweighing that of Europe. It commands global attention both for its economic growth as well as security challenges. Whereas the economic growth is led by India and China, a large number of the economic giants belong to eastern flank in terms of centres of trade, investment, financial reserves, natural resources, manufacturing hubs, service providers, science and technology development, IT sector development and expanding markets with the major regional players, such as, China, Japan, South Korea, Singapore, Malaysia. ASEAN and APEC together represent economic sample of the region. West and Central Asia is anybody's envy due to its riches in oil and natural gas commodities, basic to any industrialized country's economic sustenance and growth.

Centrality of Asia in International Terrorism

Since the contemporary wave of Islamist terrorism began in Middle-East in 1968 with the Popular Front for the Liberation of Palestine's

(PFLP) acts of plane hijackings and kidnappings, terrorism attained both global and regional dynamics with Asia being a centre of gravity in terms of perpetrating and perpetuating the acts of terror. It is a region that poses serious security challenges to the region as well as to the world at large. The dramatic history of Asia in recent years is marked by the factors like—the fall of Soviet Union, emergence of five relatively unstable Central Asian Republics, nuclearization of South Asia, 9/11 attacks and subsequent regime changes wrought by the American power in Afghanistan and Iraq and developments in Iran and North Korea having potential for creation of new war zones in the continent. It is the main theatre of pervasive and seemingly unending 'war on terror' as well as persisting and protracted political and ethnic conflicts and insurgencies having a spill-over effect on not only the neighboring regions but also in other parts of the world. Of late, religious fundamentalism and the terrorism driven by it have become symbols associated with this region. Asia poses serious challenge to global security for being a continent where most of the emerging as well as aspiring nuclear weapon states are located. Then, there is the most haunting specter of poverty, disease, democratic denial, distortion and abuse of human rights, existence of failed and failing states, spread of HIV/AIDS and low levels of growth.

If one goes by *Country Reports on Terrorism* (previously called "Patterns of Global Terrorism"), a number of Asian countries have been designated as state sponsors of terrorism. According to the 2010 Report, out of Cuba, Sudan, North Korea, Iran and Syria, the last three belong to the Asian region with Iran topping the list and the three out of these five being the countries with predominantly Muslim population. Iran has been there on the terror sponsor states list since 1984. Iraq was removed from the list in 1982 to make it eligible for US military technology while it was fighting in Iran-Iraq war, but it was put back on the list in 1990 for its invasion of Kuwait. Since US invasion of Iraq in 2003, it was officially removed once again in September 2004. In recent times Asia has taken the lead in world incidences of terrorism. By 2000, Asia already accounted for 75 percent of all terrorism casualties worldwide [Patterns of Global Terrorism, 2000/2001]. From Afghanistan and Pakistan on the West to Bangladesh and Burma on the East, India is at the geographical center of a belt of terrorist, insurgent and separatist violence [Chellaney: 2001/02]. Internally, India is faced with terrorist violence in Jammu and Kashmir, in the North-East and the South. Militant organizations have links to external agencies, and these links can be surprisingly long. For example, one of the terrorists freed by India in the Kandahar hijacking of a commercial jetliner in 1999 was suspected to have financed a hijacker in the September 11 bombing of the World Trade Towers [CNN, "Suspected Hijack Bankroller Freed by India in '99," http://www.cnn.com/2001/US/10/05/inv.terror.investigation/index.html].

To quote Rohan Gunaratana, "Two momentous events in 1979—the Islamic Revolution in Iran and Soviet invasion of Afghanistan—marked the rise of new wave of Islamist movements in Middle-East, Asia, Africa, Caucasus and Balkans and also in Western Europe" [Gunaratana:2002]. It was amidst a highly volatile and violent phase of international politics in Asia that the groups like al-Qaeda came into being and the terrorists like Osama bin-Laden conceptualized *jihad* against Soviet troops. United States also emerged as a major player in the region in its anti-Soviet bid and cultivated the *mujahideen* who after Soviet withdrawal from Afghanistan helped launch a number of Islamist movements at various locations where Muslims were perceived to be suffering injustice at the hand of non-Muslims.

Geographical Profile of Asia

1. Central Asia: The five 'stans', namely: Kazakhastan, Kyrgyzstan, Tazakhistan, Turkmenistan and Uzbekistan—all of these became independent republics after they broke away from the former USSR. Rich in gas, these nations are bound together by the harshness of their landscape and a shared history [Clough: 2008]. In centuries past, the area was traversed by traders seeking silk and spices, as well as by mighty conquerors, such as Attila, Mahmud, and Genghis Khan. Though some scholars would like to include Afghanistan also in this region due to religious-cultural affinity and geographical proximity, but the recent inclusion of Afghanistan on 3rd April, 2007 into South Asian Association for Regional Co-operation (SAARC) as its 8th member in its 14th Summit held in New Delhi confirms it as a state in South Asian region.

2. East Asia: It is the modern term for traditional European term 'Far East'. It is comprised of the countries: China, Hong Kong (SAR of China), Macau (SAR of China), Taiwan, Japan, North Korea, South Korea and Mangolia. Densely populated East Asia is home to a wide array of languages and religious beliefs as well as challenges.

While China's vast population is predominantly rural, industrial pollution is also a major problem. Though Japan is a leading manufacturer, trader, and financier, it still faces high unemployment and low economic growth. Despite maintaining the world's fourth largest army, North Korea faces severe shortages of food and electricity for its population. But there are bright notes for East Asia: China has the fastest-growing economy in the world, and South Korea is the world's 12th largest trading nation. There are predictions that these nations will play a major part in the 21st century.

3. Middle-East or in the United Nations' terminology—south-west Asia (including some parts of North Africa also)—is home to most ancient human civilizations like: Egyptian, Assyrian, Babylonian, Persian, Greeks and Roman—all of whom left their marks on the region. UN includes Turkey, which is partially located in Europe, and South Caucasus (Armenia, Azerbaijan and Georgia) in Western Asia though they have socio-political ties with Europe. The geographic home for Islam, Christianity, and Judaism, the Middle-East has been the fighting ground for major religious battles for centuries, especially since the Ottoman Empire was dismantled after World War I. The major countries of the region are: Armenia, Azerbaijan, Bahrain, Egypt, Georgia (partially located in Europe), Iran, Iraq, Israel, Jordan, Kuwait, Lebanon, Oman, Qatar, Saudi Arabia, Syria, Turkey (partially in Europe), Turkmenistan (parts of it), United Arab Emirates (UAE), Yemen, Sinai (Eastern Egypt), Cyprus, Gaza strip and West Bank. The list is indicative of the region's centrality in being the main theatre for the forces of Islamic fundamentalism to conveniently unleash terrorist movements from here. The region attracts big international players due to it being the reservoir of the largest share of an important natural energy resource: OIL. The controversial territories of West Bank and Gaza Strip and the state of Israel form part of this region, popularly called West Asia.

The region is important for being the cradle of Islam from where the Prophet Mohammad preached the revelations of Quran that spread to other lands also. Two holiest shrines of Islam, Mecca and Medina, are located in the region and has the oldest institutions of Islamic learning. Due to these religious and cultural factors, most of the interpretations of Islam filter through Arab lens—a factor resisted by the non-Arab Muslims throughout the world. Being dominated by authoritarian political systems, the Arabian Middle-East generates extremist political movements with message often articulated in the language of political Islam. Even some South Asian countries with majority Muslim populations or with significantly large Muslim populations were greatly influenced by Wahhabi ideology of the Arabs whose ideological successors in Pakistan launched Taliban in Afghanistan responsible for much of Islamist terrorism after the end of cold war.

The region gains strategic significance and has become the centre of global interest and attention as three of the world's major religions—the monotheist traditions of Judaism, Christianity, and Islam—were all born in the Middle-East and are all inextricably linked to one another. Christianity was born from within the Jewish tradition, and Islam developed from both Christianity and Judaism. While there have been differences among these religions, there was a rich cultural interchange between Jews, Christians, and Muslims that took place in Islamic Spain and other places over centuries. Because Judaism, Christianity, and Islam—recognize Abraham as their first prophet, they are also called the Abrahamic religions. According to PBS website, while there was always a small community of Jews in historic Palestine, in 73 C.E., the Roman Empire dispersed the Jews after an insurrection against Roman authority. Most Jews then lived in Diaspora, as minorities in their communities, until the founding of the state of Israel in 1948. When Jews from all over the world came to settle in modern Israel, they found that various subcultures had developed in different areas with distinctive histories, languages, religious practices, customs, and cuisine.

[visit:http://www.pbs.org/wgbh/globalconnections/mideast/themes/religion/index.html]

Jews believe in one God and his prophets, with special respect for Moses as the prophet to whom God gave the law. They, however, do not believe in the prophets after the Jewish prophets, including Jesus and Muhammad. Therefore, they do not subscribe to the idea that Jesus was the Messiah and the son of God, nor do they believe in the teachings of Islam. Christianity started as an offshoot of Judaism in the first century C.E. Until the emperor Constantine converted to Christianity in 324 C.E., early Christian communities were often persecuted. It was then that the Roman Empire became the Holy Roman Empire, and its capital relocated from Rome to Constantinople (formerly Byzantium and now Istanbul). The development of Christian groups derived from major and minor splits. The Orthodox Church and its patriarch split away from the

Roman Catholic Church and the Pope in 1054 C.E. because of political and doctrinal differences. In the 16th century, Martin Luther, upset at the corruption of the Catholic papacy, spearheaded a reformation movement that led to the development of Protestantism. Christian missionaries proselytize all over the world, and there are large populations of Christians on every continent on earth, although the forms of Christianity practiced vary.

Michael L. Ross in his *Foreign Affairs* article (May/June 2008) titled: "Blood Barrels: Why Oil Wealth Fuels Conflict" says that though the world has grown much more peaceful over the past 15 years, the oil wealth of oil-rich countries often wreaks havoc on a country's economy and politics, helps fund insurgents, and aggravates ethnic grievances. And with oil ever more in demand, the problems it spawns are likely to spread further. Today, with violence falling in general, oil-producing states make up a growing fraction of the world's conflict-ridden countries. They now host about a third of the world's civil wars, both large and small, up from one-fifth in 1992. According to some, the US-led invasion of Iraq shows that oil breeds conflict between countries, but the more widespread problem is that it breeds conflict within them. More than a dozen countries in Africa, the Caspian basin, and Southeast Asia have recently become, or will soon become, significant oil and gas exporters in addition with the traditional oil and gas supplier countries of Persian Gulf. Some of these countries, including Chad, East Timor, and Myanmar, have already suffered internal strife. Most of the rest are poor, undemocratic, and badly governed, which means that they are likely to experience violence as well. On top of that, record oil prices will yield the kind of economic windfalls that typically produce further unrest.

*4. **Southeast Asia:*** The region is comprised of the countries that are geographically south of China, east of India and north of Australia and has cultural influence either from China or from India. The countries falling in this region are: Brunei, Myanmar, Cambodia, East Timor (Timor-Leste), Indonesia, Laos, Malaysia, Philippines, Singapore, Thailand and Vietnam. Out of these, Myanmar, Cambodia, Laos, Thailand and Vietnam constitute Indo-China. The Golden Triangle region, known for illicit drug trade also lies in this region. The dominant religion of people of this region is Islam followed by Christianity. Prior to the 13th century, Buddhism and Hinduism were the main religions in Southeast Asia. Western influence started to enter in the 1500s, with the arrival of the Portuguese and Spanish in Moluccas and the Philippines. Countries in Southeast Asia practice many different religions. *Mainland* SEA countries, that is, Myanmar, Thailand, Laos, Cambodia, and Vietnam, practice predominantly Buddhism. Singapore is also predominantly Buddhist. In the Malay Archipelago, people living in Malaysia, Indonesia and Brunei practice mainly Islam. Christianity is predominant in the Philippines, Eastern Indonesia and East Timor. The Philippines has the largest Roman Catholic population followed by Vietnam, both ex-colonies of European powers. Religions and peoples are diverse in Southeast Asia and not one country is culturally homogeneous. In the world's most populous Muslim nation, Indonesia, Hinduism is dominant on islands such as Bali. Pockets of Hindu population can also be found around Southeast Asia in Singapore, Malaysia, etc. [Wikipedia search]

5. South Asia: This region comprises of Afghanistan, Bangla Desh, Bhutan, India, Maldives, Nepal, Pakistan and Sri Lanka. **South Asia**, also known as **Southern Asia**, is a southern geopolitical region of the Asian continent comprising territories on and in proximity to the Indian sub-continent. It is surrounded by (from west to east) Western Asia, Central Asia, Eastern Asia and Southeastern Asia. The peoples of the region possess several distinguishing features that set them apart anthropologically from the rest of Asia; the dominant peoples and cultures are Indo-Aryan and Dravidian and have a great affinity with the Iranian Plateau and the Caucasus. Persian, Arab and Turkish cultural traditions from the west also form an integral part of Islamic South Asian culture, but have been radically adapted to form a Muslim culture distinct from what is found in the Middle-East, e.g. pilgrimage to *dargahs*. South Asia ranks among the world's most densely populated regions. About 1.6 billion people live here—about one-fourth of all the people in the world. The region's population density of 305 persons per square kilometer is more than seven times the world average. The region has a long history. The region was at its most prosperous before the 18th century, when the Mughal Empire held sway in the north; European colonialism led to its expansion in the region, by Portugal and Holland, and later Britain and to a lesser degree France. Most of the region gained independence from Europe by the late 1940s. (Wikipedia search).

South Asia is the quintessential "bad neighborhood." Overpopulated, poor and poorly governed, it has messy borders and a messier history of conflict, as well as an incendiary mix of strong ethnic identities and diverse religious communities, many of which are concentrated within exclusionary ghettoes. Islamist extremism has flourished in this intemperate soil, and it is here that the world's first *global* Islamist terrorist movement was bred and nurtured, and from where it was exported—first into the immediate neighborhood, and then across the continents, until it finally struck the heart of "fortress America" on September 11, 2001 [Sahni: 2005].

In a nutshell, Asia is home to many failing and failed states. Such states provide operational bases and safe havens for local and international terrorists who, taking advantage of the porous borders of such states and absence or near absence of law-enforcement agencies, get opportunity to move men, weapons and money around the globe or to smuggle precious resources, such as, diamonds and narcotics that help fund their terrorist operations. Failed states are also the fertile grounds for terrorist recruitment where poor and disillusioned youth easily harbor religious or ethnic grievances. It is common knowledge today that much of world's illegal drug supply comes from a failed state—Afghanistan—that has suffered decades of foreign occupation, instability, internal strife and almost no development of any kind. Such states affect the neighboring and far-flung states also by boosting immigration pressures and these developments, in turn, create a lucrative traffic in flesh trade and illegal workers filling the war chests of terrorists and criminals. Strategic and well thought-out use of 'religious hurt' in such situations is a sure recipe for creating violent situations. Sadly, many Muslim majority states in various sub-regions of Asia are suffering from this syndrome: low levels of development or underdevelopment, extremely low level of GDP and per capita income, absence of democracy and political stability, high eventuality of rioting and lawlessness and virtual absence of any control of government on its territory and people—leave aside the 'good governance'—and too much interference from outside. Afghanistan, Hezbollah and Yemen are some such examples in Asia while Sudan, Algeria, Sierra Leone, Somalia, Haiti are only a few examples of terrorists' breeding grounds in African continent.

This brief profile of the region would help in understanding the deep-rooted problem of religious fundamentalism and international terrorism as a number of states falling within Asia are either the perpetrators or the victims of terrorist violence. The pertinent question here is: whether terrorism is a product of religious fundamentalism given the presence of Muslim dominance in many of these states which suffer from low levels of development and inter and intra-state conflicts, or is it the presence of outside players who, driven by vested interests, have stakes in the region and try to manipulate and politicize the existing situation in the manner that give rise to forces of strife and conflict who then use terrorism as a strategic device to counter the political fundamentalism of some 'hegemonic' states in the region. In order to find out answers to these questions, let us undertake a region-wise survey of the socio-political and economic situation existing in these states individually and as a region or sub-region that play a catalytic role in exacerbating terrorism having international consequences in terms of adverse effect on peace and security.

Patterns/Profile of Religious Fundamentalism and Terrorism in sub-regions of Asia

Cental Asia

After the collapse of the USSR in 1991, Islamic fundamentalism made its appearance in the Central Asian region through Pakistani organizations such as the Tablighi Jamaat, the Harkat-ul-Mujahideen (HUM) and the Harkat-ul-Jihad-al-Islami (HUJI) and through the Hizbul Tehrir (HT), which came from the UK via Pakistan. The Pakistani organizations spread the Wahabi ideology through *madarssas* and mosques, which led to the Central Asian Republics (CARs) becoming a hotbed of *jihadi* extremist/terrorist activities in pursuit of the objective of an Islamic Caliphate. *Jihadi* terrorism in the CARs has a purely religious dimension. It is seen as a struggle against their own leaders, who are trying to suppress them, and not against foreign occupying forces. Long years of communist rule drove the religious faith of the people underground, but it could not be suppressed or eradicated completely. The success of the Afghan *mujahideen* against the Soviet troops through the 1980s demonstrated the power and influence of Islam in motivating people to make sacrifices to defend their religion.

The collapse of the USSR has led to a resurgence of Islam in the CARs. The newly born Islam here has assumed an increasingly political color. But, the ruling dispensations in the countries of the region see in political Islam a threat to their power. No liberal political movement has been allowed to come up to confront democratically in town halls and other public fora. The policies of the remnants of the Communist era continue to hold sway. The result: the mosques have become not only places of worship, but also places of confrontation with the authorities.

These five Central Asian Republics, despite their significantly large Muslim populations, opted to remain secular in outlook. However, some Islamic organizations and preachers from Gulf gained ground in this region because of ethnic and regional proximity and made the region a fertile ground for Islamic revivalism. High rates of unemployment and widespread governmental corruption made the task easy for radical Islamists who were working overtime to establish *madarassas* and mosques and were smuggling weapons into the region with Saudi funds. Any adverse activity in one country was bound to have a spill over effect on the neighboring one. It can be argued that due to the failure of the newly independent states of Central Asia to alleviate poverty, provide adequate educational and employment opportunities for their citizens and raise their standards of living, the region easily fell prey to the designs of Islamic fundamentalists. Their proximity with Afghanistan facilitated al-Qaeda and Taliban to use this region as a safe haven and turn it into a base for a new wave of post-Taliban Islamism.

The situation in the CARs is a matter of common concern to India, China, the US and Russia. There are cross-border linkages between the Taliban, al-Qaeda, the Pakistani *jihadi* terrorist organizations and the Uighur extremists on the one hand and with the Chechens on the other. The possibility of the Ferghana Valley becoming one day a new hub of international *jihadi* terrorism cannot be ruled out. Also, the region might tap the expertise of its own scientist community and of elsewhere too to develop a WMD capability. The region being location of vast energy resources attracts the dangers of the terrorists acquiring control of them one day. It is important for these five countries to get together, exchange ideas on the situation in the region and work towards a common counter-terrorism approach.

East Asia

China is home to a large population of adherents of Islam. According to the *CIA World Fact Book,* about 1-2 percent of the total population in China are Muslims, while US Department of State's *International*

Religious Freedom Report shows that Muslims constitute about 1.5 percent (or a population of 19,594,707) of the Chinese population. According to the 2000 census, the largest of the ten traditionally Muslim ethnic groups in China are the Hui (9.8 million in year 2000 census, or 48 percent of the officially tabulated number of Muslims). The other nine, in descending order of size, are Uyghurs (8.4 million, 41 percent), Kazakhs (1.25 million, 6.1 percent), Dongxiang (514 thousand, 2.5 percent), Kirghyz (161 thousand), Salar (105 thousand), Tajiks (41 thousand), Bonan (17 thousand), Uzbek (12 thousand), and Tatar (5 thousand). However, individual members of traditionally Muslim ethnic groups may profess other religions or none at all, while Muslim believers may also be found among traditionally non-Muslim groups (one example being the Kache, who are ethnically Tibetans). Muslims live predominantly in the areas that border Central Asia, Tibet and Mongolia, i.e Xinjiang, Gansu and Oinghai, which is know as the **Quran Belt**.

Amongst these, the demand by some elements of the Uyghur ethnic minority in China for a separate state of East Turkestan has a long history pre-dating 9/11. Since 11 September, however, the Chinese government has insisted that these separatists are no different from other Islamic terrorists that are the focus of the global 'War against Terror'. An examination of the situation in the region through relevant sources suggests that Uyghur terrorism in western China reached its height in the early and mid-1990s. Xinjiang, the homeland of the Uyghurs, has no less than eight international borders including borders with Pakistan, Afghanistan and India: the countries so vigorously the hubs of terrorism. Given the cross-border character of modern terrorism, this geographical factor gives events in Xinjiang regional significance and hence the violence in Xinjiang. However, despite the porous nature of the international borders of Xinjiang, Uyghur extremism in western China poses no immediate threat to South Asia [see: Vicziany: 2003]. The main impact of the Uyghur separatist movement on South Asia is that it has enabled the Chinese government to argue that it shares a common concern with the United States and India in the War against Terror.

The government of Peoples' Republic of China considers these separatists, those who are fighting for an independent state in the northwestern province of Xinjiang which the separatists call "Eastern Turkestan" as part of a network of international Islamic terror, with funding from the Middle-East, training in Pakistan, and combat experience in Chechnya and Afghanistan. The scholars like Chien-Peng Chung differ and feel that the latest wave of Uighur separatism has been inspired not by Osama bin-Laden but by the unraveling of the Soviet Union, as militants seek to emulate the independence gained by some Muslim communities in Central Asia [see: Chung: *Foreign Affairs*: 2002]. For a decade now, Xinjiang has been rocked by demonstrations, bombings, and political assassinations. Amidst allegations and counter-allegations on both sides of this movement, members of a shadowy "Eastern Turkestan Islamic Movement" have obtained funds and training from al-Qaeda. Uncertain future of Uighur separatist movement notwithstanding, current situation is that there are a number of Islamist groups fighting for independence, many of them having developed extensive presence in Pakistan, Kazakhastan, Kyrgyzstan and Germany where the funds are raised. The terrorist attacks launched by Uighurs even in Beijing confirm to al-Qaeda modus operandi of targeting the centres instead of peripheries. The influx of Chinese Muslims to Pakistan and Afghanistan in 1990s for indoctrination and training and also their participation in Afghan *jihad* and later to use these techniques in China after 9/11 reveals how far the terror network has spread its wings in far flung regions of Asia.

Middle-East: (Southwest Asia including some parts of North Africa)

In last 60 years since 1948, a number of conflicts have taken place in Middle-East, such as, Arab-Israel Conflict, Jordan-Syria tensions, 1970 Black September massacre, Lebanese civil war, Libya-Egypt conflict, the 1980 Iran-Iraq war, Second Gulf War (Iraq-Kuwait), UN-Iraq conflict, general conflicts of US and its allies and then the latest being US invasion of Iraq in 2003.

The Arab-Israel conflicts include: 1948 Arab-Israel war, 1956 Suez Crisis, 1967 Six-Day War, 1970 War of Attrition, 1973 Yom Kippur War, 1982 Lebanon War, 1987-90 First *Intifada,* 1982-2000 South Lebanon Conflict, 2000 al-Aqsa *Intifada*, 2006 Lebanon-Israel war. This statistics of wars and conflicts in the region clearly presents a picture of the region being severely infested with religious radicalism, violence and terrorism operating not only within the region but also globally and having connections with global terror networks. If one looks at the wars from Mohammed's conquests since 610 AD till modern times, we would find that Muslims have been engaged in wars since the birth of Islam as a religion in the year 622

CE.. In addition with the Muslim-dominated region of Middle-East, Muslims are involved in wars and conflicts elsewhere also, such as, war in Somalia, Darfur Conflict, Kashmir issue between India and Pakistan, Iraq war, Afghanistan and Chechnya—to quote only a few. Add to this the ongoing conflicts in Western Sahara, Morocco, Chad, Libya and Algeria where Muslims are part of the conflict. A brief stocktaking of the countries of this region associated in some significant way with terrorist activities is very much in place here.

After 9/11, a debate has broken out in ***Egypt*** over the country's now most 'famous' and embarrassing export: Islamic militancy. The question is why and how the country produced the anti-Western religious fanatics at the center of the Sept. 11 terrorist attacks and at the core of the al-Qaeda network which, it seems all but certain, carried out the atrocity. American investigators think one of the men who plunged hijacked aircraft into the World Trade Center towers on Sept. 11 was an Egyptian. The main deputies of Osama bin-Laden, the leader of al-Qaeda, are Egyptian. One of them, Dr. Ayman Al-Zawahiri, is the leader of Egypt's militant *Jihad* group, and has been mentioned as the "brains" behind al-Qaeda's operations. And seven of 22 people on a Federal Bureau of Investigation "most wanted" terrorist list are Egyptian.

So it is wondered why the United States' "moderate" Arab allies: Egypt and Saudi Arabia—the latter being responsible for supplying a striking number of suspected participants in the Sept. 11 attacks—have spawned these people. The United States and its allies have accused Egypt and Saudi Arabia of running dysfunctional, undemocratic states that simultaneously abuse the human rights of suspected militants and quietly permit religious fanaticism to spread throughout society. Egypt and Saudi Arabia's critics have charged that many Islamist radicals, squeezed by security forces rather than neutralized by pluralistic policies, left Egypt in the 1990s to join bin-Laden in Afghanistan. *Newsweek*'s Fareed Zakaria, one of the first to dig the knife in, said the Bush administration needed to prod Egypt and Saudi Arabia to "actively fight the virulent currents that are capturing Arab culture" because both countries had "resisted economic and political modernization." After September11 attacks, Egyptian President Hosni Mubarak repeatedly stressed that a resolution of Israel-Palestine issue would take the wind out of radical Islamists' sails. Domestically, Egypt also blames much of radicalism on the oldest, biggest and most influential Islamist group—Muslim

Brotherhood—as promoting radical ideologies which is viewed by the Arab societies and the West as something outside Islam [see: worldpress.org/mid-east/329.cfm]. In an article in the *Terror Finance Blog*, "Egypt's role in the destabilization of the Middle-East", it is argued that at the time the international community imposes economic sanctions against the terrorist government of Hamas, Egypt facilitates the transfer of millions of dollars in cash to support and sustain it.

In late 2006, Hamas-appointed Prime Minister Ismail Haniyeh returned to Gaza with US$35 million in cash. On January 5, 2007, Haniyeh declared, and was allowed by the Egyptian authorities to carry USD 20 millions he was given by the Saudis to support the Hamas resistance. Moreover, Mubarak is alleged to fuel terrorism in the Gaza Strip—the source of all Middle-East terrorist organizations. Egypt allows "the systematic smuggling of surface-to-surface, anti-tank and anti-aircraft missiles, explosives and even terrorists into the largest terrorist base in the Mideast, Gaza," says Yoram Ettinger.

While the economic condition of the majority of the Palestinians in the Gaza Strip is worsening, the cash smuggled by the Hamas leaders is used to pay their operatives and members in the security services, to ensure their loyalty. By allowing cash and weapons for Hamas, Egypt supported a terrorist organization which is a branch of the outlawed Egyptian Muslim Brotherhood. Through his actions, Mubarak almost repeated the actions the Saudis took in the 1990's to keep bin-Laden out of the Kingdom. They too, supplied a radical element with money and weapons to fight others elsewhere. But the Saudi plan succeeded only partially; although bin-Laden fought the Russians and attacked the US, his followers keep their strife against the Saudi Royals alive. Indeed, a strong Hamas—the Palestinian branch of the MB, helped embolden their Egyptian brothers. It is suggested in the article that if President Bush is serious about his "war against the radicals, he should have added Egypt to that list" [see:www.terrorismfinance.org/the-terror-finance-blog/2007/01/mubaraks.willfu.html].

Such views apart, it is also true that a moderate Egypt is key to peace and stability in the Middle-East. After 9/11, US-Egypt relations are in trouble. Though Egypt has stood firm with US in latter's war on terror, a section of US Congress is critical of Egypt's role in promoting terrorism and its relations with Israel. For the past several decades, Egypt has been a battleground between moderate and radical

Islamic forces. The role of Egyptian militants in al-Qaeda and other radical movements is considerable. However, US should recognize the importance of the role Egypt can play in any peace process between Arab states and Israel. While no single set of voices defines modern Islam, Egypt's intelligentsia, religious hierarchy, and institutions are central in defining Islamic religious and secular perspectives and diminishing the influence of radical Islamic tendencies [*CRF Paper: 2002*].

The Egyptian government's well-known opposition to Islamist terrorism (before rise of insurgency against Mubarak in 2011) and its effective intelligence and security services made it less likely that terror groups will seek safe haven there. However, Egypt's rugged northern Sinai region remained a haven for smuggling arms and explosives into Gaza and a transit point for Gazan Palestinians trying to infiltrate Israel. In addition, Palestinian HAMAS officials have allegedly carried millions of US dollars in cash across the Gaza-Egypt border. Criminal networks that smuggle weapons and other contraband through the Sinai into Israel and the Gaza Strip may be associated with or used by terrorist groups in the region. The apparent radicalization of some Sinai Bedouin might be linked in part to these smuggling networks, to heavy-handed Egyptian efforts to dismantle them, and to the Bedouin's long-standing distrust of the central government [see: Country Reports on Terrorism, 2007].

In Egypt's post-Mubarak politics 2011 onwards, though the liberal activists were responsible for the takeoff of the revolution but, so far, Islamists and the military have been defining the landing. Sensing an opportunity, Islamists grabbed it. When, over the next ten days, the regime tested the option of using real military force to quash the protests, the protestors' most effective and best organized manpower came from two sources—first, the Muslim Brotherhood and its Islamist allies and second, the well-oiled machines of soccer fan clubs and soccer team security forces. Only the former, of course, had a strategic political agenda, and from the moment the Islamists committed themselves to the fight, their goal has been to capture, exploit and inherit the revolution. For the Muslim Brotherhood, the demise of the Mubarak regime provided an opening to regularize their role in Egyptian political life after decades on the political margins or in the political shadows. That is to say that the Brotherhood recognized that a post-Mubarak Egypt might provide an opportunity for it to resume legitimate political activity after decades of an uneasy relationship with the country's military leadership [Satloff: 2011].

Iran, Lebanon and Hezbollah and Syria

The **Islamic Republic of Iran, Lebanon** and a highly dreadful and resourceful resistance movement of Lebanese Shias—**Hezbollah**—and all these three along with Syria form the most serious terrorist network in the Middle-East. According to US State Department's April 2007 *Country Reports on Terrorism*, Iran remained the most active state sponsor of terrorism. Its Islamic Revolutionary Guard Corps (IRGC) and Ministry of Intelligence and Security (MOIS) were directly involved in the planning and support of terrorist acts and continue to exhort a variety of groups, especially Palestinian groups with leadership cadres in Syria and Lebanese Hizballah, to use terrorism in pursuit of their goals.

al-Qaeda, till Osama was there, was a well placed organization to threaten global security, which it might continue in near future also. It thrived on failed and failing states, such as, **Syria** where extremist Sunni groups have long operated, particularly in the country's second-largest city, Tripoli, which was controlled by a Sunni fundamentalist group during much of the 1980s, before Syria cracked down. The United Nations force stationed in Lebanon are under constant threat, since *the jihadists* consider it to be another crusading army in the Muslim world.

Gaza, which was earlier captured by Israel in 1967 was transferred to Palestinian National Authority in 1993, is another prime candidate for al-Qaeda operatives. It is already divided between Hamas and Fatah, and there is evidence that a small al-Qaeda apparatus is forming there. Israeli security sources have expressed growing alarm about this new al-Qaeda presence on their doorstep, though the Hamas leader Ismail Haniya denied any such presence in an interview with CNN in December 2010.. Yemen, bin-Laden's ancestral homeland, may also make an appealing base. In November 2006, the group al-Qaeda of Jihad Organization in the Land of Yemen claimed credit for attacking oil facilities in the Hadramawt region "as directed by our leader and commander Sheik Osama bin-Laden ... [and in order] to target the Western economy and stop the robbery [and] the looting of Muslim resources."

According to United States' strategic analysts, **Iran** has been maintaining a high-profile role in encouraging anti-Israeli terrorist activity rhetorically, operationally, and financially. Supreme Leader Khamenei and President Ahmadinejad praise Palestinian terrorist operations and provided Lebanese Hizballah and Palestinian terrorist groups—notably HAMAS, Palestinian Islamic Jihad, the al-Aqsa Martyrs Brigades, and the Popular Front for the Liberation of Palestine-General Command—with extensive funding, training, and weapons. Iran continues to play a destabilizing role in Iraq, which appears to be inconsistent with its stated objectives regarding stability in Iraq. Iran provided guidance and training to select Iraqi Shia political groups and weapons and training to Shia militant groups to enable anti-Coalition attacks, has been providing them the capability to build IEDs with explosively formed projectiles. The Revolutionary Guard, along with Lebanese Hizballah, implemented training programs for Iraqi militants in the construction and use of sophisticated IED technology. These individuals then passed on this training to additional militants in Iraq.

Iran has also been alleged for remaining unwilling to bring to justice senior AQ members it detained in 2003, and it refused to publicly identify these senior members in its custody. Iran has repeatedly resisted numerous calls to transfer custody of its AQ detainees to their countries of origin or third countries for interrogation or trial. Iran also continued to fail to control the activities of some al-Qaeda members who fled to Iran following the fall of the Taliban regime in Afghanistan. But such allegations and charges need further authentification.

One more controversy surrounding this Islamic state in the Middle-East is regarding Iran's alleged nuclear activities inviting vehement criticism from US and its European allies. Launched in 1950s with the help of United States, the Iranian nuclear program was temporarily disbanded after the Islamic Revolution of 1979 but then revived again with less Western assistance. Iran's current effort includes several research sites, a uranium mine, a nuclear reactor, and uranium processing facilities that include a uranium enrichment plant. The Iranian government asserts that the program's goal is to develop nuclear power plants, and that it plans to use them to generate 6,000 MW of electricity by 2010. The US and some other nations' officials say the program covers an attempt to acquire nuclear weapons. Iran's officials have categorically denied these accusations. As of 2006, nuclear power does not contribute to the Iranian energy grid.

According to Gawdat Bahgat, the Director of the Center for Middle-Eastern Studies at the Indiana University of Pennsylvania state, the Iranian nuclear program is formed by three forces, namely:

perception of security threats from Pakistan, Iraq, Israel and United States; domestic economic and political dynamics and the national pride [see: Bahgat: 2006]. Having little confidence in the international community very often working under US pressures and its bitter experiences during Iran-Iraq war of 1980s and the Gulf War of 1990s, Iran was under great domestic pressure to become self-sufficient in its power to deter the military advances of other states. A signatory of Nuclear Non-Proliferation Treaty (NPT) since 1970 and therefore subject to International Atomic Energy Agency's (IAEA) regulations, Iran resumed its nuclear program after Revolution and informed IAEA about it. But, the assistance that was being given to it to enrich uranium by IAEA was terminated under US pressures and hence developing an indigenous expertise in the field became a question of national pride. Amidst allegations and counter-allegations, threats and fears and all kinds of political maneuvering and arm-twisting, the current situation is that Iranian President Ahmadinejad is firm on his country's progress in nuclear program despite UN sanction. On 23rd December 2006 the United Nations Security Council (UNSC) unanimously passed Resolution 1737, banning the supply of specific nuclear materials and technology to Iran, and freezing the assets of individuals and companies linked to Iran's nuclear program. The resolution also specifies that if Iran fails to suspend nuclear enrichment, further non-military sanctions may follow. The language of the resolution is lighter than that sought by the United States, and was changed in response to Chinese and Russian concerns. On March 24, 2007, further to Resolution 1737, the UNSC adopted Resolution 1747. Iran's nuclear program has become political issue in two ways: local and international. Iranian politicians use it as part of their populist platform, and there is international speculation about Iran's possible use of nuclear technology.

Iran is a member of the NPT, which it ratified in 1970. However, the International Atomic Energy Agency (IAEA) believes that recent Iranian non-co-operation makes it impossible to conduct adequate inspections to ensure that the technology is not being diverted for weapons use. Nevertheless, Iran has a legal right to enrich uranium for peaceful purposes under the NPT. Iran, and many other developing nations who are signatories to the NPT, believes the Western position to be hypocritical, since the NPT calls for universal nuclear disarmament. Iran also compares its treatment as a signatory to the NPT with three nations that have not ratified the NPT. Each of these nations—Israel, India and Pakistan—developed an indigenous nuclear weapons capability. However, as the NPT calls on member-states to refrain from receiving weapons technology, the weapon plans Iran received from AQ Khan place it in technical violation of its NPT obligations.

The potential reason behind US resistance to a nuclear Iran lies in Middle-Eastern geopolitics. In essence, the US feels that it must guard against even the possibility of Iran obtaining a nuclear weapon, as a nuclear-armed Iran would dramatically change the balance of power in the Middle-East, weakening US influence. It could also encourage other Middle-Eastern nations to develop nuclear weapons of their own, further reducing US influence in a critical region. Moreover, another primary US concern with Iran obtaining nuclear weapons is its belief that Iran sponsors international terrorism, which includes giving support and weapons to the Iraqi insurgency. Iran's support of Hamas and Islamic Jihad leads to US fears that Iranian nuclear weapons could find their way into the hands of Islamic militants. Over the years, Tehran has supported several militant groups in Palestine including Fatah and Islamist movement Hamas and also Palestinian Islamic Jihad. It also has daunting array of interests in Iraq with which it shares a long border and taking advantage of the current instability situation in Iraq, Tehran has flooded Iraq with many intelligence agents including some presence of Lebanese Hezbollah that serves as a proxy for Iran. It also tries to influence the Iraqi Shiite groups who have a significant presence in Iraqi government also which in the long-run might serve the strategic interests of Iran as against those of United States.

The US also maintains that Iran does not need nuclear power due to its abundant oil reserves since nuclear power is more expensive for the Iranians to generate than oil-fed power. However, it should be noted that this argument has taken a back seat as developing nations have re-invested in their civilian nuclear industries and the magazines such as *The Economist* have taken an economic stance similar to that of Iran's. The US government is concerned that Iran does not formally recognize Israel's right to exist, and some Iranian politicians have openly called for the destruction of Israel.

The controversy notwithstanding, Iran has a vast support for its nuclear program, not necessarily limited to peaceful purposes, from a number of Islamic countries. The 118 countries of Non-Alignment

Movement, in their Summit meeting in Havana on September 16, 2006 declared support for Iran's nuclear program for civilian purposes in their final written statement. It is a clear majority of the 192 countries comprising the entire United Nations, which comprise 55 percent of the world population.

According to Daniel Byman, the Director of Center for Peace and Security Studies at Georgetown University and a Senior Fellow at the Saban Center for Middle-East Policy at Brookings Institute, Iran since Islamic Revolution in 1979 has been one of world's most active sponsors of terrorism. Tehran has armed, trained, financed, inspired, organized and otherwise supported dozens of violent groups over the years not only in the Persian Gulf neighborhood but also terrorists and radicals in Lebanon, Palestinian Territories, Bosnia, Philippines and elsewhere. Yet, Iran is not likely to transfer chemical, biological, nuclear or radiological weapons to terrorists for reasons that: (a) these terrorists and radical groups can operate effectively with existing methods and weapons; (b) Iran has become cautious of its terrorist activities in recent years; and (c) escalation in its support to terrorists would invite US wrath and international opprobrium. So, Tehran is not likely to pass chemical, biological and nuclear weapons to terrorists [see: Byman: Deadly Connections: 2005; *Iran, Terrorism and Weapons of Mass Destruction*: 2005].

As for **Hezbollah**, it is a radical Shiite Moslem organization that operates in Lebanon. A key pillar of its extremist ideology is its call for *jihad* (holy war) against Zionists and supporters of Israel, particularly US, to ultimately foment an Islamic revolution so as to translate its ambition of turning Lebanon into an Islamic republic like Iran into reality. Hezbollah's ideology comes from Iran; moreover, it receives military, logistical and economic support from both Iran and Syria. Iran, which is accused of completely opposed to all efforts to achieve peace with Israel and which is suspected to have supported terrorism to subvert such efforts, as well as seeking to develop its own non-conventional weapons capability, backs Hezbollah to the hilt. Syria, which dominates Lebanon and occupies parts of it, uses Hezbollah as a belligerent instrument against Israel, in order to advance Syria's political and strategic objectives in its own conflict with Israel.. Syria also actively supports a number of Palestinian terrorist organizations, including Hamas and Islamic Jihad, and hosts terrorist headquarters and training camps in the Damascus area. Lebanon, too, supports Hezbollah, allowing it to operate against Israel from within Lebanese territory, in direct contradiction to international law [Israel, the Conflict and Peace: 2003].

Iraq being a secular state did not have any revealing connections with Islamic terrorist groups in the region or outside it. However, the American speculations and unfounded allegations that Saddam Hussein possessed weapons of mass destruction (WMD) and the subsequent turn of events after United States' preemptive action in Iraq in 2003 has created a whole lot of media industry and academic studies on the subject having both the supporters and opponents of US action in Iraq. The illegal and immoral elimination of Saddam Hussein and the mess it created due to outside intervention is an important aspect of Iraq imbroglio. The question of Iraq's link to terrorism grew more urgent with Saddam's suspected determination to develop weapons of mass destruction (WMD), which Bush administration officials feared he might share with terrorists who could launch devastating attacks against the United States. But there is no hard evidence linking Saddam to the 9/11 attacks. In fact, as many commentators have also noted, Baghdad failed to express sympathy for the United States after the attacks and led to heightened US suspicion against Iraq. It is quite unlikely that the Islamist al-Qaeda and Saddam's secular dictatorship would be allies.

The question why US waged a war on Iraq in March 2003 haunts all those who are concerned with analysis of international politics and those who were associated with Bush administration. Initially the decision was made to remove Saddam Hussein projected by the administration as a dictator building and hiding weapons that could give him a leverage to dominate Middle-East. According to President Bush, the Iraq invasion was also a part of a larger war on terrorism, despite the lack of support from western allies like France and Germany. In fact, no WMD have so far been recovered from Iraqi soil. However, despite tall claims and three elections since 2005 for preliminary government, for making a constitution and for a new government, insurgency remains a hard reality raising doubts about the success of this ambitious and unjustified American project in Iraq. If at all Saddam supported terrorist groups, it were the secular terrorist groups such as various splinter nationalist groups of Palestine rather than the Islamist groups such as al-Qaeda.

In fact, US was supporting Saddam Hussein as a regional counter-weight to Ayatollah Khomeini when Iraq launched a long brutal war against Iran in 1980. The relations soured only when in

1990, Iraq invaded the neighboring, oil-rich state of Kuwait. Though a US-led war ousted Saddam from Kuwait, UN was maneuvered to impose sanctions on Iraq, including imposing international inspection to watch that the Iraqi dictator does not pursue a policy of building WMD.

United Arab Emirates (UAE): Though by-and-large liberal in its political and cultural dispensations, UAE had some connections with al-Qaeda and was amongst the first three countries to recognize Taliban after its capture of Kabul and a large part of Afghanistan. It has also been the major center for al-Qaeda's financial operations. It has been alleged as having served as an important base for financing many Islamist organizations including al-Qaeda; it was important centre for channeling explosives and other materials to Afghanistan that were purchased from various locations including Europe. Yet, despite its 'some' connections with terrorist groups, knowingly or unknowingly, UAE has fully co-operated with US in latter's fight against terrorism.

Saudi Arabia, a highly fundamentalist Islamic state and a closely policed state shows no or little evidence of it sponsoring terrorism from its soil or harboring terrorists for the reason that since long, it is under the protectorate of United States complying fully with it in all political, military and economic affairs though it is a country where the craziest religious diktats find their origin. Though Saudi Arabia does not directly support terrorism and terrorist groups, the US government has expressed concerns about Saudi financial ties to terrorism. Islamic networks originating in Saudi Arabia reportedly provide financial backing to terrorist groups that operate in Middle-East and around the world. The fact that many militant groups are mosque-based makes crackdowns difficult, but Saudi leaders now understand the need to control the militant Islamist elements in the country. In order to fight terrorism, the need to control the mosques and the clerics that incite violence has been acknowledged. The efforts in this direction are: holding the first-ever International Conference on Counter-Terrorism in February 2005 and flagging-off a public relations campaign to discourage religious radicalism and terrorism.

After 9/11 at least, Saudi government is completely in co-operation with Washington and no particular terrorist group belongs to it. However, being home to the two most revered mosques in the Islamic world, its actions and behaviour are closely watched by all Muslims around the world and therefore the radical and rigid

Islamists could not tolerate Saudi Arabian Royal family seeking American security umbrella and coming under US protectorate in the aftermath of second gulf war and the presence of US troops on the soil where two holiest places of Islam are located. Despite claims of not harboring any anti-American terrorist on its soil, surprisingly fifteen out of nineteen 9/11 hijackers were Saudis. Osama bin Laden, the mastermind behind attacks on America and the virtual creator of al-Qaeda was also a Saudi national and is known to have enjoyed great Royal favors before turning against the US and the Royal family of Saud. Though Saudi Arabia is a tightly governed country and the kind and quality of liberties and freedoms enjoyed in a liberal democracy by any normal person there, citizen and non-citizen, are virtually absent here, yet al-Qaeda succeeded in creating a large support base here conducting small-scale operations against American and British targets, particularly attacking those who were associated either with Saudi oil industry or belonged to US and allied forces engaged in Operation Southern Watch, the air coalition over Iraq. Khobar Tower attacks are only one among many such examples. Other than this direct or indirect terrorist connection of Saudi Arabia, it has also remained one of the most well-known places from where finances are supplied to terrorists via investments and charities with various rich *sheikhs* donating heavily for Islamic cause.

According to a Reuter's report published in *Hindustan Times* in April, 2007, Saudi government foiled an al-Qaeda linked plot of some Saudi terrorists, more than 170 of them, who had plans to attack oil facilities and military bases. These included some trainee pilots who were preparing for suicide operations and many others were being trained in use of weapons, some were sent to other countries to study aviation in preparation to use them to carry-out terrorist operations inside the Kingdom. Since Saudi Arabia is the world's top oil exporter, supplying about 7 million barrels a day to the world markets and holds nearly a quarter of the world's oil reserves, it remains on the centre of terrorists' radar. An analysis of the report makes it clear that this location in the Asian continent remains the target as well as breeding ground of the undercurrents of radical Islamic ideology and anger due, to a large extent, to the interference and policies of the West in the region. [see: "Saudi Terror Plot Foiled", *Hindustan Times*, April 28, 2007]

The US fears surrounding Saudi Arabia's connections with terrorism are also based on 1996 terrorist attacks on Saudi military housing complex that killed 19 US military personnel and wounded 500 people including 372 Americans, a series of bombings in 2003, an attack on US Consulate in Jeddah and two car bombings in Riyadh in December 2004 and a February 2006 attack on an oil complex—all these confirm the links between al-Qaeda and Saudi radicals. Though neither petty nor organized crime are serious problems in Saudi Arabia, the country's quest both for modernity and religious purity have given rise to a kind of unrest amongst its people, particularly the youth, who are unable to express their predicaments and internal conflict through democratic means, which are totally absent in an absolute monarchy. [see: Country Profile: Saudi Arabia: 2006].

Al-Rashid in his book [see: Al-Rashid: 2006] mentions about a working alliance between the Saudi dynasty and the Wahhabi religious establishment. According to the arrangement, the state has let the ulama (Islamic clergy) dictate the details of social and personal conduct, while the ulama has proclaimed the religious imperative of obeying the ruler. This religiously sanctioned justification for authoritarian rule came to be contested, especially in the 1990s, by a group of clerics proclaiming the need for a *sahwa* (awakening), but they largely tamed their ideological challenge after 9/11, renouncing violence and restricting their demands to socially very conservative reforms rather than regime change. Now, both the *sahwa* movement and the state have confronted hard-core Saudi *jihadists*, who are in exile or are imprisoned. These Saudi ideologues and ideologies can best be understood within their distinctive Saudi-Wahhabi historical context.

Yemen is a Muslim state where not more that 35 percent of its territory is under any influential control of its government and thus it serves as a favorable breeding ground and ideal base for terrorists, their training and for providing them safe havens. Throughout 1990s, Yemen was an important base for al-Qaeda network, particularly to carry out operations in Sub-Saharan and North Africa.

Israel and Occupied Territories

The Palestinian-Israeli conflict, originally a native-settler conflict, is one of the most durable and intractable conflicts since World War II. Various peace process between the Palestinians and Israelis have not yet delivered positive results. Under these conditions, the dangers of a spill over effect of this ongoing violence add to the rising wave of Islamist terrorism all over the globe. Historically, in 1922 after the collapse of the Ottoman Empire that ruled the region of Palestine for four centuries (1517-1917), the British Mandate of Palestine was created which was contested by the Zionist movement. Given the dispute over Palestine, the United Nations Partition Plan in 1947 proposed to partition the mandated territory between the Arabs and the Jewish state, with Jerusalem and surrounding areas to remain under international regime. Under the Plan, the Gaza Strip and West Bank were to form part of the Arab States, which the Arab leaders rejected.

The Palestinians in fact are resisting the 'occupation' of Gaza and West Bank by Israel which as of today is understood to be the root cause of much of Islamic radicalism-generated terrorism in the Middle-East as well as in the world. The success of War on Terror largely depends on a serious address of this outstanding issue. 'End the Occupation' is the most ubiquitous line of argument today among Palestinian spokesmen, who somehow justify terrorism as a political instrument. Palestinian groups that support and carry out politically-motivated violent acts have included: Hamas, Palestinian Islamic Jihad, Fatah's Al-Aqsa Martyrs Brigade, the Popular Front for the Liberation of Palestine (PFLP), the Popular Front for the Liberation of Palestine—General Command (PFLP—GC), the Democratic Front for the Liberation of Palestine, and the Abu Nidal Organization. A United States Congress decision in 1987 determined that the Palestine Liberation Organization (PLO) was also a terrorist organization which was reversed after the 1993 Oslo Accords. The *al-Aqsa Intifada*, the second major wave of violence between Palestinians and Israelis after 1967 is considered by Palestinians as a war of national liberation against foreign occupation, whereas many Israelis consider it to be a terrorist campaign. Like in much of the cases of political violence anywhere, the perpetrators would always say that their attacks are justified, while the states targeted feel otherwise. But regardless of the moral, political or tactical justifications, these attacks are defined as acts of terrorism when they are indiscriminate or directed at non-combatants according to all academic definitions of "terrorism", and also in accordance with the definitions used by the United Nations.

In July 2000, a Middle-East Peace Summit was held at Camp David, hosted by US President Bill Clinton and attended by Palestinian Authority Chairman Yasser Arafat and Israel's Prime

Minister Ehud Barak. The Israel's official version is that "during the summit, Israel expressed its willingness to make far-reaching and unprecedented compromises in order to arrive at a workable, enduring agreement. But Yasser Arafat chose to break-off the negotiations without even offering any proposals of his own. Consequently, the summit adjourned without any outcome [Israel Minisrty of Foreign Affairs: 2003]. It is true that both Palestinians and Israelis of 'Occupied Territories' live in a daily fear of being murdered by suicide bombers, car-bombs and gunmen, who in most parts are Palestinian militants known for using these violent techniques. The international mainstream media has been overly pro-Israel presenting it in a favorable light while treating Arab side in a negative manner.

The Palestinian militant groups are also found to have forged ties with al-Qaeda and many of their members went to Afghanistan to help the latter in its fight against the Soviet occupation. There are charities in Sudan and Pakistan that were funding al-Qaeda and Hamas and Palestinian Islamic Jihad. If one believes the version of the Israeli security and intelligence set-up, al-Qaeda had infiltrated the Occupied Territories and was all set to mount operations against Israel [see: Gunaratana: 2002].

Undoubtedly, Palestine deserves a state of its own. But techniques like hijackings and suicide bombings and the 'cult of martyrdom' are perhaps the products of Palestinian terrorist groups that display reverence for bombers who set examples for others to emulate them as the 'martyrs'.

South Asia

Terrorism has been a ubiquitous phenomenon in South Asian countries much before it hit the United States in a grievous and dramatic fashion on September 11, 2001. These countries have been the victims of one or more forms of terrorism: as a part of the al-Qaeda network; arising out of religious fundamentalism; brought about by structural factors such as extreme and large scale poverty, glaring inequality; prolonged and gross forms of injustice; and as a result of the general collapse of law and order.

The challenges faced and the strategies evolved by the South Asian states in responding to these challenges are inevitably to be in accordance with the nature of terrorism faced in each of the South Asian countries as each has its own specific aspects and character while being interlinked at the same time. South Asia remains mired in violence and terrorism due to the existence of many **flash points such as Kashmir** tangle between India and Pakistan, **ethnic crisis between Tamils and Sinhalese in Sri Lanka** and terrorism unleashed by Liberation Tigers of Tamil Elam (LTTE); the ongoing mess in **Afghanistan** that started three decades ago and the other countries also facing problem of sorts within their territories. The complex concerns of state failure in almost every political entity on India's periphery—Afghanistan, Pakistan, Nepal, Bangla Desh, Sri Lanka and Myanmar—and the terrorist movements of different varieties going on in almost all these countries of the region make South Asia a highly fragile region in terms of peace and security. The existence of Islamist terrorism in Afghanistan and Pakistan gives the region a specificity having implications for international peace and security as well.

South Asian states of Pakistan and Afghanistan have been hubs of terrorist activities since the period of Soviet invasion of Afghanistan in 1979 and the subsequent American interest and involvement in this region in the framework of cold war. When the cold war ended, South Asia was towards the lower end of America's interests and priorities and with the ouster of Soviet troops, US also abandoned Afghanistan, making a very supportive infrastructure available to future terrorists in the form of Taliban and their Arab guests: Osama bin-Laden and his al-Qaeda network. Taliban was another factor in Afghan politics that had immense support and role of Pakistan in its creation and sustenance in the form of getting not only the funds diverted from elsewhere but also the recruits for the Taliban who were trained in *jihad* in *madrassas* (Islamic schools) in Pakistan. Pakistan was also amongst the first three countries, along with Saudi Arabia and United Arab Emirates, to recognize Taliban the moment Kabul fell into its hands on September 26, 1996. Pakistan's regular armed troops, often disguised as Taliban members, were fighting against Northern Alliance while the financial and other logistic support was coming from Saudi Arabia also. Since the cold war days, Pakistan was an old and a trusted ally of US. But its involvement and support to terrorist activities of Taliban and al-Qaeda on one hand, and offering to become a frontrunner in America's War on Terror after 9/11 terrorist attack on the other shows a contradictory role of Pakistan in South Asian affairs. As puts Gunaratana, "Without the strategic advice, logistics and manpower of Pakistan, the Taliban could never have brought nearly 90 percent of Afghan territory under its control" [Gunaratana: 2002: p. 43].

Thus, whereas the international terrorist environment in South Asia concentrates mainly on Afghanistan and Pakistan, the other countries in the region such as, Nepal, Sri Lanka, Bhutan and Bangladesh and India have some connections as well, though these countries are only marginally related to international dimension of terrorism. Another feature of South Asian terror profile is that the terrorist outfits in the neighboring Southwest Asia have their reach and effect in South Asian countries. Within South Asia, there are interlinkages of various terrorist groups and organizations. Al-Qaeda, its Taliban supporters, indigenous Pakistani terrorist groups and *jehadi* militants of Kashmir—all have links and remain active in the entire sub-continent, [see: Kronstadt: 2003]. Keeping these factors in view, let us take a country-wise survey of terrorism in South Asia:

> A strange feature of terrorism in Asia is that some of the radical Islamist groups engaged in violence and terrorism, divided either on Shia-Sunni lines or on some other ideological plane, are paradoxically found to be working in close association with each other when it comes to irritate the common enemy, i.e., US and the West. For example, the Shiite Islamists of Iran and Lebanon, including Hezbollah, have been found to be working in close association with largely Sunni Taliban and al-Qaeda, thus overcoming sectarianism amongst the Muslims. These developments unfolding in the Asian theatre surprised even the CIA officials who were shocked to find Shia and Sunni terrorists working together in developing agent-handling system, conducting long-range operations and the Hezbollah and Iranian experts training al-Qaeda recruits in suicide bombing techniques. The Hezbollah-al-Qaeda strategic partnership paved the way for further ideological tolerance and inter-group cohesion as a means of confronting a common enemy. Its signs could be witnessed in the whole of Asia including the Middle-East and South-East Asia. Even a 'secular' Iraq under Saddam Hussein was giving aid to those Islamist terrorist groups that were targeting western interests. Iraqi intelligence is believed to have supplied arms, ammunitions, money and explosives to the countries ranging from Thailand to Philippines and Sri Lanka with intentions to attack US and allied targets though there is no strong evidence to directly link Iraq to spectacular acts like 9/11. Palestinian Islamic Jihad and Hamas are also found to have association with other dreaded terrorist groups existing in Asia.

Of the various ideological streams that currently inspire and provoke political violence and terrorism in South Asia, the most destabilizing and lethal, and the one with the greatest extra-regional impact, is Islamist terrorism. A multiplicity of sub-sets and a complex, sometimes conflicting scheme of inter-linkages, has been documented in connection with the extended range of Islamist terrorist groups operating in the region.

Pakistan

Various shades of radical political Islam define the Pakistani identity and nation, and have long been the state's principal tool of internal political mobilization and of external projection in an extraordinary and audacious enterprise of strategic over-extension. Crucially, the footprint of almost every major act of international Islamist terrorism, for some time before 9/11 and continuously thereafter, invariably passes through Pakistan. Pakistani power elite, located in the regressive military-mullah-feudal combine, is yet to abandon terrorism as a tactical and strategic tool to secure what it perceives as the country's quest for 'strategic depth' in the region. This remains the case despite the increasing 'blowback' of Islamist terrorist violence within the country, and the progressive erosion of the Army's status and control in expanding areas of the country. While the Pakistani Army has taken selective action against particular groups of Islamist terrorists—particularly those who have turned against the state, the fact remains that Pakistan continues to support and encourage the activities of a wide range of terrorist and Islamist extremist organizations. This is particularly the case with organizations that are active in Afghanistan and in India [Pakistan Terrorist Groups: An Overview, *South Asia Terrorism Portal,* Institute for Conflict Management, New Delhi].

In South Asia, Pakistan is the first and foremost country harboring much of terrorist groups and exporting much of terrorism in neighboring countries. It is in the badlands of Pakistan where a number of terrorist training schools and *madrassas* for religious indoctrination are very well entrenched. Even al-Qaeda's single most important refuge after Afghanistan has been Pakistan despite persistent pressures from US to track down terrorist networks. Speaking on October 7, 1994 on Pakistan's links with fundamentalism and international terrorism, Peter Duetch of Florida commented in US House of Representatives that "I am shocked to see reports detailing the extensive involvement of the Islamic Republic of

Pakistan in supporting Islamic fundamentalist terror groups in Afghanistan and India. I have seen Peter Arnett's excellent documentary "Terror Nation? US Creation?" shown on CNN last month. The film provides a graphic account of the links between the Islamic Republic of Pakistan and the fundamentalist regime of Gulbuddin Hekmatyar. I was disturbed to note that some Afghan groups that have had close affiliation with Pakistani Intelligence are believed to have been involved in the New York World Trade Center bombings". In his address Peter Duetch mentioned about the holy-war headquarters in Peshawar, Pakistan, the bustling base of operations for the Afghan resistance. There is also a mention in the speech about Pakistan showering Afghan terrorists with US-provided weapons and Pakistan's military leaders' involvement in their pursuit of an Islamic nuclear bomb and export of fundamentalism into India. He also mentioned about a report published in *Washington Post* of September12 1994 on how Pakistan's military has been involved in the drug trade. Pakistan's army and particularly its intelligence agency—ISI—(the equivalent of the CIA) are immensely powerful and are known for pursuing their own agenda. Over the years, civilian political leaders have accused the military (which has run Pakistan for more than half of its 64 years of independence) of developing the country's nuclear technology and arming insurgents in India and other countries without civilian knowledge or approval and sometimes in direct violation of civilian orders. Historically, the army's chief of staff has been the most powerful person in the country.

The **monstrous ISI**, after the withdrawal of Soviet troops, began smuggling heroin into the Western countries, using its big profits to sustain itself and wielded so much power that it became almost a state-within-state. Shortly after 9/11, it came to be known that an agent of ISI had wired $100,000 to Mohammad Atta's bank account in Florida at the instructions of the then Director of ISI, General Mahmoud Ahmad [*ABC News*, September 30, 2001; *Wall Street Journal*, October 10, 2001; *AFP*, October 10, 2001]. Out of 13 incidents of hijacking of Indian airplanes, seven were carried out by groups having links with Pakistan's ISI [Raman B: 2000]. Writing on Pakistan's *Jihadi* culture, Jessica Stern at Harvard University's Kennedy School of Government quotes US State Department's report that South Asia has replaced the Middle-East as the leading locus of terrorism in the world and wrote in *Foreign Affairs*, "Although much has been written about religious militants in the Middle-East and

Afghanistan, little is known in the West about those in Pakistan—perhaps because they operate mainly in Kashmir and, for now at least, do not threaten security outside South Asia. General Pervez Musharraf, Pakistan's military ruler, calls them "freedom fighters" and admonishes the West not to confuse *jihad* with terrorism. Musharraf is right about the distinction—the *jihad* doctrine delineates acceptable war behavior and explicitly outlaws terrorism—but he is wrong about the militant groups' activities. Both sides of the war in Kashmir—the Indian army and the Pakistani "*mujahideen*"—are targeting and killing thousands of civilians, violating both the Islamic "just war" tradition and "international law" [Stern, Jessica: *Foreign Affairs*: 2000]. Pakistan's reasons to support the so-called *mujahideen* might be: the Pakistani military's determination to pay India back for allegedly fomenting separatism in what was once East Pakistan and in 1971 became Bangladesh and India having an edge over Pakistan in population, economic strength and military might and therefore the Pakistani government supports the irregulars as a relatively cheap way to keep Indian forces tied and assisting militants active in Indian part of Kashmir. But in doing so, though Pakistan might be succeeding in executing its agenda in Kashmir, it however loses international reputation and credibility, which is already at very low ebb. By facilitating the activities of the irregulars in Kashmir, the Pakistani government is inadvertently promoting internal sectarianism, supporting international terrorists, weakening the prospect for peace in Kashmir, damaging Pakistan's international image and spreading a narrow and violent version of Islam throughout the region. As writes Gunaratana, successive Pakistani governments used the *jihadi* training and operational infrastructure on the Pakistan-Afghanistan border to arm, train and finance up to two dozen Kashmiri groups [Gunaratana: 2002: p. 167]. These groups have often been used by Pakistani intelligence apparatus as proxy military forces to undermine Indian control of Kashmir. Pakistan's religious fundamentalist orientations are also due to the fact that it is country founded on an avowedly religious principle. Its militant and armed Sunni organization—Sipah-e-Sahaba Pakistan and its underground splinter group—Lashkar-e-Jhangvi, openly support attacks on Shia mosques and also carry out anti-Iran activities through Afghanistan and also have gained popularity in their gruesome acts of violence against Indian security forces in Kashmir. Even in the destruction of the 6th century old, the priceless, pre-Islamic Bamiyan Buddha in

Afghanistan in 2001, though the Taliban were the executors, in fact the masterminds were Arabs and Pakistanis. The dreadful act invited the ire of India, China, Japan, Russia and some Central Asian Republics. Even UNESCO pledged to rebuild these age-old world heritage statues.

An ally in war against terrorism, Pakistan had provided a safe haven to the 'most wanted' Osama bin-Laden who was hiding somewhere in Pakistan when in a US military operation in May 2011, he was located at a compound in Abbottabad, Pakistan and was killed. US intelligence agencies are convinced that **A.Q. Khan network** supplied North Korea with technology to pursue its nuclear program in gross violation of bilateral assurances to US. Pakistan's actions also have facilitated a signatory to the Nuclear Non-proliferation Treaty (North Korea is a signatory) to violate its commitments. Pakistan is "double-dealing" with the US,—working with it in war against terror and maintaining ties with North Korea—making it debatable amongst US strategists whether to reward it for the former or to punish it for the latter. The dilemma of US administration was that it could not afford to lose Pakistan's cooperation. As a neighbor of Afghanistan, Pakistan served as a launching pad for military action against Taliban and to rout out other terror networks active there. Additionally, Pakistan is the most politically unstable of all nuclear powers, economically bankrupt, and a home to rising Islamic fundamentalism. In these circumstances, Pakistan had to be co-opted despite its bleak records. [see: Carbough, 2007].

Questioning American stand on always giving a clean chit to a rogue state, Jessica Stern asks in *New York Times*—"It's true the Pakistanis have helped us to capture some of the leading al-Qaeda figures, but you also have to wonder: Why do we find them all in Pakistan?" The mastermind of the first attack on the WTC was Ramzi Yousef, a Pakistani national with links to the Pakistani government. The attacks on the US embassies in Africa were masterminded by terrorists based in Pakistan. The finances used by Mohammad Atta (the ring-leader of the hijackers that attacked the WTC-Pentagon) were wired by terrorists with links to Pakistan's Inter Service Intelligence—an agency with deep links to the Taliban and al-Qaeda. All the 19 terrorists, as well as Zacarias Moussaoui (the suspected 20th hijacker) and Richard Reid (the shoe-bomber) are known to have spent time training in Pakistan in institutions funded by the Pakistani intelligence [see: Thundyil: 2004].

As says Stephen Cohen, an expert on Pakistan, that Pakistan is at a critical juncture, which is not new for the state as it has had four wars, many small military clashes with India, four coups and a collapsed economy several times, yet each time whenever it was to be declared a 'failed state, it bounced back—albeit with a weakened economy, a more fragmented political order, less security in comparison to is powerful neighbor and disturbing demographic and educational trends [see: Cohen: 2004: p. 267]. Though there might be some truth in the usage of the cliché of three A's—Allah, Army and America—as the forces saving Pakistan from each turmoil, Pakistani Establishment needs to do a lot if it wants to avoid a situation which might turn Pakistan into a rogue state—one with a weak or malevolent government that supported terrorism and possessed weapons of mass destruction.

Pakistan is not a victim of terrorism; rather it is the victim of its own sponsored *jihad* which it would like to make a permanent part of its strategy as the army has a vested interest in its continuance. Though it chose to join US in WOT out of compulsion, its army, ISI and many varieties of Pakistani Islamists, from moderate to radical, were highly critical of this choice. Islamists, increasingly better organized, have strengthened their hold on Pakistani areas bordering Afghanistan making them a safe haven for the Taliban. After Osama's killing by US in May 2011, and later developments that deteriorated diplomatic relations between the two countries, the US might lose Pakistani cooperation against al-Qaeda and Taliban and other extremist groups as much of the remaining al-Qaeda terrorists and Haqqani network are believed to be hiding in Pakistan [Shah: The *Guardian*: 2011].

Kashmir: A brief mention of Kashmir and Pakistan's role therein is quite imminent here to understand the nuances attached with regional aspect of Islamist terrorism in which politics of all the key players remains a determining factor in sustenance and escalation of terrorism. The continued stand-off between India and the Islamic Republic of Pakistan over Kashmir is responsible for much of terrorism and instability in the region also having negative socio-economic fallout not only between these two but also affecting other countries of the region too. For Islamabad, the liberation of Kashmir is a sacred mission, the only task unfulfilled since Muhammad Ali Jinnah's days. Moreover, a crisis in Kashmir constitutes an excellent outlet for the frustration at home, an instrument for the mobilization of the masses, as well as gaining the support of the Islamist parties and primarily their loyalists in the military and the Inter Services Intelligence (ISI). Back in the 1970s, Pakistan started to train Sikhs and other Indian separatist movements as part of Zulfiqar Ali Bhutto's strategy, the then Pakistan's Prime Minister and started sponsoring terrorism and subversion as an instrument to substitute for the lack of strategic depth. (see: www.Kashmir-information.com).

In Kashmir, the Pakistan-backed terrorists continue daily provocation against India, and an increasingly frustrated Indian government feels that it has no recourse short of full-scale war. As one of the world's longest-suffering victims of terrorism, India had high hopes for the US-led campaign against global terrorists. Islamists carry weight in Pakistan and are said to be beyond government control. In addition, the army is an irresistible temptation for all governments in Pakistan, both military and civil, to use terrorists in its campaign against India in the state of Jammu and Kashmir. As a result, Pakistan has sought to let what India calls cross-border terrorism in Kashmir continue.

As for the lack of international interest and support in Kashmir, before 1995 the West was not even prepared to concede that there was terrorism there. Only after the kidnapping of some American and West European tourists in 1995 by the Harkat-ul-Mujahideen (HUM) under the name of Al Faran was there a willingness to recognize that there was terrorism in J&K. Pakistan has been supporting all these terrorist movements—whether in the North-East or Punjab or J&K—but with nuances. In the case of J&K, it talks openly of its right to extend political, moral and diplomatic support to the terrorists, whom it projects as freedom-fighters, because it considers the territory as rightfully belonging to it. The Kargil fiasco in 1999 exposed Pakistan's overtures in the valley and created domestic political problems leading to a military coup that handed over the reigns of power in General Musharraf's hands. In fact, it was Musharraf who along with three other army Generals masterminded not only the invasion of Kargil but also supervised the establishment of terrorist training camps for HuM. The **rationale behind US tolerating this military ruler** was that: (1) US did not want to alienate a nuclear weapon Islamic state; and (2) it wanted to get continued support of ISI to capture or kill bin-Laden. Since 9/11 Musharraf was walking a political tightrope because his country had a huge constituency of radical Islamists who were known for being engaged in terrorist activity. So, his dilemma was to keep the fundamentalists on his side to gain support within the country, but at the same time face pressure from the United States and the West to crack down on these groups and individuals.

The *US State Department report "Patterns of Global Terrorism" 1995* says, "There are credible reports of official Pakistan support for militants fighting in Kashmir..." After the Soviet withdrawal, the *mujahideens* from Afghanistan were diverted to Kashmir theatre, which exploded from 1989 onwards. As compared to native Kashmiri groups, the Afghan Arabs who had training under al-Qaeda and Pakistan's *Harkat-ul-Jihad-e-Islami,* were highly motivated and fanatical and came to Kashmir with a one-point agenda of 'liberating' Kashmir and their Muslim brethren there from the clutches of 'infidels'. In Kashmir there was an all-pervading Pakistan, Taliban and al-Qaeda factor that made and continues to make terrorism in the valley so dreadful and so 'Islamist'.

The existence of terror camps on Pakistan soil have been regularly confirmed by various sources. Satellite imagery from the FBI and data produced by India's Research and Analysis Wing (RAW) clearly suggest the existence of many terrorist camps in Pakistan. A terrorist outfit, the Jammu Kashmir Liberation Front (JKLF) openly admitted that more than 3,000 militants from various nationalities were being trained in Pakistani terrorist training camps. Other non-partisan resources also confirmed that Pakistan's military and ISI both include personnel who sympathize with and help Islamic terrorists adding that "ISI has provided covert but well-documented support to terrorist groups active in Kashmir, including the al-Qaeda and its affiliate Jaish-e-Mohammed".

Until Pakistan became a key ally in the War on Terrorism, the US Secretary of State included Pakistan on the 1993 list of countries, which repeatedly provide support for acts of international terrorism. In fact, it is clear that Pakistan has been playing both sides in the fight against terror, on the one hand helping to curtail it while secretly stoking terrorism. Even the noted Pakistani journalist, Ahmed Rashid has accused Pakistan's ISI of providing help to the Taliban. According to one author, Daniel Byman, "Pakistan is probably today's most active sponsor of terrorism." Such role of Pakistan is confirmed by a CRS report also [Kronstadt: 2003].

Some of the radical Islamist outfits using terrorist tactics in Kashmir and having their origins in the subcontinent are as follows:

The ***Harkat-ul-Mujahideen (HuM)*** is a Pakistan-based terrorist outfit. It was formerly known as *Harkat-ul-Ansar* till it was banned by United States in 1997 following reports that it was linked with Osama bin Laden and al-Qaeda. Following the September 11, 2001 terrorist attacks the *HuM* came under scrutiny of the US government for its extensive links with Osama bin-Laden. Therefore, on September 25, US President George W. Bush signed an order officially banning the outfit. It's General Secretary, Maulana Masood Azhar, and its J&K unit Chief, Sajjad Afghani, were arrested in Srinagar by Indian Security Forces in 1994. Massood Azhar was later bargained for release by Taliban for freeing the passengers of a hijacked Indian Airlines plane IC 814. The HuM's operational capabilities have been severely curtailed by the formation of the Jaish-e-Mohammed (JeM) by Masood Azhar, immediately after his release.

The ***Harkat-ul-Jihad-i-Islami (HuJI)*** was launched in 1980 as part of the Pakistan-based *Jihad* network fighting troops of the then Soviet Union who were deployed in Afghanistan. It has undergone several changes during the past two decades and spread its area of activities to Jammu and Kashmir, other parts of India and Bangla Desh. *HuJI* cadres are suspected to be involved in the terrorist attack on the Kolkata office of the United States Information Service (USIS) on January 22, 2002. Security agencies have also reported that HuJI cadres are linked to the perpetrators of the train arson at Godhra in Gujarat on February 27, 2002. It functions under the patronage of the Inter Services Intelligence (ISI) and has formed alliances with the proscribed Students Islamic Movement of India (SIMI) and other Islamist extremists to spread violence in India.

Lashkar-e-Toiba (LeT) (also known as Jama'at-ud-Da'awa) was formed in 1990 in the Kunar province of Afghanistan and is based in Muridke near Lahore in Pakistan and is headed by Hafiz Muhammad Saeed. Its first presence in Jammu and Kashmir was recorded in 1993 when 12 Pakistani and Afghan mercenaries infiltrated across the Line of Control (LoC) in tandem with the Islami Inquilabi Mahaz, a terrorist outfit then active in the Poonch district of J&K. The LeT is outlawed in India under the Unlawful Activities (Prevention) Act. It was included in the Terrorist Exclusion List by the US Government on December 5, 2001. The US administration designated the Lashkar-e-Toiba as a FTO (Foreign Terrorist Organization) on December 26, 2001. It is also a banned organization in Britain since March 30, 2001. The group was proscribed by the United Nations in May 2005. The military regime of Gen. Pervez Musharraf banned the Lashkar-e-Toiba in Pakistan on January 12, 2002. In addition with challenging the sovereignty of Indian state in Jammu & Kashmir, LeT aims at the restoration of Islamic rule over all parts of India and is also active in Chechnya and other parts of Central Asia.

The Jaish-e-Mohammed (JeM) or the Army of the Prophet has been held responsible for the December 13, 2001 terrorist attack on the Indian Parliament in New Delhi. The outfit has been banned by the Indian government under provisions of the Prevention of Terrorism Act (POTA) on October 25, 2001. The US Secretary of State, Colin Powell, in a notification on December 26, 2001, designated the outfit as a foreign terrorist organization. Like the Lashkar-e-Toiba, the JeM too is an outfit formed, controlled and manned by Pakistan. The outfit was launched on January 31, 2000, by Maulana Masood Azhar in Karachi after he was released from an Indian jail during the terrorists for hostage swap of December 31, 1999. Its aim is to use violence to force a withdrawal of Indian security forces from J&K. The outfit claims that each of its offices in Pakistan would serve as schools of *jihad.* In its fight against India, he added that the outfit would not only "liberate" Kashmir, but also would take control of the Babri Masjid in Ayodhya, Amritsar and Delhi.

The Jamiat-ul-Mujahideen (JuM) was formed in 1990 and its followers are mostly Kashmiris from the Ahle Sunnat (Deoband) school. Unlike the other terrorist organizations who draw support along sectarian lines, the outfit recruits terrorists from all sects. It is opposed to the All Parties Hurriyat Conference (APHC) which is described by its leaders as a big danger for the terrorist movement in Jammu and Kashmir.

Al-Qaeda (Arabic for "The Base") emerged out of Afghanistan in the late 1980s as the world's first international terrorist organization to advance the cause of radical Islam. Its reach is global. In the Islamic world, its task is to purify societies and governments according to a strict interpretation of the Qu'ran, or Koran. In the non-Islamic world, its task is to compel governments there to withdraw their culture, influence, and military ties from the Islamic world. And except for the country that harbors its headquarters, no Islamic government supports its existence or its terrorist operations. It is a rogue network. As its name implies, it is a base for the support of other Islamic terrorist groups, such as the *Harakat ul-Mujahideen*, which operates out of the Pakistani controlled segment of Kashmir and which struggles to rid all of Kashmir of Indian "occupation." It is believed to provide these fraternal groups with money, contacts, and trained recruits in their particular battles to advance strict Islam. *Al-Qaeda* also has its own international network of cells under its command. It is this network that threatens world peace and security.

Osama bin-Laden, the mentor and father of al-Qaeda was determined to remove any non-Muslim penetration of the Islamic world, such as the presence of American military forces in Saudi Arabia. He had vowed that Israel must be destroyed and worked hard to ensure that all regimes become and remain true to strict Islamic principles. Rashid Ahmed says that every assumption made about the terrorist organization today is wrong as al-Qaeda remains the most dangerous international security threat to both the Western and Islamic worlds. Though some terrorism experts might think that al-Qaeda was reduced to a "soft" organization after America's war on terror, its members pursued terror projects relentlessly. The group's second-in-command, Ayman al-Zawahiri, issued 15 major speeches in 2006 on audio or videotape. These discourses dealt with such subjects as how al-Qaeda should approach Iraq following the departure of US troops; how the conflict in Somalia should be waged; and the need for a new terrorism offensive in Europe. [Ahmed: 2007].

The 9/11 Commission found no credible evidence of any operational connection between al-Qaeda and Iraq before the attacks. However, in post-9/11 years, al-Qaeda continued to expand its original core with clearly pronounced objectives. It has proven quite skillful at propaganda, churning out promotional DVDs that swayed opinion in the Muslim world. Before the September 11 tragedy, al-Qaeda maintained a presence only in Afghanistan and Pakistan. Afterwards, the organization emerged as a major threat in both these countries, but it also has established itself in Iraq, Somalia, Saudi Arabia and Sudan. Turning to Afghanistan, al-Qaeda and its Taliban allies restored combat capabilities that were destroyed by the US-led anti-terrorist coalition in late 2001.

In Iraq, al-Qaeda strategy was to isolate US forces by driving out all other foreign forces with systematic terrorist attacks, most notably the bombings of the United Nations headquarters and the Jordanian embassy in Baghdad in the summer of 2003. More important, he focused on the fault line in Iraqi society—the divide between Sunnis and Shiites—with the goal of precipitating a civil war. In February 2006, it attacked one of the country's most sacred Shiite sites, the Golden Mosque in Samarra. Most of all, al-Qaeda in Iraq has continued to orchestrate massacres against Shiites in Baghdad. It is also known to have developed closer ties to Kashmiri terrorist groups, such as Lashkar-e-Taiba and Jaish-e-Muhammad. From these accounts, it is clear that al-Qaeda today is a global operation—with a well-oiled propaganda machine based in Pakistan, a secondary but independent base in Iraq, and an expanding reach in Europe. Its leadership is intact except bin-Laden. Its Taliban allies are making a comeback in Afghanistan, and it is certain to get a big boost there if NATO pulls out.

The role of ISI in the murder of *Wall Street Journal* reporter, Daniel Pearl (in 2002) who was kidnapped and later murdered while investigating in Pakistan over involvement of the role of Pakistan's secret Intelligence in trying to blow up an American airliner clearly reveals the reach of ISI in terrorist acts. Pearl's kidnappers were not just ordinary terrorists as was suggested by their demands, especially their demand that the United States sell F-16 fighters to Pakistan, reflecting the desires of Pakistan's military and the ISI and giving some indications that kidnappers were connected to the ISI. After it was learned that Pearl had been murdered, it was also learned that Saeed, the ISI agent who had wired $100,000 to Mohamed Atta, had been involved in the kidnapping. In a January 2011 report released by the Center for Public Integrity, it has been confirmed that there was seamless cooperation between Pakistani terrorist groups such as Lashkar-e-Janghvi (LeJ), which took the lead in the kidnapping, and al-Qaeda. Members of other Pakistani terrorist groups such as Harkat-ul-Jihad-e-Islami (HUJI), Harkat ul-Mujahideen (HUM), and Sipah-e-Sahaba Pakistan (SSP) were also involved in Pearl's kidnapping and murder. [http://www.longwarjournal.org/archives/2011/01/new_investigation_in.php#ixzz3qH42wgo5]. Such reports confirm how dreadful role Pakistan has been playing in instigating, helping and executing international terrorism.

While moderate sections of Pakistani society have generally been marginalized, religious parties and their causes flourished. The military followed pro-US policies but the compulsions of domestic legitimacy resulted in an alliance of expediency with the religious sector. As a result of the military's unwillingness to extricate itself from domestic politics, the religious right, *jihad* and Islamisation became acceptable currency in political life, threatening regional peace and fundamental political, economic and social rights of Pakistanis.

Pakistan's tribal belt, including Pushtun-majority Federally Administered Tribal Areas (FATA) covering seven districts bordering on south-eastern Afghanistan, have become fertile grounds for terrorists and terrorism to flourish with the support and active involvement of Pakistani militants which negatively impacted US and NATO efforts to stabilize Afghanistan. Due to an approach alternating between excessive force and appeasement of militants, the Musharraf government in 2006 released militants, returned their weapons, disbanded security check-posts and agreed to allow foreign terrorists to stay if they give up violence. While the army virtually retreated to barracks, the Accord with militants in fact facilitated the growth of militancy and attacks in Afghanistan on US and NATO troops by giving pro-Taliban militants a free hand to recruit, train and rearm [see: International Crisis Group Asia Report No. 125, December, 2006]. State failure and lack of governance has gone in favor of militancy and militants. In FATA, the Pakistan government, as a matter of policy, has embarked on a path of appeasing the militants allowing them to establish parallel, Taliban-style police and court system in both parts of Waziristan. One of the poorest regions of Pakistan, FATA is economically and developmentally neglected for decades and being on the border of Afghanistan, serves as a convenient transit route for smuggling and trafficking in drugs and weapons, necessary for facilitating the sustenance of terrorist activities. Along with the dangers inherent in Pakistani scientist A. Q. Khan's nuclear black market, an absent democracy, weak civilian institutions and a history of dysfunctional civil-military relations have added to the problems for whoever has to deal with Pakistan.

Existence of Madrassahs: is a disturbing factor in the context of terror connection of Pakistan. These religious schools educate millions of students in the Muslim world and have been blamed for all sorts of ills since the attacks of September 11, 2001[1]. It is an Arabic word for any type of school, secular or religious (of any religion) and is variously transliterated as *madrasah, madarasaa, medresa, madrassa, madraza, madarsa,* etc. *Madrasa Islamia* translates as 'Islamic school', *Madrasa deeneya* translates as 'religious school' and *Madrasa khasa* translates as 'private school'. *Madrassah* networks serve multiple purposes—ideological indoctrination, focal point creation, and leadership training—and are an important policy issue and unraveling terrorist social networks is the key to a successful counterterrorism policy in the whole of Asia. They have been denounced as dens of terror, hatcheries for suicide bombers, and repositories of medievalism. As Samina Ahmed and Andrew Stroehlein of the International Crisis Group wrote in *The Washington Post* after 2005 July's London bombings, "Jihadi extremism is still propagated at radical *Madrassahs* in Pakistan. ... And now, it seems, the hatred these *Madrassahs* breed is spilling blood in Western cities as well".

Although the *madrassas* were created to fill the hole left by the state in educating young people free of charge, some became recruiting centers for terrorists, as most of the financing for the institutions came from terrorist groups and not from the government. There was also a great dearth of well-rounded education in these institutions, as their graduates were only good for Mosque services, and not other fields of life. Thus, social and economic factors played a great role in helping to spread intolerance. The word Taliban itself means "students", with "Talib" (singular) meaning a student. Saudi Arabia has funneled billions of dollars into West Asia, Pakistan and the rest of the world for over three decades for the propagation of puritan Islam in *madarssas.* This has made it easier for young minds to accept the cult of violence and be prepared and ready to kill in the name of religion. There has been a lethal mix of Saudi money and Pakistani manpower supplies to *jehad.* Saudi funding through various trusts like the Al-Haramain Islamic Foundation and the Al Rashid Trust, have helped finance *madarssas* and mosques. Saudi financial contribution to the making of the Pakistani nuclear bomb and contribution to the Afghan *jihad* has emboldened Pakistani adventurism [Wikram Sood, "Global Terrorism and Responses", *Indian Defense Review*, Vol. 21.2, March, 2008].

Pakistan-Bangla Desh factor: Pakistan has also played an active and prominent role in the growth of Islamist radicalism and terrorism in and from **Bangladesh**, another theatre of urgent concern in the South Asian region. There is increasing evidence, moreover, of a rising trend in operational cooperation between Pakistani and Bangladeshi intelligence agencies and *jihadi* organizations, particularly in their efforts to target India. A number of terrorist attacks and arrests in different parts of India, including the suicide attack at Hyderabad on October 12, 2005, the attack at the Indian Institute of Science in Bangalore on December 28, 2005, the *Diwali* bombings in Delhi on October 29, 2005, and the Mumbai serial bombings of July 11,

2006, among others, have exposed evidence of joint Pakistan-Bangladesh operations and terrorist modules. This pattern of collaboration and networking compounds the dangers within the region, and acts as a force multiplier for Islamist terrorist organizations seeking to project their capacities internationally [Sahni, Ajai, "The War on Terror: Assessing US Policy Alternatives on Pakistan", *Faultlines*, Vol. 18, January 2007].

With these developments and track record of Islamic fundamentalism, it can be argued that while in 2001, there was no civilizational conflict between Islam and the West and the September 11 tragedy was generally viewed as the work of a fanatical band of terrorists, today, the danger of civilizational conflict might become real if not arrested. And the threat doesn't just involve the potential clash between the Western and Islamic worlds; there additionally exists a danger of a clash within the Islamic world, pitting Sunni against Shi'a.

Afghanistan

Afghanistan, or as it is officially called the Islamic Republic of Afghanistan, is a landlocked country located approximately in the centre of Asia and its placement is such that it is variously taken as being located in Central Asia, South Asia and middle-East. It has religious, ethno-linguistic and geographic links with its neighbors such as, Pakistan, Iran, Turkmenistan, Uzbekistan and Tajikistan and China. Being a crossroad between East and West, it has an important geostrategical location, connecting South, Central and Southwest Asia. Though it has a long history of many upheavals, in August 1919 the country regained its full independence from Great Britain. Since the late 1970s, Afghanistan has suffered continuous and brutal civil wars which included foreign interventions also, such as, the Soviet invasion of 1979 and the recent US-led invasion by NATO in 2001. Thus seeing Afghanistan in a historical setting, one would find that the pride of the nation has been repeatedly breached by foreign involvements, mostly by alien, non-Islamic nations. But, if one learns lessons drawn from history, one would agree with ***Milton Bearden's assertion that Afghanistan is the graveyard of great empires*** as the country has witnessed the traverse of the world's great armies on campaigns of conquest to and from South and Central Asia. All eventually ran into trouble in their encounters with the unruly Afghan tribals.

From Alexander the Great's conquests in 327 BC. to Genghis Khan's and the great Mughal emperors' efforts to enter Asia through Afghanistan up to the nineteenth century contest between the United Kingdom and Russia for control of Central Asia and India, the tribal, backward Afghanistan has remained a focal point of entry and conquest. However, all these were disasters for the conquerors. The most recent being the failed adventure of the big Soviet empire of late 20th century and now the United States seems to be meeting the same fate. Engagement in the new war on terrorism poses severe challenges for the United States as rooting out bin-Laden's network also required military success in a country that the Soviet Union could not conquer in ten years of trying, as well as support from unstable surrounding nations. Washington may be tempted to try to oust the Taliban regime, but doing so could rekindle Afghanistan's brutal civil war. The United States must proceed with caution—or end up on the ash heap of Afghan history.

The ten-year long Soviet occupation of Afghanistan resulted in the killings of at least 600,000 to 2 million Afghan civilians. Over five million Afghans fled their country to Pakistan, Iran and other parts of the world. Faced with mounting international pressure and great number of casualties on both sides, the Soviets withdrew in 1989. It was seen as an ideological victory in the US, which had backed the *Mujahideen* through three US presidential administrations in order to counter-Soviet influence in the vicinity of the oil-rich Persian Gulf.

But, following the removal of the Soviet forces, the US and its allies lost interest in Afghanistan and did little to help rebuild the war-ravaged country or influence events there. Because of the fighting, a number of elites and intellectuals fled to take refuge abroad. This led to a ***leadership vacuum in Afghanistan.*** Fighting continued among the victorious Mujahideen factions, which gave rise to a state of ***warlordism***. The most serious fighting during this period occurred in 1994, when over 10,000 people were killed in Kabul alone. The conflict amongst various factions of Afghan society soon degenerated into fighting along regional, tribal and ethnic lines, resulting from a breakdown of state institutions and widespread arming of the population whose legacy can be traced into super-power rivalry in the state in the last decade of the cold war era. It was at this time that the Taliban developed as a politico-religious force, eventually seizing Kabul in 1996. By the end of 2000 the Taliban had captured

95 percent of the country. During the Taliban's seven-year rule, much of the population experienced restrictions on their freedom and violations of their human rights. Women were banned from jobs, girls forbidden to attend schools or universities. Those who resisted were punished instantly. Communists were systematically eradicated and thieves were punished by amputating one of their hands or feet. Meanwhile, the Taliban managed to nearly eradicate the majority of the opium production by 2001.

As a consequence of these developments after the end of cold war, the Asian state of Afghanistan replaced once confirmed terrorist training hub—Lebanon—as the major centre of international terrorist training location and by October 2001, nearly forty terrorist groups were operating there. The lack of a far-reaching US policy in Asia after the *mujahideens* defeated the Soviet forces created a Frankenstein out of the local *mujahideen* fighters, which America had to rue later in September 2001 and thereafter. In 1996, a heavy shroud was placed on the people of Afghanistan when Taliban captured Kabul, distorted the peaceful and sacred scriptures of Prophet Mohammed and turned them into the rule books of terror, enforced through a religious police a perverse rendition of Islam which gruesomely joins the constant and faithful prayer with barbaric practices of torture, beating, rape and execution of the simple, gullible people of Afghanistan, both men and women. Well before September 11 attacks, Taliban was engaged in widespread ethnic cleansing, littering Afghanistan with mass graves of ethnic and religious minorities.

Despite repeated requests from a victim of terrorism—India—in whose northern-west, Kashmir, Pakistan was sponsoring terrorism by using the trained and well-equipped former *mujahideens* who had perfect *jihadi* training and vast operational infrastructure, United States kept turning a blind eye till the shoe started pinching itself with a shocking surprise. The Taliban, itself notorious for its fundamentalist frenzy that it let loose on the simple people of Afghanistan, after bringing a large part of the country under its control, joined hands with al-Qaeda in their joint fight against the US and the West and unhesitatingly offered its land and resources to turn Afghanistan into a safe haven and breeding ground for terrorists. This nexus shaped the future course of dreadful terrorist events and brought Asia on the forefront making it 'the most dangerous place on the earth to live on' a fear expressed by Bill Clinton, the former US President during his March 2000 visit of the sub-continent. This nexus kept getting enlarged as a vast majority of radically indoctrinated Asian Muslims from the near and far flung countries, such as, Indonesia, Malaysia, Singapore, Philippines, newly independent Central Asian Republics along with Chinese, Pakistanis and Bangladeshis kept pouring in Afghanistan or joined terrorist groups in their own countries having connections with al-Qaeda.

Narco-terrorism is another serious problem in Afghanistan, particularly it reached alarming proportions during years after 9/11 and NATO military action there. Writing in *International Herald Tribune* on the ever deteriorating conditions in Afghanistan, Bruce Riedel and Karl Inderfurth argued that since the violence including murders and suicide attacks kept increasing in Afghanistan with each passing day in the years of war on terrorism, development work got a setback and therefore they stressed on the need for NATO to join in the fight against Afghanistan's exploding **drug trade**. According to them, Afghanistan was in danger of becoming a full-fledged narco-state. The opium harvest rose by almost 60 percent in 2006, accounting for about 92 percent of the world's supply. Between a third and a half of Afghanistan's economy is dependent on the illegal drug trade. Drug proceeds were supporting the Taliban and helping fuel the growing insurgency. US and NATO-led forces were reluctant to take part in combating the drug trafficking. But it became clear that since the Afghan Army, police and counter-narcotic forces were not adequate to the job, NATO must have assumed a counter-drug mission. So long as the Taliban has had a safe haven in Pakistan, it could continue their insurgency indefinitely. Therefore, 1,600 mile frontier with Pakistan has to be stabilized. India provided tremendous aid to Afghanistan in its bid to fight religious extremism so that Afghanistan does not fall to extremist forces again, and at least a decade long foreign military presence is required to control this drug trade [Riedel: 2007]. Ashley Tellis, an expert on South Asian strategic affairs, maintains that the state building process in Afghanistan, undertaken by US since 2001, is far from complete as the Central governmental authority does not extend meaningfully throughout the country, challenges of providing law and order have not been met, warlords have yet to be integrated into the political process, raising a new national army, elimination of Taliban and breaking al-Qaeda networks are all outstanding issues in Afghanistan's reconstruction which is frustrated every now and then by the resurgence of Taliban violence in different parts of the country.

The al-Qaeda leadership did not anticipate the rapid collapse of the Taliban regime in Afghanistan in the fall of 2001. Up to that point, Afghanistan had been a fertile breeding ground for the organization. According to some estimates, al-Qaeda had trained up to 60,000 *jihadists* there. Al-Qaeda leaders welcomed the invasion by US and coalition forces on the assumption that they would quickly get mired in conflict, as the Soviets had two decades earlier. Al-Qaeda and the Taliban thought they had decapitated the Afghan opposition and severely hampered its ability to fight by assassinating the Northern Alliance commander Ahmed Shah Masoud two days before 9/11. But in December 2001, Mullah Muhammad Omar, the Taliban leader and self-proclaimed "commander of the faithful," to whom bin-Laden had sworn allegiance, lost Kandahar, the capital of the Taliban's fiefdom. The Taliban had already lost considerable support among Afghans by the time of the invasion because of their draconian implementation of fundamentalist Islamic law and their harsh crackdown on poppy cultivation, the mainstay of the Afghan economy. But the key to their defeat was the defection of Pakistan. According to Ahmed Rashid, an expert on the Taliban, **up to 60,000 Pakistani volunteers had served in the Taliban militia before 9/11**, alongside dozens of active-duty Pakistani army advisers and even small Pakistani army commando units. When these experts left as Pakistan chose to be a part of 'War on Terror', the Taliban lost their conventional military capability and political patronage, and al-Qaeda lost a safe haven for its operational planning, training, and propaganda efforts. However, by 2006 the Taliban recovered sufficiently to launch a major offensive in Afghanistan and even attempted to retake Kandahar. New tactics imported from Iraq, such as suicide bombings and the use of improvised explosive devices, became commonplace in Afghanistan. Writing in *Parameters: US Army War College Quarterly* (Vol. 32, No. 2, 2002), William Hawkins sees a silver lining in all the dark picture of Afghanistan and says that we are lucky that Osama chose Afghanistan to locate in a country without ballistic missiles, without weapons of mass destruction, or for that matter, even without a conventional army of any size. The restoration of peace and order is basic to peace in the terror-infested world.

Bangladesh

According to South Asian Terrorism Portal (SATP) of the Institute for Conflict Management, New Delhi, the chief terrorist outfits in Bangla Desh are Harkat-ul-Jehad-al-Islami (HuJI), Jagarata Muslim Janata Bangladesh (JMJB), Jama'atul Mujahideen Bangladesh (JMB), Purba Bangla Communist Party (PBCP) with Islami Chhatra Shibir (ICS) as an extremist organization. The Jihad Movement in Bangladesh was one of the original signatories of bin-Laden's 1998 declaration of war on the West. In 2006, as bitter feuding between the two main political parties was increasingly ripping the country apart, there were growing indications that Bangladeshi fundamentalist groups were becoming radicalized.

With 147 million people, largely Muslim Bangladesh has substantial Hindu and Christian minorities and is nominally a secular democracy. But, the then ruling Bangladesh Nationalist Party (BNP) had struck a Faustian bargain with the fundamentalist party Jamaat-e-Islami years ago in order to win power. In return, the government generally looked the other way as the Jamaat systematically filled sensitive civil service, police, intelligence and military posts with its sympathizers and Jamaat-sponsored guerrilla squads patterned after the Taliban operated with increasing impunity in many rural and urban areas. Even the coveted post of industries minister was given to Matiur Rahman Nizami, a high-ranking Jamaat official who had helped promote the growth of a Jamaat economic empire that embraced banking, insurance, trucking, pharmaceutical manufacturing, department stores, newspapers and TV stations. A study by a leading Bangladeshi economist showed that the "fundamentalist sector of the economy" earns annual profits of some $1.2 billion [Harrison: 2006].

As for the mushrooming of *Madarssahs* is concerned, these began to proliferate in a big way both in Bangladesh as well inside India along the border areas. The estimates were that in Bangladesh alone, there were about 64,000 madrassas and the students who graduate from these religious seminaries are often lured by fundamentalists and fall prey to the attraction of Jihad. Matter of concern is the *madrassahs* proliferating along the Bangladesh border some of which harbor fundamentalist as well as insurgents/militant groups crossing over. *Madarssahs* in Chittagong Hill Tract have been used by Harkat-ul-Jihad for arms training.

Inside Bangladesh there are a number of training, liaison camps and safe houses for the underground insurgents of the Northeast India like ULFA, Muslim United Liberation Tigers of Assam, NDFB,

NSCN (IM), PLA, KYKL, etc. ULFA leaders are known to be permanently residing in Dhaka and Chittagong. They run lucrative business in Bangladesh and live a luxurious life. ULFA is also suspected to have had a working relationship with the Bangladesh Government. Indian intelligence reports have indicated that there are at least 190 militant camps inside Bangladesh which train and house Northeast militants with the help of ISI and al-Qaeda.

Through Bangladesh, the ULFA has made contacts with arms dealers in Cambodia and Thailand. Cox's Bazaar became a transit route of weapons. Bangladesh also served as a place for currency conversion and flowing finances out of country. As a result of harboring insurgents, Bangladesh has been flooded with small arms specially the Chittagong Hill Tracts. These weapons have been procured from insurgents groups in Myanmar and are brought by small speed boats to Chittagong ports or to unmonitored Cox's Bazaar's port and further purchased by other insurgent groups of North-East. The presence of number of terrorist groups like Harkat-ul-Jihad Al-Islami, Islami Liberation Tiger of Bangladesh, Jamaat-e-Islami, Islami Chatra Shibir, Arakan Rohingya National Organization, etc. in Bangladesh have made their impact in North-East insurgency in India. Linkages have been established between Harkat-ul-Jihad, al-Islam and al-Qaeda.

According to Jane's Terrorism Watch Report, Bangladesh is one of the poorest countries on earth, on the brink of being a failed state, and that makes it a perfect target for al-Qaeda and its ever-expanding network of Islamic extremist organizations. Virtually unnoticed by the world at large, Bangladesh is being dragged into the global war on terrorists by becoming a sanctuary for them. Neighboring India, which has had turbulent relations with Bangladesh since it gained independence from Pakistan in 1971, alleges that there are 195 camps in Bangladesh where guerrillas seeking autonomy or independent statehood in north-eastern India are being trained. Various governments in Bangladesh have repeatedly denied support to anti-Indian militants or to Islamic organizations, some of them linked to al-Qaeda. The US and its Western allies are gradually waking up to the potentially explosive situation developing in Bangladesh, which former prime minister Sheikh Hasina, leader of the Awami League, the main opposition party, calls the "Talibanisation" of Bangladeshi society ["Terrorism in Bangla Desh", *Jane's Terrorism Watch Report,* 27 January, 2005; see also: Bangla Desh Assessment: 2008, *South Asia Terrorism Portal,* Institute for Conflict Management, Delhi].

The Anti-Terrorism Ordinance, 2008 passed by the Government of Bangla Desh enables it to combat religious terrorism in the country more effectively. Though some human right groups condemn the ordinance which might be used for widespread persecutions, it empowers the Central Bank to freeze the accounts of a suspected terror financier. The judges now can impose death penalty for terror financing and staging murder to create panic and jeopardize the country's sovereignty. Special tribunals will be constituted, under the new Ordinance, to deal with such offenses. Anyone resorting to murder, kidnapping or damaging property to create panic among the people and jeopardize the country's security by using explosives, arms and chemicals, will be charged with committing terrorist offense. These will provide a deterrent to the terrorist outfits that are funded from terrorist cocoons like the Middle-East and not surprisingly even some western countries. Thus, although considering the demographic composition of its population Bangladesh is a Muslim majority country, it by and large enjoys much liberal legal and cultural environment since its inception.

Nepal: The birthplace of Buddha, an icon of world peace, the only Hindu state in the world, this Himalayan Kingdom, Nepal, was both a zone of peace and a favorite resort for young travelers from across the world before the genesis of the People's War of the Maoists began in 1996. Following the June 1, 2001 massacre of the King Birendra and his immediate family, the US intervened militarily in Nepal and sought to brand the Maoist revolutionaries as 'terrorists'—denying them not only legitimacy but also placing them outside the scope of universally recognized international law relating to armed conflict. But the events since 1994 have witnessed the gradual acceptance of Nepal's revolutionaries by its neighbors such as India and abandonment of the 'terrorist' terminology for these revolutionaries [Mage: 2007]. The 2006 collapse of the monarchy and historic peace agreement with the Maoists create hopes that overdue action can now be taken to address the needs of the poor which have been neglected over the last decade.

One year after the historic April 2006 uprising, Nepal has seen the end of 11 years of armed insurgency, the inclusion of the Maoists in the parliament and government, and the promulgation of an interim constitution. Despite the positive momentum toward peace, Nepal's transition has been crippled by demonstrations organized by an increasing number and variety of groups seeking national attention. Unfortunately, the demonstrations have often turned violent, resulting in loss of life and extensive property damage. Nepal's capital, Kathmandu, has been the epicenter of these protests, and students have been among the most active participants.

In a 2005 report of the International Crisis Group on Nepal's Maoists, it is argued that in less than ten years, the Maoist insurgency has transformed Nepal. The Communist Party of Nepal (Maoist) has spread armed conflict across the country and reshaped its political environment irrevocably. The Maoists are at heart a political party. They have developed military capacity but it is subordinated to political control. They use terror tactics and coercion but they are not simply terrorists. They maintain links to other communist revolutionary groups on the subcontinent but they are neither Khmer Rouge clones nor is their campaign part of any global terrorism. Maoist strategy is of a protracted people's war, both political and military—and the two cannot be separated [see: *International Crisis Report*, October 27, 2005]. But, by far Nepal does not look to be on the terrorist radar despite the ongoing civil war and insurgency in the country.

Sri Lanka: Sri Lanka gained independence from Britain in 1948 as the Dominion of Ceylon, although the British Royal Navy retained a base there until 1956. In 1972, the country became a republic, adopting the name Sri Lanka. Since this time, the country has experienced two major conflicts—a civil war and a Marxist uprising.

The Sri Lankan Civil War lasted from 1983 to 2009. The strategists on Sri Lanka opine that acts of "terrorism" had been committed by all sides during the civil war, but although all parties in the conflict had resorted to the use of these tactics, in terms of scale, duration, and sheer number of victims, the Sri Lankan state was particularly culpable. This was echoed by the Secretary of the Movement for Development and Democratic Rights, a Non-governmental organization (NGO), which further claimed that the Sri Lankan state viewed killing as an essential political tool. This had originally prompted the demand for a separate state for minority Tamils called Tamil Eelam in the north of the country. Assaults on Tamils for ethnic reasons have been alleged, and the experience of state terrorism by the people of Jaffna has been alleged to have been instrumental in persuading the United National Party to increase their hostilities there [Sri Lanka and State Terrorism: Wikipedia search].

As for the Tamil Tigers' share of violence goes, the war, over the

Tigers' claim to a Tamil-minority "homeland" in the north and east of Sri Lanka, claimed some 65,000 lives between its start in 1983 and the ceasefire in 2002. In the Batticaloa district of eastern Sri Lanka some 160,000 people were displaced in a few months alone. The Tamil Tigers are indeed a repellent partner in any peace negotiation. But they are the only one on offer, thanks to their ruthless elimination of most opposition, to their skill in exploiting Tamil grievances at feeling second-class Sri Lankans, and to the Sri Lankan army's backhandedness and violence, which has driven many Tamils into the Tigers' protection.

According to Prof. Peter Chalk of Rand Corporation, the LTTE threatens not only the domestic stability of Sri Lanka and India but also the security of the international system as a whole. In a recent study done for the Canadian Security Intelligence Service and published as Commentary No. 77, Prof. Chalk has said that the activities of the LTTE "octopus" include movement of refugees, arms transfers, terrorism, money laundering and drug-trafficking, all "by-products" of the militant group's insurgency in Sri Lanka. Activities of this sort conducted by various terrorist groups have repeatedly demonstrated the potential to disrupt national and international security in the post-cold war period, especially in the globalized world of today.

India: Terrorism in India has been multidimensional having religious, ethnic, secessionist/separatist contexts. The states that have suffered most due to terrorist activities and violence are **northern states** of Jammu & Kashmir (J&K) and Punjab, **North-Eastern states** of Assam, Meghalaya, Manipur, Mizoram, Arunanchal Pradesh, Tripura and Nagaland and the **Central States** of Bihar and West Bengal (Naxalism and Maoism) where except for J&K, the terror spell has been sporadic and of low to high intensity depending on the different phases. In August 2008, National Security Advisor M.K. Narayanan said that there are as many as 800 terrorist cells operating in the country [*Times of India*, 12 August, 2008]. Most of the terror activities relate to cross-border terrorism sponsored by neighboring Pakistan in which though the Kashmir valley is largely and regularly hit, the other parts of the country also become victim. The Sikh terrorism over a separatist demand for a state of Punjab was active in the decades of 1980s and early 1990s, and it is almost non-existent now. Terrorism in North-East also has had its highs and lows. But of all terrorisms, the one sponsored by the Islamist terror groups in

Pakistan is most dreadful, resourceful and sustained. The largest number of *jihadi* terrorist organizations from a single country are from Pakistan. They are the HUM, the Harkat-ul-Jihad-al-Islami (HUJI), the Lashkar-e-Toiba (LET), the JEM and the Lashkar-e-Jhangvi (LEJ). The LEJ is an anti-Shia terrorist organisation, which is not active in India. The remaining four are active in Indian territory—in Jammu & Kashmir and outside [Raman, B.: 2006]. While al-Qaeda as an organization is not yet active in Indian territory, its influence on the evolution and spread of *jihadi* terrorism in Indian is palpable. These *jihadis* operate in J&K through their own Pakistani cadres as well as through Kashmiri Muslims recruited and trained in camps in Pakistani territory. They operate in other parts of India through Indian Muslims recruited with the help of organizations such as the Students' Islamic Movement of India (SIMI) as well as through sympathetic members of the Indian Muslim Diaspora in the Gulf recruited directly by these organisations. The Kashmiris recruited by them are taken to Pakistan for training overland across the Line of Control (LOC) [Raman, B.: 2006]. In a nutshell, India is a victim of international terrorism and lacks a coherent policy to deal with these terrorist organizations so as to check its sway in various parts of the country.

Terrorism in Southeast Asia

Islam has always had a political component of it in Southeast Asia since it reached the region between twelfth to thirteenth century AD and superimposed itself on the local political culture. However, after the arrival of Europeans to the region, Islam became an important locus of anti-colonial struggle and also of the nation-building later on, though over time it got marginalised by the development-oriented secular states of the region reducing political space for Islam [Abuza: 2003, p. 33].

Southeast Asia has frequently had episodes of political violence, which have either been explicitly or implicitly linked to 'terrorism'. These have included activities by Communist groups (for example, in Malaysia, Thailand and the Philippines) and activities directed against ruling Communist regimes (such as the bomb attacks in Laos in 2000-2001). Terrorism has also at times been state-sponsored or condoned by states, as in the case of Christian anti-separatist groups in the southern Philippines (opposing Muslim secessionists) and militias in East Timor, Papua and other parts of Indonesia. Due to the ties of radical Islamist groups in Southeast Asia with al-Qaeda terrorist networks since mid-1990s and the region fast becoming a base for the past, current and future al-Qaeda operations, the international community, particularly US feels concerned. The concern was reinforced, as the members of Jemaah Islamiyah (JI) are known to have assisted two of September 11 hijackers who have also confessed to have plans to carry out attacks against western targets. October 12, 2002 Bali bombings, September 9, 2004 attacks on Australian Embassy in Jakarta confirm the fact that JI remains capable of carrying out large-scale plots against western targets.

The **Bali bombings on October 12, 2002** and then on October 1, 2005 have directed attention to the issues that Southeast Asia is a region conducive to the activities of both indigenous and international terrorist groups, and that elements of the two have been closely interlinked. The bombings have also emphasized that at a time when extensive attention is being directed towards the problem of Iraq, the 'war on terror' is far from won. However, much before the Bali episode, developments in Afghanistan since 1980s have fostered and introduced new elements into the terrorism in the region. Several factors have been important, such as:

(a) The Afghan experience has been central to the recent development of more radical Islamic groups in Southeast Asia. The region is home to an estimated 230 million Muslims (20 percent of the world total), though majority of them have moderate and tolerant views and have been willing to coexist with other religious groups and secular institutions. However, the more strident and anti-pluralist vein of Islam propounded from areas such as Saudi Arabia, Pakistan and Afghanistan has had an appeal to a minority. This appeal was deepened by the process of international recruitment for the anti-Soviet resistance in Afghanistan and many leaders of Southeast Asia's radical Islamic groups served or trained there. In addition, many Southeast Asians have studied in *madrassahs*—either in Southeast Asian countries or overseas, for example in Pakistan—where a more strict interpretation of Islam is taught.

(b) The advent of a large-scale and radical Islamic resistance in Afghanistan also added a new dimension to Southeast Asian Islamic separatism, whereby local groups could now gain added inspiration, assistance and funding from prestigious and well-financed international movements. Southeast Asia up to the 1980s already had several regionalist and separatist Islamist movements, such as in southern Thailand and in the southern Philippines. Of these two, Afghan resistance had a major impact on the southern Philippines, where Islamic radicals have found added sources of support and finance.

(c) **Socio-economic factors** have fostered further support for radical Islamic groups in the region and efforts to attain autonomy by Islamic movements have been intensified such as in the southern Philippines. The Asian financial crisis of 1997 did put pressures on regional governments and spending on crucial areas such as education. This factor contributed to the attraction towards the religious schools. Furthermore, well-funded Islamic radical movements have been able to offer financial support both to their active members and their families in the event of a member's death in combat. This has had considerable appeal to those in outlying and economically disadvantaged areas.

(d) Moreover, the factors like the **porous borders, weak immigration controls and ineffective administration** have significantly contributed to the rise of various Islamist movements in the region. Malaysia, for example, did not require visas till recently by anyone from those Muslim countries, which are members of the Organization of Islamic Countries (OIC). The Philippines has a weak immigration system and it is easy for foreigners to marry Filipino citizens and effectively change their identity. Administrative requirements can also be circumvented through corruption.

(e) There are long-standing economic and trade links between Southeast Asia, Middle-Eastern and South Asian countries, many of them outside normal financial channels and not readily monitored by governments. These have facilitated fund transfers from the Middle-East and South Asia to radical groups in the region. Criminal activity, including **drugs trafficking**, is widespread in the region and can assist resource-movement by radical groups. The region also has large supplies of weapons, both indigenously produced and imported.

According to Rohan Gunaratna, Southeast Asia has become an

important arena for international terrorism, notably al-Qaeda, which has a highly decentralized and elusive transnational terrorist network that is difficult to identify or combat. Though al-Qaeda has been damaged by Coalition operations, the organization may well retain two thirds of its core leadership and the great majority of the approximately 20,000 activists who were trained in its Afghan camps after 1996. Gunaratna argues that their leaders are handpicked, mostly educated in the Middle-East, speak Arabic unlike the vast majority of Asian Muslims, and are of a radical bent. Al-Qaeda's Asian core is handpicked from several hundred *jihadi* volunteers who fought in Afghanistan, including, *inter alia,* Central Asians, Chinese, Pakistanis, Bangladeshis, Indonesians, Malaysians, Singaporeans and Filipinos [see: Gunaratana: 2002].

According to CRS Report on *Terrorism in Southeast Asia,* the emergence of radical Islamist movements in the region can be attributed to several phenomena such as: reaction to globalization—which has been particularly associated to United States in the minds of the regional elite—, frustration with repression by secular regimes, the desire to create a pan-Islamic Southeast Asia, reaction to the Israeli occupation of Gaza Strip and the arrival of the thousands of terrorist veterans after fighting and winning in Afghanistan. Forging of connections between al-Qaeda and domestic radical Islamist groups in Southeast Asia is part of this trend.

In Southeast Asia, al-Qaeda's activities appear to have been concentrated in the Philippines, Malaysia, Singapore and Indonesia which established contacts in the region from 1988 and established a logistics base in the Philippines in the early 1990s. This base included use of Islamic charities and specially established businesses to channel funds. While many of its activities remain to be revealed, it has been alleged that al-Qaeda developed ambitious plans to conduct a series of bombings of airlines in the region, as several cases have illustrated, such as, a Pakistan-born al-Qaeda member Ramzi Yousef had plans to destroy a Philippines Airlines flight from Manila to Tokyo with a bomb in November 1994. He also was involved in planning a series of bombings and assassinations before his plans were uncovered after bomb components set-off a fire in his Manila apartment in February 1995. Al-Qaeda went on, however, to establish extensive links with two radical groups, Abu Sayyaf Group (ASG) and the Moro Islamic Liberation Front (MILF), both based in the southern Philippines. Another activist, Fathur Rohman al-Ghozi, born in Indonesia and trained by al-Qaeda, operated extensively in the Philippines and organized a series of bombings in Manila on 30 December 2000, which killed 22 and injured over 100. Another major plan was uncovered by Singapore's Internal Security Department in January 2002 in which it was found that the radical Islamic group Jemaah Islamiyah (JI) planned to develop large truck bombs to attack targets in Singapore. These included US personnel and installations including the US Embassy and possibly the adjacent Australian High Commission.

Arrests in Singapore and later Malaysia and the Philippines since late 2001 suggest that Jemmah Islamiah (JI) has had an extensive capacity to organize a large and possibly still largely unknown network of 'sleeper cells', both logistical and operational. It appears to operate in loose association with several other groups including the Moro Islamic Liberation Front (MILF) and, according to Rohan Gunaratna, together they have been 'part of a regional terrorist network operating under the aegis of al-Qaeda. JI is conceivably al-Qaeda's instrument connecting mainstream and renegade terrorists and guerrilla elements in the region.' [see Gunaratna: 2002, p. 193].

Other than these well-known groups, a number of other Islamist groups, nearly sixty in number, also function either independently or in league with some well-known Islamist groups. In Indonesia, for example, a number of Islamist groups are working overtime to disrupt the country's stability by aggressively campaigning for the enforcement of the Islamic law—*Sharia* and try to politicize and radicalize the moderate Indonesian Muslims who are traditionally more understanding and are willing to accommodate the scientific views in the interpretation of Koran that is compatible with democracy, human rights and social justice.

Though there is some instability in Southern Thailand due to a separatist struggle going on there and the Islamic character of this struggle can be felt, they cannot be termed as Islamist movements. Nevertheless, Thailand is serving as a safe haven, logistic base and weapon procurement center for over a decade, but that is due to widespread corruption among some segments of military and not because of terrorist linkages there. Southern island of Mindanao in Philippines is centre for sectarian violence between the majority Catholics and Muslims and the denial of a rightful place for the latter has given rise to a number of Islamist movements there. Moro Islamic Liberation Front (MILF) and Abu Sayyaf (ASG) have been two prominent groups having an objective to create a separate, independent state for Muslims in the Moro region in Southern Philippines.

Major radical terrorist groups active in at least four countries: Indonesia, Malaysia, Thailand and the Philippines and having links with al-Qaeda and with other movements in the region are:

Jemmah Islamiyah (JI) was founded in the mid-1990s in Indonesia, the world's fourth largest and the most populous Muslim country, with the grand aim of establishing an independent Islamic state encompassing Indonesia, Malaysia and the southern islands of the Philippines. Intelligence officials, notably from Singapore, have investigated the group since it came to wide attention in January 2002. JI has also been implicated in a number of bombings including those in Manila in December 2000. JI is alleged to be led by a radical Indonesian cleric, Abu Bakar Bashiyar, who runs a Muslim boarding school in Solo, central Java. The plot to stage bombings in Singapore was allegedly organized by a deputy to Bashiyar, Riduan Isamuddin, also known as *Hambali*, whose current location is unknown. JI is thought to be associated with other groups including the *Kumpulan Mujahideen* Malaysia (KMM). Singapore has alleged that JI received some funding from al-Qaeda over three years. Singapore officials have also claimed that *Hambali* has been seeking to coordinate the activities of JI with Muslim radicals in Thailand and separatists in the southern Philippines, especially the Moro Islamic Liberation Front (MILF) into an alliance called *Rabatitul Mujaihidin.* A peculiarity of JI is that it is the only Islamist organization in the world to recruit women. Though Islamist extremism is not popular amongst the educated and widely read/widely traveled moderate Singaporeans, the JI Singapore is dominated by foreign extremist ideologues mainly from Malaysia and Indonesia who subscribe to an anti-American, anti-western agenda [Gunaratana: 2002: pp. 187-88]. As says Gunaratana, the common denominator among those arrested either from al-Qaeda or JI is that they are driven neither by poverty nor by lack of education, but by a shared religious ideology that depicts the US as the enemy of Islam and the belief that Allah would reward them for waging a global *jihad.*

Lashkar Jihad (LJ) was established as the paramilitary wing of Forum Komunikasi Ahlus Sunnah wal Jammah (Communications Forum of the Followers of the *Sunnah*), established in Jogjakarta in early 1998. The LJ was formed on 30 January 2000 in response to religious violence in Maluku. LJ claims a three part mission—social work, Muslim education and a 'security mission' and it has had over 10,000 members, some of whom have been active in eastern Indonesia in communal violence. *Lashkar Jihad* has gained support from Indonesia's armed forces (TNI) and has also been able to embezzle money from the military (over $US 9 million). Its founder claims to have rejected approaches from al-Qaeda but supported the September 11 attacks in the US. In mid-October, Laskar Jihad announced that it had been disbanded but the veracity of this claim has yet to be determined decisively.

The Front Pembela Islam (FPI) is another Indonesian radical Islamic group, formed in August 1998 and now claims branches in 22 provinces. Based in Jakarta, the FPI is led by Habib Muhammad Riziek Syihab, a religious teacher who was educated in Saudi Arabia. Its stated goal is the full implementation of Islamic *Sharia* law, although it supports Indonesia's present constitution and avoids calling for an Islamic state. The FPI has a paramilitary wing called Laskar Pembela Islam and is well known for organizing raids on bars, massage parlors and gaming halls. In late 2001, it took the lead in threatening to sweep Americans out of Indonesia because of the US operations in Afghanistan, although the threat was not in fact carried out.

Abu Sayyaf Group ASG (The bearers of the Sword) is an outgrowth of the long-term struggle for autonomy in the southern Philippines, is opposed to any accommodation with the Christians and believes that violent action is the only solution. Abu Sayyaf has mounted terror and criminal attacks since 1991 and directed a wave of such attacks against Christian civilians in 1993. Its founder Abdurajak Janjalani was a veteran of the Afghanistan conflict who had brought back with him enthusiastic followers of radical Islamic ideology. It has strong al-Qaeda links—Osama bin-Laden is reported to have sent the Pakistani terrorist Ramzi Yousef, who is responsible for at least three operations, namely—bombing of the World Trade Center in 1993, New York Landmarks (1993) and an aircraft over Japan in 1994, for training with Abu Sayyaf and al-Qaeda has given financial assistance. Abu Sayyaf has also gained extensive revenue from kidnapping—including a $25 million payment from Libya to free hostages in March 2000. The group has recently suffered from serious internal divisions and its factions—whose interests appear to be primarily criminal—are now thought to have possibly about 500 members.

The Moro Islamic Liberation Front (MILF) has disclaimed connections with al-Qaeda but hundreds of its members are reported to have trained with al-Qaeda in Afghanistan. MILF split from the Moro National Liberation Front (MNLF) which had led a struggle for autonomy for Muslim areas of the southern Philippines from 1972. The MNLF suffered a series of setbacks in the 1990s, with a number of leaders either defecting to the government or joining MILF. Its aim is an independent Moro Muslim state and by the 1990s it had become the primary Moro rebel movement. The movement has been able to gain funds from sympathetic Islamic organization abroad, including in Malaysia, Pakistan and the Middle-East. It has had up to 35,000 members and has trained members of other groups, including JI. Major problems confront the goal of an Islamic Moro state, not least the fact that long-term immigration into the southern Philippines means that non-Muslims outnumber Muslims in most provinces of Mindanao.

The New Peoples' Army (NPA) was declared to be a terrorist organization by the US government in August 2002. It is believed to be the military wing of the Communist Party of the Philippines (CPP) having radical orientation. Although primarily a rural based group, the NPA has an active urban infrastructure to conduct terrorism as it also uses city-based assassination squads. Drawing most of its funds from contributions by supporters in the Philippines, Europe and elsewhere and from 'revolutionary taxes' levied on businesses, NPA opposes any US military presence in Philippines.

Thus, the terrorist groups in Southeast Asia are well established and have linkages with groups such as al-Qaeda, with an objective to build a pan-Islamic network in the region. Since the 9/11 attacks on US, efforts have increased to coordinate regional and international action against terrorism in the region. The Association of Southeast Asian Nations (ASEAN) announced increased cooperation at its annual ministerial meetings in July-August 2002. During these meetings, the United States also signed an anti-terrorism agreement with ASEAN to combat terrorist activities and related transnational organized crime. Agencies will also focus on combating the financing of terrorism and countering money laundering. Increased efforts are needed to pursue law-enforcement measures by the states in the region [see also: Parliament of Australia websites, reviewed last on April 11, 2003].

However, many governments view increased American pressure

and military presence in their region with ambivalence because of the political sensitivity of the issue both with mainstream Islamic and secular nationalist groups. Indonesia and Malaysia are majority Muslim states while Philippines and Thailand have sizable, and historically alienated and separatist-minded Muslim minorities [see: CRS Report for Congress, *Terrorism in Southeast Asia*, February 7, 2005].

Caucasus: (South and North)

After the collapse of Soviet Union, many breakaway republics got independence and others like Chechnya tried hard to get it though its rebellion was suppressed with a heavy hand by Russian military. It was a time when, amidst the highly volatile and confused situation in the region, al-Qaeda and other terrorist organizations tried to gain a foothold giving Muslims of this region a common platform to co-operate with each other in their fight not only against Russia but also against the other non-Muslim rulers of the region, such as, Christian Armenia.

The South Caucasus region consisted of ***Azerbaijan, Armenia and Georgia***, a cluster of states in the former Soviet Union forms part of larger Asia and suffers from instability in terms of numbers, intensity and length of ethnic and civil conflicts. Other emerging and full-blown security problems include crime, corruption and terrorism, proliferation of weapons of mass destruction and the trafficking in narcotics. The neighboring Iran, Russia and Turkey, having some kind of ethnic ties with these states, have security concerns in the region and the US and other EU states also try to pursue their different interests in the region. The region seems to have become a playground for outside powers. Add to this the tensions amongst these three on various accounts. The existence of various terrorist groups and of Islamic militants cannot be ruled out though the neighboring Central Asia is more prone to terror activities than the South Caucasus. All the three accuse each other for sponsoring terrorism in other's territory. Though before 9/11, Azerbaijan was serving as a conduit to terrorists in Chechen's war against Russia, after 9/11 however, it controlled *mujahideens* to operate from its territory which prompted al-Qaeda to criticize Azerbaijan for alignment of a Muslim country with the infidels. A CRS report in 2006 says that as part of the US Global War on Terror, the US military in 2002 began providing equipment and training for Georgia's military and security forces.

Azerbaijani and Georgian troops participate in stabilization efforts in Afghanistan and Iraq, and Armenian personnel serve in Iraq. Some observers argue that developments in the South Caucasus are largely marginal to global anti-terrorism and to US interests in general. They urge great caution in adopting policies that will heavily involve the United States in a region beset by ethnic and civil conflicts. Other observers believe that US policy now requires more active engagement in the region. They urge greater US aid and conflict resolution efforts to contain warfare, crime, smuggling, and Islamic extremism and to bolster the independence of the states. Some argue that energy resources in the Caspian region are central US strategic interests [CRS report: December 8, 2006].

Armenia's geographic location, porous borders, and loose visa regime presents opportunities for traffickers of illicit materials, persons, and finances. Its reliance on ties with neighboring Iran have dampened Armenian criticism of Iranian extremism and led to closer trade relations between the two countries. Diplomatic and trade relations with Iran are seen as a geographic and strategic necessity for the landlocked country, in light of closed borders with Turkey and Azerbaijan, and the perceived risk of instability in Georgia. President Kocharian spoke out against the possibility of international sanctions against Iran [See: Country Reports on Terrorism: April 2007].

Chechnya, located in the Northern Caucasus, is a federal republic of Russia. Following the bloody First Chechen war with Russia, which included a mass exodus of non-Chechen minorities, the republic gained a *de facto* sovereignty, although till January 2000, only the Afghan Taliban government had recognized it. Russian federal control was restored after the Second Chechen War. Since then there was a systematic reconstruction and rebuilding process, though unrest remains an issue [*Wikipedia* search]. Chechen rebels had some kind of association with al-Qaeda also right from 1994 and thousands of Afghan guerrillas were fighting the Russians in Chechnya during their first and second war with Russia. Al-Qaeda is believed to have provided training, ideological support and finances to Chechen *mujahidin.* The intelligence agencies of Saudi Arabia, Lebanon and Iran were secretly assisting Chechen rebels giving some hard time to Russian soldiers. During the first Chechen war, Ukraine served as a prime transit point for *mujahideen* fighters and weapon smuggling. [see: Gunaratana: 2002: p. 135].

Horn of Africa

Though Asia remains the main thrust of this study, a quick look at the African region, where a number of failed and failing states are located, will not be out of order. According to a special report of United States Institute of Peace (USIP) issued in January 2004, the Horn of Africa—Kenya, Ethiopia, Djibouti, Somalia, Eritrea, and Sudan—remains a major source of terrorism. Following the 9-11 attacks, the Horn has come under increased scrutiny as a strategic focal point in the war against terrorism. The area is of some strategic importance as it links the East with the West through the Red Sea and contains countries such as Sudan which is major oil producers attracting interest and external influence which the Islamists seek to counter and hence the importance of the region for being vulnerable to terrorist activities. According to the Failed States Index Ranking 2006 prepared by *Foreign Policy* magazine, Sudan tops the list of failed states while Somalia stands on the sixth and Ethiopia on the 26th place, Kenya on 33rd and Eritrea stands on the 54th rank.

In May 2003, the ***Kenyan*** government admitted that a key member of the al-Qaeda terror network was plotting an attack on western targets, confirming al-Qaeda's firm local presence. In fact, Kenya is both a soft target and a source of terrorism. Evidence unveiled during the trial in New York of four men linked to the bombing of American embassies in East Africa in 1998 revealed a terror network that had flourished in Kenya, taking advantage of lax immigration and security laws. The core leadership of the Kenyan cell consisted primarily of citizens of the Gulf states, Somalia, Pakistan, and the Comoro Islands who had assimilated into local cultures along the Indian Ocean seaboard. They, in turn, gradually recruited local Kenyans, particularly from the coast. Due to the corruption endemic in the immigration system, foreign residents of the Kenyan cell obtained citizenship and set-up small businesses and Muslim non-governmental organizations (NGOs).

Ethiopian Muslims have not been receptive to Islamic fundamentalism and they lack centralized power. They tend to identify first with their ethnic kin. Muslims and Christians are geographically intermixed throughout most of the country. Islam in Ethiopia has been benign during the past century. Yet, the potential for conflict is present though so far, Ethiopia appears to have remained free of terrorist attacks instigated by al-Qaeda and other Middle-East terrorist groups. Export of Islamic radicalism from

Muslim Somalia and Muslim Sudan, with both of whom Ethiopia shares a porous and corrupt 1000-miles-long frontier, is bound to impact it as it itself has a predominantly large Muslim population.

Djibouti's importance to terrorists derives from its transit capabilities rather than its potential as a base for international terrorist organizations. Events since 1999, however, may have increased Djibouti's attractiveness to international terrorists. Whether or which terrorist groups operate in Djibouti is difficult to assess. However, the groundwork to support such groups has expanded: in the last five years a new Islamic Center has been opened, supported by Saudi Arabia; reportedly, the number of mosques has grown from approximately 35 to over 100, financed and often staffed from outside Djibouti. The young cleric in the most important mosque in Djibouti City reportedly was trained at the Cairo theological center known for its radical Islamist doctrine. Economic links to individuals in the United Arab Emirates (U.A.E.) and Somalia are close as are communal links to families in Yemen.

Though so far, ***Somalia*** has not been the site of significant terrorist activity, nor has it been the site of terrorist attacks, nor it has hosted terrorist training bases; it has not proven to be a profitable recruiting ground for al-Qaeda; few foreign terrorists appear to have used the lawless country for safe haven; and no portion of the country has fallen under the direct political control of a radical Islamist group since 1996, yet Somalia has played a role in Islamist terrorism, albeit a specialized one. It has served primarily as a short-term transit point for movement of men and materiel through the porous and corrupt border between Somalia into Kenya, which has been a preferred site of terrorist attacks. Somalia remains high on the list of potential terrorist safe havens. This is due to a confluence of factors, such as, Somalia is a collapsed state, where terrorists can operate beyond the rule of law; the country's largest radical Islamist group, al Ittihad al Islamiya (AIAI), is very active; Somalia's commercial activity in remittances (*hawilaad*), transit trade, and smuggling and its long coastline and porous borders facilitate unrecorded movement of cash, people, and material and because of high levels of insecurity, the western physical presence is dramatically reduced in the country. According to Indian Intelligence interrogation of some Pakistani terrorists, a number of Arab *Mujahideens* who fought in Afghanistan moved to Somalia after Soviet withdrawal and joined AIAI.

Since 9/11, ***Eritrea*** has been included on the list of countries harboring terrorists, due to allegations that the Eritrean Islamic Jihad Movement (EIJM) was linked to al-Qaeda. Eritrea's inclusion in the "coalition of the willing" threatens to widen the gap between moderate and radical Eritrean Muslims due to the regime's use of the "war against terrorism" to eliminate all dissent.

The government of ***Sudan*** has a long history of harboring terrorist organizations and radical Islamic groups. It is the only sub-Saharan African government on the US State Sponsors of Terrorism List and the only one that has officially provided support and safe haven for terrorist organizations. The National Islamic Front (NIF), which changed its name to the National Congress Party in 1999, seized power in a military coup in 1989 and promoted an aggressive Islamist agenda. The new government sought to create an Islamic state in Sudan and to establish the country as the capital of the militant Islamist world. To pursue these goals, the NIF developed close ties with radical Islamist groups and terrorist organizations, including Osama bin-Laden and the origins of his al-Qaeda organization. Sudan openly harbored bin-Laden and al-Qaeda from 1991 to 1996 and though Osama and his cohorts left Khartoum in May 1996, the links between Sudanese regime and Osama continued, the latter enjoying the patronage of the former. For years, the Sudanese government has used its territory to provide safe haven, training bases, and staging areas to numerous terrorist organizations, including al-Qaeda, Egyptian Islamic Jihad (EIJ), Hezbollah, Hamas, Palestinian Islamic Jihad (PIJ), Abu Nidal, and Gama'at al Islamiyya. Operatives not only moved freely in and out of Sudan, but also established offices, businesses, and logistical bases for operations. Training camps were opened in the east of the country, which sent fighters from Lebanon, Afghanistan, and Algeria across the border into neighboring Eritrea and Ethiopia. Though Eritrea continues to accuse Khartoum of supporting Eritrean Islamic Jihad, the Sudanese regime has scaled down its exportation of radicalism in an attempt to improve relations with its neighbors and its standing in the international community. Under international pressure, Sudan expelled bin-Laden in 1996. Khartoum accelerated the moderation of its foreign policy and distancing from terrorist organizations after the September 11 terrorist attacks. Currently, the government of Sudan stands at a crossroads. It is attempting to move in a new direction through serious peace negotiations with the Sudan People's Liberation Army (SPLA) and improved relations with the United

States, but those efforts are being hindered by high-ranking officials who remain committed to the radical Islamist agenda.

An effective US response to terrorist threats in the Horn of Africa must include increased and targeted foreign aid, improved regional intelligence capabilities, and increased pressure on exogenous forces (especially Saudi Arabia) that stoke the flames of radicalism through Muslim "charities" and religious training programs [see: USIP Special Report No. 113: 2004].

Sub-Saharan Africa

Consisting of 42 countries in **Central** (Democratic Republic of Congo, Republic of Congo, Central African Republic, Rwanda, Burundi), **East** (Sudan, Kenya, Tanzania, Uganda, Djibouti, Eritrea, Ethiopia, Somalia), **Southern** (Angola, Botswana, Lesotho, Malawi, Mozambique, Namibia, South Africa, Swaziland, Zambia, Zimbabwe) and **Western** Africa (Benin, Burkina Faso, Cameroon, Chad, Cote d'lvoire, Equatorial Guinea, Gabon, Gambia, Ghana, Guinea Bissau, Liberia, Mali, Mauritius, Niger, Nigeria, Senegal, Sierra Leonne, Togo), sub-Saharan Africa is the poorest region in the world, still suffering from the legacies of native corruption, inter-ethnic conflict, overall ignorance of the indigenous populations, violence and political strife. The region contains many of the least developed countries in the world.

These countries have some terrorist activities and connections due perhaps to the many permeating weaknesses from which they suffer. Given that significant portions of the countries in this region remain either ungoverned or poorly governed, terrorist groups are bound to utilize the opportunities to their advantage. However, despite this dismal internal scenario and existence of the conditions of crime and conflict at the local level, it is unlikely that the region becomes a major supplier of international terrorists due to the reason that there are profound differences in Islam as practiced in Africa than that is practiced in Middle-East. Despite large Muslim populations here, the terrorists do not have supportive regimes such as Taliban in Afghanistan. But because of the poor or no government control in the region, terrorists find the environment inviting to hide fugitives and weapons. But again, given the absence of technologically advanced communicative infrastructure facilitating links with the outside world, Sub-Saharan Africa might not be an attractive proposition for substantial terrorist groups and networks. Majority of terrorism in Africa is waged by local populations against their own governments that lack legitimacy and is unleashed by poorly trained and ill-equipped fighters [see: National Intelligence Council's Discussion paper—Mapping Sub-Saharan Africa's Future].

Funding International Terrorism: Asian Context

(Hawala, Drug-trafficking, contributions by charities and religious donors, corruption)

The issue of financing terrorism has always been on the agenda of the nations victim of terrorism. The subject acquired seriousness and currency after the terrorist attacks of 9/11 on US. Since then, combating financing of terrorism has become an important component of nations' overall strategy to combat terrorism in all its forms. Middle-East and South Asia are specifically targeted as main hubs from where most of the direct or indirect terrorist-funding emanates, be it through *hawala*[2] transactions (underground banking system), charities or registered or unregistered money services. Saudi Arabian charities, which were prime sponsors of terrorist groups around the world, are now under much tighter controls albeit there is still a lot to do in the Middle-East and South Asia. Terrorist groups like al-Qaeda are on the run albeit they are also innovating—in making/moving monies and in hiring of their key operatives—the new terrorist is a western educated middle class technology savvy person and the source for getting information is the internet.

A pertinent question raised by Ehrenfeld Rachel in "Funding Terrorism" (2002) is explained thus: "In the turmoil following the initial bombing of the World Trade Center on February 26, 1993, few noticed that the first man arrested, Mohammed Salameh, the poor, unemployed, and illegal immigrant, offered five million dollars for bail. Where could he possibly get this kind of money? The judge refused bail, but was the source of Salameh's offer the same as that which funded the eight men, arrested shortly afterwards, who had planned to blow up Manhattan's tunnels and bridges and assassinate public officials?" Were the same money sources behind the final attack on the World Trade Center on September 11? The question speaks volumes about the linkages of money-laundering channels, illegal drug trade and corruption at high levels with terror networks.

Quite frequently the **charitable organizations** are used to collect and transfer illegal funds through alternative payment systems.

However, those funds are hardly enough to pay for the training camps, recruitment, conventional and unconventional weapons, travel, safe houses, propaganda, and all that is needed to carry out international terrorism activities. Yet, very little is said about the main sources of the very large funds required by the terrorists. Corruption that makes the illegal drug trade possible facilitates and in turn feeds the umbilical cord that is the terrorist lifeline, i.e., money. The illegal trade is their most vital and venomous source of funds. No other commodity, legal or not, exists on the market today that is better than illegal drugs that generates extremely high and fast returns.

According to Congressional Research Service (CRS) Report on terrorist financing issues, though the Saudi government *per se* is committed to cooperate with US in fighting terrorist financing, and is itself a victim of terrorism and thus shares the US interest in combating it, Saudi leaders acknowledge providing financial support for Islamic and Palestinian causes, but maintain that no official Saudi support goes to any terrorist organizations, such as Hamas. The report also states, however, that Saudi Arabia "was a place where al-Qaeda raised money directly from individuals and through charities" and indicates that "charities with significant Saudi government sponsorship" may have diverted funding to al-Qaeda. US officials remain concerned that Saudis continue to fund al-Qaeda and other terrorist organizations [Saudi Arabia: Terrorist Financing Issues, *CRS Report*, September 14, 2007].

In a report in the *online US News* (December 15, 2003) by David E. Kaplan, it is argued that "Over the past 25 years, the desert kingdom has been the single greatest force in spreading Islamic fundamentalism, while its huge, unregulated charities funneled hundreds of millions of dollars to jihad groups and al-Qaeda cells around the world." What was the source of the funding? Oil. A December 2003 report of the US government's General Accounting Office report in Nov. 2003 highlights the ways in which international terrorists make widespread use of alternative and informal mechanisms to raise, move or secure their funds. This makes it extremely difficult for counter-terrorist and law-enforcement officials to monitor flow of funds that terrorist groups need to sustain themselves and to execute their violent activities. The overlapping jurisdictions of different governments complicate the situation further that helps terrorists more than the counter-terror agencies. In a

December 2003 National Bureau of Asian Research (NBR) analysis, Abuza has detailed a variety of methods used by the Indonesia-based Jemmah Islamiyah (JI) terrorist group to raise and transfer funds through Islamic charities based in Saudi Arabia and how JI secures pledges from its members and transfers legitimate donations from mosques to its own coffers and also how it uses *hawala* system (underground banking) and personal couriers to transfer funds and various other things across the borders almost without trace and how some of its cells resort to petty crimes like bank robberies as in the context of Southeast Asia [Abuza: 2003].

Role of Drug Trafficking in Financing Terrorism in Asia

Narco-terrorism is a subset of terrorism, in which terrorist groups, or associated individuals, participate directly or indirectly in the cultivation, manufacture, transportation, or distribution of controlled substances, (such as poppy, cocaine, opium and other drug products like these) and the monies derived from these activities. Further, narco-terrorism may be characterized by the participation of groups or associated individuals in taxing, providing security for, or otherwise aiding or abetting drug trafficking endeavors in an effort to further or fund terrorist activities. Drug organizations fuel some terror networks. These drug organizations come from locations as far apart from each other as Afghanistan and Colombia, but all use violence in order to achieve their goals. The powerful international drug trafficking organizations operate around the world and employ thousands of individuals to transport and distribute drugs to communities world-wide. Some of these groups have never hesitated to use violence and terror to advance their interests, all to the detriment of law-abiding citizens.

The link between drugs and terrorism is not a new phenomenon. Globalization has changed the face of both the legitimate and illegitimate enterprise. Criminals by exploiting advances in technology, finance, communications and transportation in pursuit of their illegal endeavors have become a new force to be reckoned with in the category of criminal entrepreneurs.

Narco-terrorism is a term coined by former President Fernando Belaunde Terry of Peru in 1983 when describing terrorist-type attacks against his nation's anti-narcotics police. In the original context, narco-terrorism is understood to mean the attempts of narcotic-traffickers to influence the policies of a government or a society through violence and intimidation and to hinder the enforcement of the law and the administration of justice by the systematic threat or use of such violence. Pablo Escobar's (a Columbian druglord) ruthless violence in his dealings with the Colombian government is probably one of the best known and best documented examples of narco-terrorism. After 9/11, the term has come into a wider usage due to the US government's policies related to its 'War on Drugs'. The Bush administration continued funding Plan Colombia, which intended to eradicate drug crops and to act against drug lords accused of engaging in narco-terrorism. The US government is funding large-scale drug eradication campaigns and supporting Colombian military operations, seeking the extradition of notorious drug traffickers. Although al-Qaeda is often said to finance its activities through drug trafficking, the 9/11 Commission Report noted that "while the drug trade was a source of income for the Taliban, it did not serve the same purpose for al-Qaeda, and there is no reliable evidence that bin-Laden was involved in or made his money through drug trafficking." The organization gains most of its finances through donations, particularly those by wealthy Saudi individuals.

The production of and illegal trade in poppy and cocaine remains one of the most reliable sources of income throughout the world with its demand ever expanding and the money begotten from its trade sustaining the illegal, clandestine activities of terrorist organizations. According to an estimate from the State Department Office for International Narcotics Matters, the production of 1 kilo of cocaine costs about $3,000. The wholesale price of one kilo is about $20,000. Cocaine production takes as long as the leaves grow, and in Colombia, Peru, and especially Bolivia, they grow fast. Opium can be cultivated twice a year in Asia and the Middle-East and at least three to four times in Mexico and Colombia because of the tropical climate. The production of heroin is more expensive than cocaine. It costs about $4,000 to $5,000 to produce a kilo of heroin. That kilo will sell for $250,000 to $300,000 wholesale. The drugs are delivered only after the money for them is paid in full. Losing drugs in shipment does not affect the dealers because they have already been paid. Largely, the drug syndicates keep 80 percent of the revenues. According to figures released by the State Department (2001), there was over 5,000 metric tons of cocaine sold during 2001 at a street value of at least fifty billion dollars. Five hundred metric tons of heroin generated at least thirty billion dollars on the street during that same timeframe.

The Golden Triangle of Southeast Asia (Burma, Vietnam, Laos, Thailand) lies at the heart of the global heroin trade, accounting for roughly 60 percent of all illicit opium production. Narcotics from this part of the world have had an insidious, corrosive, far reaching and, at times, highly destabilizing impact. In particular, they have been linked to an explosion of AIDS, social instability, a lack of economic performance, official corruption and the growing force of organized crime. These effects have been felt, in one form or another throughout Southeast Asia, North America, Australasia, and Northeast Asia. Dealing with the threat posed by the Golden Triangle's heroin trade will require a fully inclusive strategy that emphasizes both supply disruption and demand reduction. While certain Southeast Asian states have pledged to intensify the scope and effectiveness of their drug policies, official apathy is widespread and continues to militate against effective counter-measures. [Chalk: 2000; for new drug-trafficking routes in Southeast Asia, see: *Jane's Intelligence Review*, July 1, 2002; also see: www.geopium.org].

In ***South Asia***, the Islamic State of Afghanistan is a major source country for the cultivation, processing and trafficking of opiate and cannabis products. Afghanistan produced over 70 percent of the world's supply of illicit opium in 2000. Morphine base, heroin and hashish produced in Afghanistan are trafficked worldwide. According to the official US Government estimates for 2001, Afghanistan produced an estimated 74 metric tons of opium from 1,685 hectares of land under opium poppy cultivation. This is a significant decrease from the 3,656 metric tons of opium produced from 64,510 hectares of land under opium poppy cultivation in 2000. Sadly, due to the warfare-induced decimation of the country's economic infrastructure, narcotics are a major source of income in Afghanistan. Though Afghanistan is a party to the 1988 UN Drug Convention but the country lacks the governmental resources to implement its obligations. Some intelligence sources confirmed a connection between Afghanistan's former ruling Taliban and international terrorist Osama bin-Laden and the al-Qa'ida organization. Drug Enforcement Administration (DEA), a component of the US Department of Justice, has received multi-source information that bin-Laden has been involved in the financing and facilitation of heroin trafficking activities. While the activities of the two entities do not always follow the same course, it also remains a fact that drugs and terror frequently share the common ground of geography, money, and violence. In this respect, the very sanctuary previously enjoyed by bin-Laden was based on the existence of the Taliban's drug state, whose economy was exceptionally dependent on opium.

Estimated Afghan Opium Production: Metric Tons

	2001	2000	1999	1998	1997	1996
USG	74	3,656	2,861	2,340	2,184	2,099
UNDCP	185	3,276	4,581	2,102	2,804	2,248

USG=United States Government; UNDCP=United Nations office on Drugs and Crime Prevention.

Opium prices in Afghanistan currently range from nine to eleven (9-11) times higher than what it used to be in 2000. Cultivation and production estimates may differ widely, but even under the most conservative estimate, Afghanistan has the capability to return as one of the largest opium producers in the world. The head of Afghanistan's provisional government, Hamid Karzai, supports the eradication of opium poppy cultivation and renewed the Taliban's ban on poppy cultivation and drug production in January 2002, and called upon the international community to support his efforts. The US Government and the UNDCP estimate that the area currently under cultivation could reach up to 65,000 hectares, potentially producing up to 2,700 metric tons of opium. Drug Enforcement Administration (DEA) sources have reported the observation of numerous inactive laboratory sites in Afghanistan and Pakistan, a number of significant opium dealers, large stockpiles of opium, and active opium markets in Jalalabad and Ghani Khel. The laboratories known to this point are concentrated in the regions bordering the Northwest Border Province of Pakistan, especially in Nangarhar, Laghman, and Konar Provinces in the Konduz and Badakhshan Provinces. The nexus between drugs and terrorism extends to other areas of Central and Southeast Asia as well.

According to the UN Office on Drugs and Crime, UNODC's World Drug Report, 2007, the total potential value of Afghanistan's 2006 opium harvest accruing to farmers, laboratory owners and

Afghan traffickers reached about $US3.1 billion. In addition, it is reported that in 2004, some 400 tons of cocaine was exported from one Latin American country, with an estimated domestic value of US$ 2 billion. How much of this money is used for perpetrating acts of terrorism? Estimates vary. But even a small percentage would be more than sufficient for some individuals or groups to plan, finance and carry out terrorist acts. Indeed drug trafficking has provided funding for insurgency and those who use terrorist violence in various regions throughout the world, including in transit regions. In some cases, drugs have even been the currency used in the commission of terrorist attacks, as was the case in the Madrid bombings.

There is evidence that the Taliban condoned and profited from the drug trade when it was in power and provided sanctuary to and received military assistance from terrorist groups in Afghanistan. Taliban taxes on opium harvests, heroin production, and drug shipments helped finance its military operations against rival factions. These taxes also bestowed legitimacy on Afghan drug traffickers. However, after the Taliban were forced out of power, they, and other groups seeking to undermine the regime of President Karzai, used drug trafficking to arm their militia and mount operations against the government, US forces, and international organizations in Afghanistan. The enormous profits gained through drug trafficking have been turned to strengthening warlords, corrupting local officials, and fomenting terror and instability throughout the country.

The current war in Afghanistan has brought attention to the Central Asian drug trade and the complex economic, political, and societal factors that contribute to its growth. Many agree that the same lawless conditions that enabled terrorist groups to operate in Afghanistan also allowed drug producers, traffickers, and organized crime to prosper there. However, the drug trade is not always an end in itself. It often serves to fund the illegal arms and activities of international criminal networks, which in turn become stronger and bolster the drug trade further. Groups such as al-Qaeda in Afghanistan use the money and connections that the drug trade provides to operate as supranational terrorist organizations. This connection between the drug trade and terrorism is often referred to as "narcoterror".

In addition to al-Qaeda, almost every group operating in Afghanistan—the Taliban, the Northern Alliance, and assorted warlords—is believed to have participated in and profited from the drug trade. In June of 2000, the Taliban banned opium production under international and UN pressure, but it allowed the Taliban to gain better control over drug production and sell-off stockpiles while at the same time driving up prices. Once the US attacked Afghanistan, the Taliban indicated that the ban would end, causing widespread fear of a flood of cheap heroin into other countries.

The Central and South Asian countries face a big drugs problem. Today, more than half the amount of opium exported from Afghanistan remains in the region. Young people increasingly turn to drugs to avoid facing poor social and economic conditions. Due to the accessibility of cheap heroin, drug users have switched from smoking opium to injecting heroin, and HIV/AIDS cases have skyrocketed accordingly. Attempts at interdiction often have no effect on the drug lords, who have access to an endless supply of both raw product and desperate people to press into service as couriers. Poppy cultivation and drug smuggling is the easiest, and sometimes only, way to make a living for a destitute population with limited economic opportunities.

Opinions are mixed as to what should be done next to combat the problem. There is international consensus that the government of Afghanistan needs to uphold the opium ban, but this may prove difficult without cooperation of Afghan society which is too terrified to counter Taliban. Suggestions about how to thwart the trade vary from bombing the poppy fields and stockpiles, or buying the crop and then burning it, to paying farmers with food not to grow poppy. Extreme measures may only shift the problem into other unstable neighboring countries, however. The region's economic dependence on drug production and the consequences for Central Asia go far deeper than one year's crop, and it remains to be seen what shape the international political and humanitarian response will take.

The "Golden Crescent" (Afghanistan, Pakistan, and Iran) and the former Soviet republics of Uzbekistan, Kyrgyzstan, Kazakhstan, Turkmenistan, and Tajikistan, play a central role in the global heroin trade. Native to Central Asia, the poppy plant thrives in the dry, warm climate and provides the raw opium from which heroin is produced. Combined with extreme poverty, endemic corruption, and powerful local warlords and cartels, the drug trade has flourished in the region, especially in Afghanistan. Its drug connections are mainly due to the devastating effects of the Soviet occupation and American-backed resistance in the 1980s, which created an economy dependent upon a foreign influx of cash and arms. When the Soviets withdrew in 1989, so did American financial and military support of opposition factions. Years of subsequent civil war only worsened living conditions in what was already a criminalized society with a destroyed infrastructure. Many Afghans, from impoverished peasants trying to live off war-ravaged land to power-hungry local warlords eager to augment their munitions, turned to the drug trade for the easy cash it provided. By the end of the 1990s, Afghanistan alone accounted for more than half of opium production worldwide, and is estimated to have been the source of over 70 percent of the world's heroin in 2000. As for Lebanon's Hizballah's involvement in narcotics trafficking is concerned, it has been limited, and the group's leaders have condemned the drug trade on religious grounds.

Money laundering is the process used by drug traffickers, terrorist organizations and others to convert bulk amounts of illicit profits into legitimate money. In Afghanistan, the unsophisticated banking system that previously existed has been damaged by years of war. Money laundering activity is completely unregulated. The drug traffickers also use *hawala* or *hundi* system, the informal banking system, extensively in the region. This system leaves no "paper trail" for investigators to follow. In South America, efforts to legitimize or "launder" drug proceeds by Colombian trafficking organizations are also subject to detection because of intense scrutiny by US law enforcement system. Colombian drug trafficking organizations have also developed a number of money laundering systems that subvert financial transaction reporting requirements. One such form of money laundering is known as the Black Market Peso Exchange (BMPE). The BMPE is a complex system currently used by drug trafficking organizations to launder billions of dollars of drug money each year.

Although the sources of funds may vary between terrorists and other criminals/drug traffickers—for example, terrorists may obtain funding from "clean" sources such as contributions to charities that are diverted, or from front-company operations, the methods used by terrorists and drug traffickers to transfer funds are similar. Illicit finances, and the means used to conceal profits and transfer funds, are of special concern. They also rely on bulk cash smuggling; multiple accounts; electronic transfers; commodities; and front organizations to raise, move, and launder money. They also use methods like passing money through businesses or humanitarian organizations that carry on substantial activities which are otherwise apparently legitimate, helping cover and launder the criminal and terrorist money. Both types of groups make use of fraudulent documents, including passports and other identification and customs documents to smuggle people, goods, and weapons.

Increasingly, terrorist and criminal organizations, which have fundamentally dissimilar motives for their crimes, may cooperate by networking or subcontracting on specific tasks when their objectives or interests intersect. For example, certain South American kidnapping gangs frequently sell custody of their victims to larger terrorist groups on what amounts to a "secondary market." Unlike other crime, however, drug trafficking often has a two-fold purpose for some terrorists. Some terrorists not only obtain operational funds through drugs, but also believe they can weaken their enemies by flooding their societies with addictive drugs. So while certain terrorist groups are increasingly involved in organized rackets in kidnapping, piracy, weapons trafficking, extortion, people smuggling, smuggling of cigarettes and other contraband, financial fraud, or other crimes, drug trafficking occupies a special position both in terms of profitability and as a perceived direct weapon used against the state. Some terrorist groups in particular use this argument to rationalize their involvement in illicit activity to their membership or support base. Just as terrorists and other criminals increasingly cooperate with one another or adopt one another's methods, law enforcement efforts against profit-driven and ideologically motivated criminals and terrorists also increasingly overlap. For example, the Bureau of International Narcotics and Law Enforcement Affairs (INL) sponsored training to foreign law enforcement authorities on bomb-blast investigations is valuable regardless of a bomber's motivation. The same is true of most other capacity-building programs, such as strengthening forensic laboratories or improving a police agency's management systems.

A research by *Federal Research Division Narcotics-Funded Terrorist/ Extremist Groups, Library of Congress* believes that since the end of cold war, funding of guerrilla and terrorist groups by ideologically motivated state sponsors significantly declined which led to their increasing reliance on drug-trafficking as a principle source of funding. As such, the drug-producing regions of the world, such as, Afghanistan and Colombia became heavily involved in drug-trade. And since the relationship between insurgent groups and drug-trafficking organizations or cartels is a mutually beneficial venture as it allows exchanges of drugs for weapons, use of the same smuggling routes, use of similar methods to conceal profits and fund-raising, e.g., informal transfer systems, use of the same resources for laundering money, use of the same corrupt government officials, and so forth. Though the difference between the two is that whereas the drug-cartels are primarily motivated by financial enrichment and do not seek public or media attention, the terrorists and extremists move according to their political goals, seek media and public attention and automatically come under the preview of law-enforcement agencies. As some insurgent groups have become increasingly involved in the drug trade, their criminal enterprises have assumed greater priority than their own ideological, political, or religious agendas. Examples include the Abu Sayyaf Group (ASG) in the southern Philippines and the Revolutionary Armed Forces of Colombia (FARC) in Colombia.

However, there are scholars who argue that the term 'narco-terrorism' appears to be too vague and counter-productive and cannot help understand either drug-trafficking or terrorism. Though a few cases have been highlighted by media as evidence of al-Qaeda tapping into the opium economy of Afghanistan, it is rather due to a normal habit of any crime-prone group to try to benefit from such a resource, particularly in a country like Afghanistan where the opium economy is estimated to equal half of the country's legitimate gross domestic product. In fact, says Pierre Arnoud Chouvy, for the term not to become hackneyed, 'narco-terrorism' should not refer to terrorist groups that have only been partially funded by illegal drugs, but rather to identify organized narcotic traffickers who seek to affect the policies of a government by terrorist means [Chouvy: 2004]. Moreover, one should also take into account the direct or indirect role of America's CIA and ISI in drug trade while reaching some important conclusions in how at times the interests of terrorists, drug-traffickers, resistance guerillas and governmental agencies converge in using the drug money. Chouvy argues that in post-Taliban Afghanistan and before 2004, US condoned opiate products in areas traditionally controlled both by United Front (Badshah Khan) and by various local commanders and warlords whose support was deemed strategically necessary to fight Taliban and al-Qaeda. As for al-Qaeda's links with drugs, the 9/11 Commission found no substantial links between the two, at least at the production stage, though it is likely to have involved in drug trade after the ouster of Taliban. Chouvy's conclusion is that while some proceeds from illicit drug trade undoubtedly contribute to partially funding some terrorist outfits, drug-trafficking is still far from being the main source of financing the global terrorism. Nevertheless, fighting drug-trafficking is important in itself for variety of reasons [Chauvy: 2010].

Thus, there exists a difference of opinion among the scholars on terrorism and drug-trafficking connections. Alex Schmid, in his "Links between Terrorism and Drug Trafficking: A Case of Narco-terrorism"? [Madrid Agenda: 2005] believes that though there is some empirical evidence on the simultaneous presence of armed conflict, including the terrorist variety, and the cultivation, processing and trafficking of narcotic drugs, even to the extent of the convergence between terrorist groups and organized crime groups, yet, while there are hundreds of terrorist organizations and at least as many drug trafficking organizations, the evidence, which is said to exist, is derived from relatively few cases. Out of nearly hundred countries involved in some way in the illicit drug trade, either in terms of cultivation, processing, trafficking, distribution, or the laundering of illicit profits, only some thirty countries have been empirically found to have a link between armed conflicts and illicit drug production and trafficking. While in most countries where drugs are produced, trafficked or consumed there is a causal link to crime, including violent crime, these are not necessarily terrorist crimes. There is often only sparse empirical evidence for some of the frequently cited cases of alleged connections between illicit drugs and terrorism. Even where these links have been established with reasonable certainty, estimates about the profits from drug trade going to terrorists vary widely. Lack of conclusive proof is often compensated by deductive reasoning such as this: with the end of the Cold War state-supported terrorism declined and terrorists had to look for alternative sources of financing and found it, *inter alia*, in the production, taxing and trafficking of illicit drugs like cannabis, heroine and cocaine [*Madrid Agenda: International Summit on Democracy, Terrorism and Security*, 8-11 March, 2005, Madrid].

Illicit drugs are, to be sure, not the only possible source of income for terrorists. The spectrum of terrorist funding is broad, as the following Table makes clear:

Principal Sources of Terrorist Financing

Domestic:	individual and corporate, voluntary contribution or coercive extortion
Diaspora-migrant communities	voluntary contribution or coercive extortion
Co-ethnic and co-religious support	donations and contributions from people with religious or ethnic affinity
State-sponsorship	patron states encouraging terrorist group to engage an inimical state
Public and private donors and individual financiers	support for terrorist-controlled welfare, social and religious organizations
Low level crime and organized crime	fraud, illegal production and smuggling of drugs, document forgery, smuggling, kidnapping for ransom, armed robbery, money-laundering, racketeering, smuggling of, and trafficking in, human beings
Investments and legitimate business	money earned (e.g. from publications) is used to acquire enterprises and engage in trade with profits being used to finance terrorism
Non-governmental organizations and community organizations	terrorist organizations set-up front organizations, which receive funds from sister NGOs in other countries or infiltrate established community organizations, which receive grants.

Source: Adapted from Rohan Gunaratna "The Lifeblood of Terrorist Organizations: Evolving Terrorist Financing Strategies", in Alex P. Schmid (Ed.) Countering Terrorism Through International Cooperation: Milan, ISPAC, 2001, pp. 182-85.

Conclusion

The issues highlighted in this Chapter show that any discussion on current phase of terrorism, a little before and after the end of cold war and also into 21st century, cannot be complete without taking stock of the Asian context. No doubt, Asia is rising on many fronts; yet it is a region marked with all sorts of negative indices too, terrorism remains one such factor. Being a region where followers of

Islam live in large numbers and where the socio-economic developmental indices have dodged a number of countries, the forces of religious fundamentalism have taken over. The highly attractive 'oil' commodity available in abundance in the region, particular in countries of Islam lures foreign players whose 'highly developed countries' status so much depends upon this energy resource. There are strategic linkages between religious extremism, international terrorism and drug business making it highly difficult for countries and alliances executing their 'war on terror' program.

NOTES AND REFERENCES

1. The word *madrassas* literally means "school" and does not imply a political or religious affiliation, radical or otherwise. In post 9/11 context, however, the word has often been used to define Islamic schools—especially in the negative context of anti-Americanism and Radical Extremism. The Yale Centre for the Study of Globalization examined bias in United States newspaper coverage of Pakistan since the 9/11 terrorist attacks, and found the term has come to contain a loaded political meaning: "When articles mentioned '*madrassas*,' readers were led to infer that all schools so-named are anti-American, anti-Western, pro-terrorist centers having less to do with teaching basic literacy and more to do with political indoctrination." While some *(madrassas)* teach a radical version of Islam, most historically have not" [Wikipedia search on *Madrassas*]
2. Hawala is an informal value transfer system. In its most basic variant, money is transferred via a network of hawala brokers and functions through transfer of money without actually moving it. In fact, a successful definition of the hawala system that is used is: 'money transfer without money movement'. A customer approaches a hawala broker in one city and gives a sum of money to be transferred to a recipient in another, usually foreign, city.

REFERENCES

Abuza, Zachary, "Funding Terrorism in Southeast Asia: The Financial Network of al-Qaeda and Jemmah Islamiyah", *National Bureau of Asian Research (NBR) Analysis*, Vol. 14, No. 5, December, 2003.

____, *Militant Islam in Southeast Asia: Crucible of Terror*, Lynne Rienner Publishers, 2003.

Ahmed, Imtiaz (ed.), *Understanding Terrorism in South Asia: Beyond Statist Discourses*, New Delhi: Manohar, 2006.

Al-Rasheed, Madawi, *Contesting the Saudi State: Islamic Voices From a New Generation*, Cambridge University Press, 2006, p. 332.

Bahgat, Gawdat, "Nuclear Proliferation: The Islamic Republic of Iran", *Iranian Studies*, Vol. 39, No. 3, September 2006.

Bajpai, K. Shankar, "Understanding India and Pakistan", *Foreign Affairs*, May/June 2003.

Byman, Daniel, *Deadly Connections: States That Sponsor Terrorism*, New York: Cambridge University Press, 2005.

_____, "Iran, Terrorism and Weapons of Mass Destruction", remarks before the Sub-Committee on Nuclear and Biological Attacks on Homeland, September 8, 2005.

Carbaugh, John E. Jr., "Pakistan—North-Korea Connections Creates Dilemma for US", Pakistan-Facts.Com, updated April 2004.

Chalk, P., "Southeast Asia and the Golden Triangle's Heroin Trade: Threat and Response", *Studies in Conflict and Terrorism*, Routledge, Volume 23, Number 2, 1 April 2000, pp. 89-106(18).

Chouvy, Pierre Arnaud, Opium: Uncovering the Politics of the Poppy, Harvard University Press, 2010.

Chouvy, Pierre Arnoud, "Drugs and Financing of Terrorism", *Terrorism Monitor*, Vol. 2, Issue 20, October, 2004.

Chellany, Brahma, "Fighting Terrorism in Southern Asia", *International Security*, Vol. 26, No. 3 (Winter 2001/02), pp. 94-116.

Chung, Chien-Peng, "China's War on Terror: September 11 and Uighur Separatism", *Foreign Affairs*, July/August, 2002.

Clough Langdon D., "Energy profile of Central Asia". In: *Encyclopedia of Earth*. Eds. Cutler J. Cleveland, Washington, D.C.: Environmental Information Coalition, National Council for Science and the Environment, September 4, 2008.

Cohen, Stephen Phillip, *The Idea of Pakistan*, Brooking Institute Press, 2004.

Cohen, Stephen Phillip, "Pakistan's South Asia Strategy", *India Abroad*, May 11, 2007.

CRF paper, "Strengthening the US-Egyptian Relations", A *CRF paper* (Council on Foreign Relations) May, 2002.

Country Profile: Saudi Arabia, *Federal Research Division, Library of Congress*, September 2006.

Country Reports on Terrorism, US Department of State, Washington, D.C.

Ehrenfeld, Rachel, "Funding Terrorism: Sources and Methods", NY: *Centre for the Study of Corruption*, 2002.

_____, *Funding Evil: How Terrorism Is Financed and How to Stop It*, Chicago: Bonus Books, 2003.

Ganguly, Rajat and Mac duff Ian, *Ethnic Conflict and Secessionism in South and Southeast Asia*, New Delhi: Sage, 2003.

Ganguly, Sumit, *Conflict Unending: India-Pakistan Tensions since 1947*, New York: Columbia University Press, 2001.

Gill, K.P.S. and Ajai Sahni, *The Global Threat of Terror: Ideological, Material and Political Linkages*, Delhi: The Institute for Conflict Management, 2002.

Harrison, Selig, "A New Hub for Terrorism", *Washington Post*, August 2, 2006.

Harrison, Selig S., "Will Pakistan Break Up?" Remarks by Selig S., Director, Asia Program, *Center for International Policy, Carnegie Endowment for International Peace*, June 9, 2009.

Hussain, Zahid, *Frontline Pakistan: The Struggle with Militant Islamism*, New York: Columbia University Press, 2007.

"Israel, Conflict and Peace", *Israel Ministry of Foreign Affairs*, 5 Nov. 2003.

Jane's Intelligence Review, July 1, 2002.

Khatri, Sridhar K. and Kueck, G., *Terrorism in South Asia: Impact on Development and Democratic Process*, Delhi: Shipra, 2003.

Kronstadt, K. Alan, "International Terrorism in South Asia", *CRS Report for Congress*, November 3, 2003.

Kux, Dennis, *The United States and Pakistan, 1947-2000: Disenchanted Allies*, Washington: Woodrow Wilson Center Press, 2001.

Mage, John, " The Nepali Revolution and International Relations", *Economic and Political Weekly*, May 19, 2007.

Manyin, Mark, "Terrorism in Southeast Asia", *CRS Report for Congress*, 27th Sept. 2005 (updated).

Markey, Daniel, "A False Choice in Pakistan", *Foreign Affairs*, July/August, 2007.

Miller, Greg, "Influx of al-Qaeda, money into Pakistan is seen", *Los Angeles Times*: May 20, 2007.

Muni, S.D. (ed.), *Responding to Terrorism in South Asia*, Delhi, Manohar, 2006.

Patterns of Global Terrorism, Department of State, Washington, D.C.

Peters, John E., *et. al., War and Escalation in South Asia*, Santa Monica: Rand, 2006.

Phadnis, Urmila, *Ethnicity and Nation Building in South Asia*, New Delhi: Sage, 2001.

Raman, B., "Pakistan and Terrorism", a Paper by *South Asia Analysis Group*, January 2, 2000.

_____, "Global Terrorism: India's Concerns, *South Asia Analysis Group, Paper No. 151:2006.*

Rashid, Ahmed, *Taliban: Islam, Oil and the New Great Game in Central Asia*, I.B. Tauris, 2002.

Riedel, Bruce and Karl Inderfurth, "NATO Must Do More in Afghanistan", *International Herald Tribune*, February 5, 2007.

Rupesinghe, Kumar, *Ethnic Conflict in South Asia: The Case of Sri Lanka and the Indian Peace-Keeping Force (IPKF).*

Sageman, Marc, *Understanding Terror Networks*. Philadelphia: University of Pennsylvania Press, 2004.

Sahni, Ajai, "The Dynamics of Islamist Terror in South Asia", *The Journal of International Security Affairs*, Fall 2005, No. 5.

Satloff, Robert, "*Egypt after the Revolt: Prospects for Post-Mubarak Egypt: An Early Assessment", The Cutting Edge*, April 18, 2011.

Shah, Saeed, *"US suspends Pakistan military aid as diplomatic relations worsen": The Guardian*, 10 July, 2011.

Sood, Vikram, " Uneasy lies the Head", *Hindustan Times*, May 28, 2007.

Stern, Jessica, "Pakistan's Jihad Culture", *Foreign Affairs*, Nov./Dec. 2000.

Stern, Jessica, *The Ultimate Terrorists*, Cambridge, Massachusetts: Harvard University Press, 2001.

Tambiah, Stanley Jeyaraja, *Sri Lanka: Ethnic Fratricide and the Dismantling of Democracy*, p. 116.

Tellis, Ashley J., *South Asian Seesaw: A New US Policy on the Subcontinent*, Washington, D.C., Carnegie Endowment for International Peace, May 2005.

Tellis, Ashley J., *India as a New Global Power: An Action Agenda for the United States*, Washington, D.C.: Carnegie Endowment for International Peace, 2005.

Thundyil, Matt, "Pakistan is Not an Ally Against Terrorism", Pakistan-Facts.Com, updated July 2004.

Upadhyay, R., "Islamic Institutions in India: Protracte Movement for Separate Muslim Identity", *South Asia Analysis Group,* Paper No. 599, February 2003.

Varshney, Ashutosh, *Ethnic Conflict and Civic Life: Hindus and Muslims in India*, New Haven, Yale University Press, 2003.

Vicziany, Marika, " State Responses to Islamic Terrorism in Western China: Impact on South Asia", *Contemporary South Asia*, Vol. 12, Issue 2, June 2003, p. 243.

Weinbaum, Marvin, *Pakistan and Afghanistan: Resistance and Reconstruction*, Westview Press, Boulder, Colorado, 1994.

Chapter

5

Conclusion: Addressing Clash of Fundamentalisms

"Everybody's worried about stopping terrorism. Well, there's a really easy way: stop participating in it." —Noam Chomsky

In this book, both 'political Islam' (and not Islam as a religion) and 'American universalism' (in all its 'Realist' dimensions: policy prescriptions and performance) have been conceptualized as 'fundamentalisms' and after a detailed survey of these fundamentalisms in various chapters of this book, it may be concluded that such a characterization is reasonable, though the conclusions are still open to debate and might have its critics. But it certainly is a dimension that carries weight in that the universalistic, exclusivist postures of both political Islam and American neo-conservatism clashed in decades of 1980s, 1990s and later in such a destructive way that a phase of international terrorism emerged on the horizon of world politics. Though the state still remained the primary actor, the international relations got both negatively and positively impacted in significant ways due to the centre-stage taken over by the non-state actors and entities in the post-cold war world. The forces of globalization, that are taken as the product of West, had a destabilizing effect on the largely insulated world of Islam resulting in the resurgence of Islam, or what may be more accurately termed as 'political Islam'. Such a post-cold war 'New World Order', a Western construct, was largely discredited by the non-state actors of Islamist fundamentalist variety who saw it as an attack on their way of thinking and living. Mark Juergensmeyer mentions about the new phenomenon of the use of religion in igniting conflict in international affairs and calls it the 'New Cold War' which in his view is replacing the bipolar powers of cold war by new economic forces, crumbling of old empires, discrediting of communism and resurgence of parochial identities based on religious and ethnic identities [Juergensmeyer: 2008].

This corrosion of fundamentalisms brought Islamists and the West, America in particular, face to face culminating into what is now called spectacular act of terrorism: 9/11. While there is a need for America to understand that merely the promotion of democracy and political tolerance cannot mitigate the fears of Islamic world, the Islamists also need to give up their fear for these 'western' characteristics: democracy, secularism, human rights and rule of law and must come forward to eradicate the ailments of their largely religion-driven societies in the ways permissible in true Islam.

Terrorism, by its very nature of singleness of purpose that helps terrorists to remain highly focused while governments confronting them have to be acting carefully on multidimensional fronts, is inherently difficult to combat. The post-cold war terrorism starting from Soviet withdrawal from Afghanistan till the September 11, 2001 attacks on US is broadly 'Islamist' in character, though neither 9/11 is the first or the last of terrorist acts nor Islam has remained or would always remain the central driver for terrorist acts. However, in the current phase of history, the two seem to be visibly co-related. But again, Islamist radicalism, as would be the case with any religion-related radicalism, is not an independent factor in itself; its use and application is dependent upon certain precipitating catalysts, both short-term and long-term, that work in favor of its provocation that ultimately gets expressed in the form of acts of terrorism and violence. If this premise were accepted, the task of devising the strategies to combat and eliminate it would become easier. In fact, terrorism caused by Islamic radicalism is manifestation of a complex web of conflicting ideologies, points of views and behaviour of different actors and groups, each claiming to be the 'most' Islamic in its position and stand on different issues of political and social behaviour within a so-called monolithic religion. As is shown in previous chapters, since Islamist terrorism is a mixture of 'many', no single approach to eradicate or remove it would be sufficient. Rather, a mix of approaches suitable to different contexts, populations, regions and countries would be appropriate, not only to address the direct causes of terrorism but also to address those semantic obstinations in the ambitious political agendas of the world's hegemon(s) that have served as catalysts. Therefore, to unearth what ails Islam and what ails the political arrogance of unipolarist hegemon would be a useful exercise.

USFP not Responding Adequately in Cold War Years and Afterwards

The tectonic events following 9/11, such as, Operation Enduring Freedom in Afghanistan, global war on terrorism and the war in Iraq and its aftermath, have dramatically affected Muslim world and their attitude towards US, resulting in the intensification of the already existing religious fundamentalism and terrorism amongst the radical Islamists whose violent activities have put the security environment of the entire world at a great risk. Therefore, any effective counter-terror effort should develop strategies towards the Muslim world that help to ameliorate the conditions that give rise to religious and political extremism.

Timothy Nefali in his book [Nefali: 2005] blames Reagan administration as well as both President Clinton and George W. Bush for not accurately responding even after bombing of Pan Am flight 103 in 1988, Bombing of Khobar Towers in 1996 and the attack on the USS Cole in 2000 as they all were gun-shy initially and had trouble in convincing audiences at home and abroad of the need to counter-attack terrorists and their bases. Moreover, Washington's foreign policy apparatus was habitual of being trapped in the conceptual framework of 1980s and therefore **continued to see terrorism as a state-sponsored activity,** by the communist rival USSR, directed at US interests abroad, having neither resources nor courage to access a safe and invulnerable US at home. However, the vulnerability of even the most powerful at the hands of a highly focused and invisible enemy changed the entire thinking in US and the result was an ambitious and preemptive posture in post-9/11, subject of much debate on the issue having both its supporters and opponents arguing fiercely for and against such a U-turn. But fatalism and preemption are certainly not the appropriate responses to terrorism. Instead, there is a need to define terrorism realistically and to develop a range of short and long-term alternatives to counter-terrorism. A measured balance between security and liberty has to be kept to ensure public that improved security might come sometimes at a cost of curbing some civil liberties temporarily. Putting the things in right perspective will help the government in mobilizing support for its counterterror policies and actions without alarming the public of a constant threat.

Bush administration failed to understand two central lessons of 9/11. ***First***, there are terrorist groups today that operate independently of state support and pose as serious a strategic threat as rogue states. ***Second***, because of radical Islamism's ideological nature, Washington's approach to Islamist terrorists must be different from its approach to rogue states. Of course, Islamist terrorists must be captured or killed and their networks disrupted. But to stop their movement's growth, their ideology must also be properly addressed.

Though military action partially succeeded in destroying al-Qaeda network and its mastermind Osama bin-Laden has also been killed in 2011, it has not eliminated the threat of terrorism, as there is possibility of local cells having taken up the struggle and also making renewed interlinkages with other terror groups operating with identical ideology. Intelligence experts need to develop better and deeper expertise to anticipate threats rather than simply react to them when they have already inflicted damage. The awareness that terrorism exists and will continue to exist and a focused commitment to combat it would help in developing right responses and policy prescriptions nationally and globally. As far as the diaspora Muslim mindset is concerned, the Muslims in US have been resistant to radicalization while those in European countries are found to refuse to assimilation. As feels Rohan Gunaratana, "The US is very good at operational counter-terrorism : killing, capturing and disrupting operations. It is very weak in winning the hearts and minds of Muslims" (Gunratana: 2003]. The US needs to restore multilateralism at the center of its counter-terror strategy. The present strategy of over-dependence on military action has worked to strain America's relations with the world, particularly the Muslim world. Though US have to pursue its national interests, but it needs to be done with a sense of shared purpose with the rest of the world. Since democracies work better and more comfortably with each other, the countries like India, which themselves are the victims of terrorism need to be co-opted in the international efforts against terrorism. Co-opting with a country like Pakistan could be beneficial for attaining short-term victory, such as disrupting Taliban and al-Qaeda networks in Afghanistan, but in the long-run what any observer of international politics sees is that a long-time sponsor of terrorism, i.e., Pakistan cannot be relied too far and too much. With growing US concerns over Pakistan due to resurgent terror networks in its tribal belt, an enhanced co-operation with a democratically rooted India in implementing anti-terror strategies can be more rewarding.

As a *long-term strategy*, building and maintaining an anti-terrorist coalition and a close international co-operation is necessary to tackle the problem politically and militarily. Equally important is the fact that while the radical Islamists are in a minority, they have developed extensive networks spanning the mainstream Muslim world as well as the countries where Muslims form the significantly large or the largest minorities while the liberal and moderate Muslims have no such networks. The result is that while the radicals succeed in amplifying their hate-message, the moderates' views are either silenced or are hijacked by threatening techniques of Islamists. **There is a need to retrieve Islam and its moderate believers from the clutches of the hijackers of religion** by providing the moderates an international platform that ensures their security as well as freedom to challenge the Islamists' misinterpretation and misrepresentation of Koran with a powerfully articulated counter-ideology that can inflict a long-term strategic damage on their machinations they use to misguide the true and moderate followers. The military, political and financial counter-terror strategies are important in their own way; yet ignoring religious, cultural, educational and social aspects could be pretty costly in the long-run.

Forced regime change need to be stopped: A flawed thesis of the proponents of the Realist School of American foreign policy that US should embark on a policy of regime change where unfriendly autocratic rulers are found to be in league with extremist elements has not and might not always deliver the intended results or might even backfire, as happened the case with US strategy in Iraq after its military action there in 2003. It is not merely a question of what strategy in Muslim world at bringing or not bringing a regime change would serve the US interests the best, there are genuine doubts about the negative effects of its universal adaptability and applicability [Rand Study: 2004]. Instead, the main thrust should be—would a strategy of regime change be serving the interests of global peace and security? The current nuclear stand-off between US and Iran and temptation to workout an option of bringing a 'democratic' regime change in a theocratic Iran is a very sensitive matter and might neither serve US interests in the region nor the interests of international peace and security unless the unbiased and widely acceptable strategy is devised to dissuade Iran from its current nuclear program. A unilateral action in this matter is bound to serve the terrorists and radical Islamists more than the other way round. Of course, an incremental democratization of the autocratic/theocratic Muslim regimes of the Middle-East would certainly help in the long-run to isolate and eliminate the radical elements that are in league with terrorists and their networks and to remove some of the structural causes of radicalism and extremism. But a sudden and abrupt change will intensify already brewing hatred against the US and the polarization along cultural/civilizational lines cannot be ruled out. The presence of US troops on Saudi soil after the Gulf war of 1991 made both US and Saudi rulers suspect in the eyes of religious radicals who, then onwards, espoused for a more extreme version of already a very rigid official ideology enforced in the country. In the case of Pakistan, a regime change will always have its problems given the presence of terrorist networks there and the possibility of nuclear weapons falling into the hands of terrorists and Islamist radicals, who are an integral part of Pakistani military set-up. Moreover, in the absence of democratic traditions and of civil society institutions in the rigidly governed Islamic countries make the task of bringing regime change very difficult, even more destabilizing at least in the short-run.

Self-restraint in the hegemonic tendencies: As regards US dominance in the world, there are both proponents and opponents of the America's unbridled power, some arguing that America's unipolar moment is already passé and others arguing that it will be able to retain this status. US policy-makers have spent the past decade debating how best to wield American power. For the rest of the world, the debate is: how best to deal with it. Notwithstanding the ironclad rule of modern international history that hegemons always provoke, and are defeated by, the counter-hegemonic balancing of other great powers, the primacist scholars claim that US hard-power capabilities are so overwhelming that other states cannot realistically hope to balance against the United States. The United States' hegemonic grand strategy has been challenged by Waltzian balance of power realists who believe that the days of US primacy are numbered and that other states have good reason to fear unbalanced US power. Amidst these claims and counterclaims is the reality that with so much power in the hands of one country, how should other nations respond? Though many governments still value US power and seek to use it to advance their own interests, yet even Washington's close allies are now looking for ways to tame the United States' might. Islamists violent reaction is one such way. If US is to maintain its credibility in the international community, it must make its dominant position acceptable to others—by using military force only sparingly, by fostering greater cooperation with key allies, and, most important of all, by rebuilding its crumbling international image. Layne Kenneth Waltz had predicted that following the Soviet Union's demise, unipolarity would quickly give way to mutipolarity by stimulating the rise if new great powers [see: Layne: 2006; Nye: 1990, 2002, 2004; Walt: 2005; Watt: 2005]. The events of 9/11 have already exposed the unsustainability of the thesis that US is insulated from challenges within and without. The fundamental dynamics of world politics is still at work and according to these dynamic, other states always have compelling reasons to offset the preponderant capabilities of the very powerful, even if that powerful is not posing them an existential threat in near future. To mitigate the threat perceptions of the rest of the world, the US should adopt a strategy of self-restraint if the threat of terrorism against it or against any other country in the world is to become a reality.

Seeing all Muslim organizations with the same lens is counterproductive: US policy-making has been handicapped by Washington's tendency to see all Islamic organizations and all Islamic movements as monolith. One such case can be cited of Muslim Brotherhood—the world's oldest, largest and most influential Islamist organization—which is condemned both by conventional opinion in the West and radical opinion in the Middle-East. American commentators have called the Muslim Brothers "radical Islamists" and "a vital component of the enemy's assault force—deeply hostile to the United States." Al-Qaeda's Ayman al-Zawahiri sneers at them for "luring thousands of young Muslim men into lines for elections ... instead of into the lines of *jihad.*" [Leiken and S. Brooke: *Foreign Affairs*: 2007]. *Jihadists* loathe the Muslim Brotherhood (known in Arabic as al-Ikhwan al-Muslimeen) for rejecting global *jihad* and embracing democracy. These positions seem to make them moderates, the very thing the United States, short on allies in the Muslim world, seeks. But the Ikhwan also assails US foreign policy, especially Washington's support for Israel, and questions linger about its actual commitment to the democratic process. In fact, the Brotherhood is a collection of national groups with differing outlooks, and the various factions disagree about how best to advance its mission. But all reject global *jihad* while embracing elections and other features of democracy. There is also a current within the Brotherhood willing to engage with the United States. In the past several decades, this current—along with the realities of practical politics—has pushed much of the Brotherhood toward moderation. Since its founding in Egypt in 1928, the Muslim Brotherhood has sought to fuse religious revival with anti-imperialism—resistance to foreign domination through the exaltation of Islam. US Policy-makers should analyze each national and local group independently and seek out those that are open to engagement. In the anxious and often fruitless search for Muslim moderates, policy-makers should recognize that the Muslim Brotherhood presents a notable opportunity to counter-balance other severely radical Islamist groups, a long-term solution to the problem of terrorism and religious fundamentalism.

On American obsession with promotion of democracy: Though some like Gregory Gause would argue that Bush administration and his supporters believed that push of democracy in Arab world will not only spread American values but also improve US security and as democracy grows in the Arab world, the region will stop generating anti-American terrorism, others like Dobriansky and Crumpton would say that **this is nothing more than a one-dimensional solution to a multidimensional problem.** President Bush underscored his views during his September 15, 2005 speech to world leaders at the UN in New York where he spoke about confronting threats directly, engaging the enemy, disrupting terrorist networks, denying enemies safe haven, building international coalitions, forging treaties that reinforce the rule of law, denying the enemy weapons of mass destruction, and changing the conditions that terrorists exploit. He mentioned about a shifting mix of international geopolitics, economics, religion, ideology, ignorance, cultural stress, and intolerant political systems that offer little room for political expression or personal freedom. Of course, tyranny does afford terrorists an advantage, but democracy is also not a guarantee to rout out terrorism. Rather, in some cases democracies tend to be vulnerable to terrorism, but tend not to produce terrorists. And some democracies are better than others in not driving people to violence. But efforts to impose democracy, a top-down approach, without giving space to local variations and cultures, are bound to be counter-productive.

A broader media access in US is required: American society desperately needs access to broader media coverage of global events. Americans live in an isolated and self-censoring society that cuts us off from reality—a deadly luxury for a superpower. They need alternative perspectives to enhance debate: let CNN International, the real BBC, al-Jazeera/English, and Indian, Russian and Chinese news coverage—all in English be available in the US. One can only benefit from broader perspectives that represent how most of the rest of the world thinks. Hard-liners in Washington may argue that "we create our own reality" but foreign acceptance of that reality has been limited in the extreme and that is problematic too.

Economic Reconstruction in Afghanistan and a balanced cultivation of Pakistan by US: The United States should supplement the military buildup by taking the lead on a major economic reconstruction program in Afghanistan. Since 2001, the international community has delivered far less aid per capita to Afghanistan, one of the world's poorest countries, than it has to recovering states such as Bosnia. The country's infrastructure must be improved in order to develop a mainstream agricultural economy that can compete with illicit poppy cultivation, which breeds crime and corruption and strengthens the *jihadi* sub-culture. In Afghanistan, if the United States and NATO demonstrate convincingly their commitment to building Afghanistan's nascent democratic institutions, supporting a civil government and maintaining a minimal presence in the country at least for the time being, the Pakistani army will have an ever greater incentive to invest in Afghanistan's stability rather than hedge against collapse or the rise of a threatening neighbor.

As for Pakistan, the United States and its partners, including NATO, need to take a firmer position with the Pakistani government to enlist its help in tracking down remaining al-Qaeda leaders and networks. So far, despite their promises to track down terrorists on their soil, most Pakistani leaders have sought to tame *jihadists* and tolerated those who harbored bin-Laden and his lieutenants, Taliban fighters and their Afghan fellow travelers, and Kashmiri terrorists. Pakistan's Inter-Services Intelligence (ISI) had extensive links with bin-Laden; he was hiding in Pakistan's military area when he was killed in an American military operation in 2011. Pakistan should no longer be rewarded for its selective counterterrorism efforts. Washington gave a massive economic aid to Pakistan since 9/11. The nexus between Pakistan and terrorism will not be broken until Pakistani officers are back in their barracks and civilian rule is restored on a more permanent basis.

Success in Pakistan's long-term struggle against extremism will eventually demand a thoroughgoing democratic transition in Islamabad. The Bush administration failed to broaden its partnership with Pakistan much beyond army headquarters; the civilian dimension of Pakistani politics should be an integral part of the counterterrorism effort. Most Pakistanis believe that Washington is all too happy to work with a pliant army puppet. Islamabad needs greater popular legitimacy in order to muster grass-roots support for the counterterrorism agenda. The United States should work to empower Pakistan's moderate civilians even as it builds trust with Pakistan's security forces. These goals are not contradictory: Washington can win the confidence of Pakistan's military establishment without accepting its exclusive political authority, and it can help empower civilian leadership without jeopardizing the army's core interests. Though it is true that Pakistan's military is the main obstacle to counter-terrorism efforts, its civilian leaders have nearly always had to negotiate a working relationship with the army in which generals retained significant decision-making power. And even if the army eventually retires from politics, it will remain an essential instrument in Pakistan's fight against terrorism. The Pakistani army has shown its willingness to partner with Islamists in order to dominate domestic politics and project regional influence whereas Pakistan's progressive parties seem to be ideologically committed in their opposition to Islamist militancy. Only a popular mobilization of Pakistan's moderates can really address the social and developmental deficiencies which ultimately cause extremism [Markey: 2007]. From Washington's side, the strategy of more carrots than sticks needs to be reversed.

US policies in Iraq: US engagement in Iraq proved more of a trap than an opportunity for the United States. In post-2003 US military action in Iraq, both al-Qaeda and Iran wanted Washington to remain bogged down in the quagmire. Al-Qaeda openly welcomed the chance to fight the United States in Iraq. US diplomacy proved certainly clumsy and counterproductive and led to disaster. Since President Barrack Obama has delivered on his election promise of withdrawing troops from Iraq, now the Iraqis need to settle their conflicts themselves. Rather than reinforce its failures, the United States should disengage from the civil war in Iraq, though it is going to be an arduous task for the Islamic State of Iraq as the Shiites and the Kurdish militias will have no compunction about instigating problems for the war-torn Iraq. Human rights abuses at Abu Ghraib and Guantánamo Bay have further sullied the United States' reputation and honor. Washington should emphasize the concrete steps the United States is taking to heal differences between Islam and the West and to bring peace to Palestine and Kashmir, among other areas.

US policies in Palestine: A positive US involvement in brokering peace in Palestine is key to diffusing much of terrorism in the world today. In the case of the Arab-Israeli conflict, this will not be easy, but neglecting the issue is no solution either. Washington should consider various ideas for getting the opposing sides back to the negotiating table, including the Baker-Hamilton proposal calling for a new international conference. ***American foreign policy in Middle-East vis-a-vis Israel needs to be reviewed***: Most American Muslims —both Arab and non-Arab—do not sanction violence, especially terrorism, but there is a widespread resentment of the perceived influence of the Jewish lobby in shaping US foreign policy and America's role in the Middle-East, especially its economic, political and military support for Israel. Although there is some degree of sympathy for Islamists' against US, most Muslims, mostly wealthy and influential, however reject the terror tactics used to advance their aims.

In fact, the continuation of Arab-Israel conflict has emboldened ever-new terrorist groups and techniques to make the region a laboratory for forging alliances with already existing groups and movements that goes in favor of proliferation of more and more terrorism across the globe—highly detrimental to world peace and security—and conducive to the radicalization of so far neutral Arab and Muslim populations who become easily vulnerable to extremism due to almost failed socio-economic conditions within their own states. They become the most potential recruits for terrorist camps and immediately buy terrorists' ideology of hatred and violence. A stable peace in the region would be possible only by an honest solution to the Palestine's problem keeping in mind the aspirations of the peoples of both sides.

A US policy of détente towards Iran is required: As for the ongoing arms-twisting of Iran and embarking on a path of the prospects of a military option there, it is suggested that to tame the growing power of Iran, Washington must eschew military options, the prospect of conditional talks, and attempts to contain the regime.

Instead, it should adopt a new policy of détente. By offering the pragmatists in Tehran a chance to resume diplomatic and economic relations with the United States, it could help them sideline the radicals and tip Iran's internal balance of power in their favor [see: Takeyh: *Foreign Affairs*, 2007]. The dispute between the United States and Iran is not merely a simple problem of disarmament. In fact, the political and strategic differences between the two countries run much deeper—and require a far more comprehensive approach. The containment option with a state like Iran that projects its influence through indirect means, such as supporting terrorism, financing proxies, and associating with foreign Shiite parties seems to be doubtful in terms of delivering positive results. Another important question is—will other states in the region be willing to help the United States to isolate Iran? The continuance of a suspicious foreign policy towards Iran since the 1979 revolution and Iranian clerics' commitment to export the Islamic revolution elsewhere created much distrust between the two. But the costly lessons learnt after war with Iraq made the clerics realize the limitations and impracticability of their dreams and Iran thereafter tried to assert itself as a regional power and stopped being a revolutionary state bent on forcibly exporting its model of government. Moreover, sanctions and other forms of US pressure have failed to prevent Iranian conduct. Worse, the Bush administration took such steps that made containment an even less effective policy. Washington's ill-advised invasion of Iraq has benefited Iran by empowering local Shiite parties sympathetic to Tehran. Long gone are the days when a powerful, Sunni-dominated Iraq could function as a counterweight to Shiite power in Iran. Iraq's Shiites are hardly homogeneous, but the leading Shiite parties in power in Baghdad—Dawa and the Supreme Council for the Islamic Revolution in Iraq—have intimate ties to Tehran. This does not mean that Iraq's new leaders are willing to subordinate their interests to those of Iran, but they are unlikely to confront the Islamic Republic at the behest of Washington.

Moreover, widespread anti-Americanism has made it harder for governments in the region to cooperate with Washington or to allow US forces on their soil. The United States may be able to keep offshore naval forces and modest bases in reliable states such as Kuwait, but it is unlikely to have a significant presence in the region, as it is too unpopular with the masses and seems too erratic to the elite. Many Persian Gulf states now have more confidence in Iran's motivations than in the United States' destabilizing designs. And so as Iran's power increases, the local sheikdoms are likely to opt for accommodating Tehran rather than confronting it [Takeyh: *Foreign Affairs*: 2007].

The situation in the CARs should be a matter of common concern to India, China, the US and Russia. There are cross-border linkages between the Taliban, al-Qaeda, the Pakistani *jihadi* terrorist organizations and the Uighur extremists on the one hand and with the Chechens on the other. The possibility of the Ferghana Valley becoming one day a new hub of international jihadi terrorism cannot be ruled out. Also, the region might tap the expertise of its own scientist community and of elsewhere too to develop a WMD capability. The region being location of vast energy resources attracts the dangers of the terrorists acquiring control of them one day. It is important for these four countries to get together, exchange ideas on the situation in the region and work towards a common counter-terrorism approach.

Confidence-building measures between the two: For this, it will be necessary to withdraw virtually all US military forces from Muslim soil where they represent a provocative element and source of anger. US troops are a lightning rod for terrorism. Let's not forget that it was US troops on Saudi soil, unwelcome to the population that lent bin-Laden his first cause in his early manifestoes against the US in the early 1990s. The US should also return counter-terrorism to the arena of intelligence and police work—where it has always belonged. The military instrument is by definition crude, ultimately ineffectual, and creates enemies faster than we can eliminate them. So, US needs to reduce its visible presence on alien soils and instead support operations, only if pressingly necessary, from regional locations. Its military should be immediately withdrawn from the thickly populated areas such as of Iraq and develop as soon as possible the indigenous forces to take up the task.

The US experiment in exercising global **hegemony, unipolarism, and American exceptionalism** must come to an end. Most of the world hates a hegemonic imperial power; most of the world prefers a multi-polar world that provides checks, balances, and options to all. Of course, the US cannot divest itself of its power, but it might find that power, consultation, and responsibilities shared with other major world states on a more collegial basis better serves US interests. "Policemen of the world" generally end up policing only those neighborhoods and communities that serve their own parochial interests, thereby diminishing any claim to "disinterested service."

Check on unwarranted outside interference needed: Moreover, it is highly incumbent on United States government to give up its revealingly ambitious plans to do unwarranted meddling in other country's affairs, particularly the Muslim countries that hold US responsible for many of their woes and also for the long-persisting and outstanding problems. Public opinion polls indicate how US actions in Middle-East and in Muslim countries of Southeast Asia and in Israel have led to a steep rise in anti-Americanism making co-operation in counter-terror measures very difficult. American policies in Muslim world are largely perceived as anti-Muslim. To regain its credibility amongst the Muslims in these countries, US should take up, in its capacity as the world leader, Palestine-Israel dispute as the central issue in its war on terror so that its other efforts in this direction do not look like a mere lip service. Also, heavily militarized approach may backfire; instead a comprehensive approach that can address real and perceived social, economic, political and ideological challenges. Involvement in the foreign lands without understanding the local, tribal or community dynamics of that particular region is to invite trouble as these communities, tribes and clans often constitute the basis for individual and group identity and serve as a very potent force in the political life of their country. US involvement in Somalia in 1990s proved to be catastrophic as it failed to understand the tribal dynamics of that country. In Afghanistan too, the similar story seems to be repeating itself.

Respect for international institutions and norms: The US must help foster institutions and international forums of cooperation that restore respect for law and international norms that have been increasingly weakened since the end of the Cold War. The resources spent on the counterproductive wars, such as, in Iraq or on building up ever better and destructive weapons can be invested on the building of hospitals, universities, schools, clinics and training programs across the Muslim world instead. Such positive gestures are pound to change the image and role of the US in the perception of Muslims worldwide. By any standards it is a bargain—and ethically right. The US needs to permit other nations and institutions to assume equal or greater partnership in resolving political problems in the Middle-East. In particular the US has lost any cachet of objectivity as an honest broker in the Israeli-Palestinian problem: Let the UN, EU, Russia, and other forces press for an urgent solution. The outlines of a final settlement are in any case well-known to all.

Balancing Realism and Idealism: Achieving a realistic peace means balancing realism and idealism in American foreign policy. Preserving and extending American ideals must remain the goal of all US policy, foreign and domestic. But till America pursues its idealistic goals through realistic means, peace will not be achieved. Whereas idealism should define its ultimate goals; realism must help it recognize the road US must travel to achieve them. As says Rudolph Guiliani, a realistic peace is not a peace to be achieved by embracing the "realist" school of foreign policy thought. That doctrine defines America's interests too narrowly and avoids attempts to reform the international system according to American values. To rely solely on this type of realism would be to cede the advantage to the US enemies in the complex war of ideas and ideals. It would also place too great a hope in the potential for diplomatic accommodation with hostile states. And it would exaggerate America's weaknesses and downplay America's strengths [Foreign Affairs: September/October, 2007].

Ailments of Islam

The radical and dogmatic interpretations of Islam pose serious challenges not only to non-Muslims, but also to the liberal, moderate Muslims, who many a times, are sued in Islamic courts for violating Islamic laws and the conservatives judges, either out of their own conviction for Islamic laws or out of fear of radicals, prescribe penalties to the 'offenders' to the extent that the liberal Muslim intellectuals are invariably tortured, executed or imprisoned and some manage to seek asylum in more liberal countries [see: Hasan Hanafi: "Alternative Conceptions of Civil Society", in—Hashmi: 2002]. In these circumstances, it very difficult for non-Muslim societies to initiate a process of peaceful democratic evolution of these societies despite the fact that some of the governments in Muslim states might be cooperating in many areas such as relating to peace and security. But when it comes to respect for human rights, particularly of women's rights, and giving up religious frenzy and adopting more modern, liberating ways of life, they stick to their religious interpretations. As the RAND study rightly argues, no non-Muslim can influence their perceptions about their religion except some enlightened Muslims who have the credibility to challenge these radicals despite some constraints and limitations.

The multiple and variegated Islamic movements going on in most of the Middle-Eastern countries seem to be essentially religious as their point of reference is Islam, but their basic thrust and power lie essentially in the political and social arena so much so that they are functionally nearly indistinguishable. In Islam, the Koran and the sayings and actions of the Prophet bear important messages for Muslim society and governance. This belief governs the thought process and psyche of most of the Islamic movements that range from the highly moderate to highly violent, such as, al-Qaeda. Some of these movements accept some form of democratic governance but others reject democracy; some are modernist, others are traditionalist; some are radical, some are conservative. But they are all grounded in Islamic history, culture, and religion, which lend them a great deal of cachet and authenticity. And they are all evolving.

Though Muslims have historical grievances against the West, they suffer from large numbers of **problems and pathologies**, a great many of which parallel problems in the broader developing world. Dictators, poorly developed economies, corruption, poor education systems, unemployment, ultra-conservative social practices especially towards women, intolerance towards diversity in certain circles, and now a mounting sense of anger toward the outside world all represent well-documented facets of the current agony of the Middle-East. Undoubtedly, the West has contributed to the emergence of many, though not all, of these phenomena, and it continues to do so at an accelerated rate, widening and deepening the East-West fissure on a daily basis. Yet the thought that the most powerful, sole global hyper-power in history might have anything to do with the trajectory of events in the region is simply inconceivable within US discourse on the problem. Acknowledgment of almost any US role in the creation of the present Middle-East morass is absent in almost all policy and think-tank analysis. The reality is that 9/11 had precedents. The issue here is not to cast blame but to analyze the nature of the problem and decide how to resolve it.

Self-introspection by Islamic world needed: As long as the Muslim world keeps blaming the outside world for their woes, without doing a ruthless **self-introspection**, there can be no likelihood of any internal reform within the Muslim societies. There will be no significant domestic movement in the Muslim World to limit or constrain radical, violent, or even terrorist movements as long as Muslims believe that these movements represent "understandable reactions" to present circumstances. Moderate voices in this superheated climate invariably lose out in the Muslim World, much as they have in the US, Israel, and now across much of Europe. Indeed, the essence of the counter-terrorist task today is not so much the killing of actual or suspected terrorists but rather there is *need to transform the Muslim environment* so that it does not that facilitates terrorism; this is the environment in which war, occupation, political frustration, political impotence, and anger have now socialized a whole new generation of potential warriors in the name of Islam [*Harvard International Review*: 2007].

Reform of radicalized madarssas: A Rand study suggests [Rabasa: 2004: p. xvii] that though a large number of social networks in Islamic societies, in the form of *madarssas* and mosques, might be providing useful social services to the community, but a careful watch would reveal and demarcate the genuine networks from the hostile ones that clandestinely promote extremism and terrorism. **There is a need to devise strategies to disrupt these networks.** The *madarssa* school system flourishing in Pakistan and in a large number of Muslim countries in South and Southeast Asia needs reform of these Islamic schools so that they fit into a broader framework of public education system that provides modern education in humanities, social sciences and pure and applied sciences and training in useful skills that make the students employable in market. Such reforms are key to break the cycle of radicalized *madrassas* that prepare Taliban and produce cannon fodder for extremists and terrorist groups. Also, Mosques that preach radical ideology under the cover of religion need to be reformed and their management put in the hands of moderates. But the question is—who will do this? An outside interference would again generate suspicion and ill-will, and therefore it needs to be avoided at all costs. Only the Muslims themselves can bring these changes.

The population explosion in most of the Muslim countries will put intense demands on their social, economic and educational systems which needs to be protected from drifting into the hands of radical Islamists who might be tempted to seize the opportunity to provide some peripheral social services to their community's youth and in turn might capitalize on their loyalties towards the religion whose engineered interpretation is easy to feed on the young minds.

The cult of martyrdom, the misuse of schools and mosques to preach hatred and violence, inspiring hero-worship for suicide bombers and psychological preparation of children and youth to follow into the foot-steps of these 'martyrs', preparing children to happily become 'holy-warriors' in the service of God and actually training them in the use of firearms are recipes for continuation of terrorist violence based on religious fundamentalism and such tendencies need to be checked by efforts at local, regional and international levels. The Palestinian Authority's manipulation of children, which has been extensively documented by the media, constitutes a reprehensible violation of every international treaty and convention meant to protect children in situations of armed conflict. The PA's heinous exploitation of children is both profoundly immoral and fundamentally illegal. Instead of instilling a culture of hatred amongst their citizens, both sides of the conflict should strive to inculcate a culture of tolerance and peaceful coexistence. A lasting peace is possible only by educating their children and youth to respect each other's rights despite cultural differences.

Need of secular moderate elements to come forward: To counter the influence of radical groups, the secular, moderate institutions and non-governmental organizations (NGOs) must take up the lead in providing assistance and resources to generate more employment locally and inculcate a vested interest in peaceful circumstances rather than in violence in the name of religion. The resourceful countries, such as United States, must assist these Muslim countries in the development of democratic and civil society institutions, on the condition of accountability by the recipients, and in ways keeping their local and community complexities in mind. This might help cool down some anger against US that has brewed up amongst Muslims in all these years due to its hegemonic, unilateralist policies in Muslim countries. There should be arrangements among the governments to remain under the supervision of some independent monitoring mechanism that can supervise the channeling of funds through countries like Saudi Arabia and some other oil-rich countries of the Gulf so that these funds do not reach undesirable elements and are properly utilized in socio-economic development of an otherwise backward region.

Muslim community should isolate extremists: It is possible only if the Muslim community itself isolates extremists who have placed the simple, ordinary Muslim under a humiliating international scanner and are wrongly or rightly branded as terrorists. Incidentally, after 9/11, in most of the cases of terrorist violence and sabotage on the international stage, perpetrators are found to be Muslims—be it Bali bombings, London Tube blasts or the most recent bomb blasts at Heathrow airport on July 2, 2007.

On Secular state: As regarding **Islamic fundamentalists' criticism of modern secular nation-state system**—a product of European political system—Hedley Bull, in his classical work in the English school of international relations, *The Anarchical Society*, argues that despite the existence of possible alternatives, the state system is the best chance of achieving order in world politics [see: Bull: 1977]. The Islamic world at present is locked in an internal struggle over how best to address and ultimately solve the problems endemic to many of its societies: namely, widespread poverty, extreme economic inequality, the prevalence of government by despotic rulers, and the inability to keep pace with emerging economies. Within Islam, there is a stream of those moderate thinkers who believe that the social, economic and political freedoms are effective means to solve these problems faced by Islamic societies. Such a view, however, is opposed by the fundamentalist branch of Islam, particularly that guided by Wahhabism which blames all of these ills on whatever modernization the Islamic world has already embraced, and advocate an unreserved rejection of the West. This rejection includes violence against Western countries and interests, and most especially violence against "impious" Muslim rulers who have adopted "Western" ways. The fundamentalists seek the establishment of states and societies based on Islamic laws and traditional mores. But if Islamic world has to take its place alongside other countries in this community of nations, both covert and overt support to Islamists and fundamentalists has to be stopped and though rejection of western values and way of life may be taken in stride by the west also without getting offended, they should try to take their countries into the world of modern science and technology to enjoy social, economic and technological development. Let it be understood both by the West and Islamists that modernization does not necessarily mean 'westernization'. In fact, the crisis concerns the choice the Islamic world faces between two diametrically opposed solutions: the violent and peaceful.

Tariq Ramdan, the grandson of Hasan-al Banna, the founder of Muslim Brotherhood, in his recent book, *In the Footsteps of Prophet* has tried to weaken the distinction between the Muslim world (dar al-Islam) and everywhere else—"the lands of war" (dar al-harb) and exhorts Muslims in the West to see themselves not as an aggrieved minority in hostile territory but as equal members of Western society, with full rights and full responsibilities. He believes that Islam is open to interpretation and can be tailored to specific circumstances. Ramdan has made some serious suggestions on reforms and also for the political integration of Muslims in the West. Although his critics would argue that Ramdan's ideas seek reform, not reformation and his urge to Muslims in the West is merely to leave the ghetto while retaining their religious identity, what he actually wants to reassure is that full faith is compatible with full social and political participation.

Addressing other Problems as long-term Mechanisms

Problem of failed and failing states need immediate addressal: The existence of a number of failed states with large swathes of territories at their borders which are virtually ungoverned or are governed minimally or where the diktats of terrorists and not of any legal, constitutional authority rule the roost is a precipitating factor providing suitable havens for terrorists and extremists. The political stability and economic development in the ungoverned areas of Pakistan, Afghanistan and Yemen, for example, would reduce the opportunities for extremism and terrorism to take roots. It is here that the international assistance under United Nations and other individual countries is required only if these work without being guided by hidden agendas and vested interests.

The problem of Ungoverned Territories: A Rand study on ungoverned territories (Rabasa: 2007) brings out the challenges that these areas pose to international security and serve as breeding grounds for terrorism and criminal activities and launching pads for attacks against the states. Ungoverned territories are unable to control the routes that drug smugglers use to transport illegal drugs. The areas/states identified in this study include the Pakistani-Afghan border region; the Arabian Peninsula; the Sulawesi-Mindanao arc in Southeast Asia; the East African corridor from Sudan and the Horn of Africa to Mozambique and Zimbabwe; West Africa from Nigeria westward; the North Caucasus; the Colombian-Venezuelan border; and the Guatemala-Chiapas (Mexico) border.

The two dimensions: ***ungovernability,*** which means that the state is unable or unwilling to perform its primary functions of governance and maintaining law and order and ***conduciveness*** to terrorist or insurgent presence means that absence of adequate governance suits terrorists as they can conveniently use available infrastructure such as transportation and communication facilities, favorable demography, presence of extremist groups and the preexisting state of violence may be exploited and become breeding grounds for terrorists. Adequacy of infrastructure also means that a terrorist group must have a basic communications and transportation network and the means of transferring funds in order to operate and access its targets. When funding from external sources is not available, terrorists need to raise money locally in order to fund their operations.

As mentioned in the Chapter on the Asian context of international terrorism, the presence of organized armed groups outside of state's direct control is the characteristic of Pakistan-Afghan border (in South Asia), Yemen (in Arabian Peninsula), Somalia and Sudan (in East Africa). These territories are also among the areas of highest concern with regard to their potential for becoming sanctuaries for international terrorist groups. Afghanistan also makes a case of ***abdicated governance.*** The central government here abdicated its responsibility of producing public goods (such as safety, law and order and social services) in areas where the war lords control the difficult terrains. One of the very negative results of such an abdication of responsibility is that security forces collude with illegal drugs and arms trade which is so basic to sustain the terrorist campaigns.

Interference by external states has also been an important characteristic of international terrorism and many Asian locations discussed in this study suffered from this syndrome. Soviet intervention in Afghanistan in 1979-89 and the American interference as a reaction to this in the cold war framework are case in point.

To eradicate this scenario of ungovernability, the international agencies, like United Nations and its various social-economic bodies and programs and various states, such as US, must use development assistance as a tool to encourage recipient governments to invest in infrastructure and institutions and should also ensure accountability of these governments to use these assistance funds for the purpose they are given so as to check the incidents or possibility of siphoning-off the money for other purposes. The regional and local cooperation might be an effective tool to arrest the tendencies of bureaucratic lethargy and corruption and also to build the capacities of local military and counter terrorism forces. An improved governance with the help of international bodies and organizations will certainly boost the functional capacity of security forces to cut the terrorist networks from their lifelines: funding sources, illegal money transactions, safe havens and arms movement.

Unwatched borders of ungoverned states are potential transit routes for their movement from one base to another and also to transport their logistics. To improve a state's ability to improve its control over its borders, vulnerable states need to be trained in detecting illegal entries, fake passports, border crossing incidents and so on. Intelligence sharing and expedited communication between intelligence agencies may prove very effective in countering terrorism.

Proper handling of underdeveloped regions required: Though the objectives like nation-building and promotion of democracy are hard to achieve but certainly not impossible if approached with care and humility. In fact, weak and failed states are the strongest sources of global disorder and terrorism. In the underdeveloped parts of the world, such as in Africa and Latin America as also in parts of Asia, the problem of development has been overlooked and if at all taken up by international bodies like United Nations, these regions have been taken only as developmental problems. They have not been looked upon as potential breeding grounds for terrorism and therefore as security risks except when countries in these regions become security threats. The economic and social fallout of globalization also need to be objectively analyzed and adequate mechanisms of accountability be worked out. A balanced policy of promotion of economic and political development through multiple multilateralisms appropriate for a globalizing world needs to be pursued.

Need to address the root causes of terrorism: Elimination of al-Qaeda, Taliban and their international networks is and should be the highest priority of the international community joining US sponsored war on terror. But, this will serve only as a short-term strategy. To root-out the menace completely and effectively, the root causes need to be addressed simultaneously. In this context, unless the international community genuinely **addresses the pending and serious issues like Palestine** and unless, instead of being in league with them for some vested interests, it exposes the corrupt and autocratic regimes of Middle-East, the propagandist ideology of the terrorist groups would remain appealing for a common, ordinary Muslim whose chances of falling into terrorists' trap will remain very high. However, an **obsession with promoting democracy at any cost and shaping the Muslim world in West's image,** without taking into account their societal and cultural complexities, will not only exacerbate the already deteriorating international political environment but would also provide fodder to mischief mongers who are always desperately looking for such situation to emerge which they can capitalize to their advantage.

***The "new" terrorism thesis tends to blame terrorism on religion*:** Two consequences follow from this: (1) Although its proponents acknowledge that many different religious doctrines have been associated with terrorism, Islam receives by far the most attention. There is thus a risk of encouraging stereotyping and prejudice. (2) This interpretation also neglects the political and social aspects of terrorism. It thus provides at best a partial picture. There is a need to ponder upon the real causes of changes that are affecting the occurrence of terrorism. Many changes have to do with the environment in which terrorism occurs. One is the set of processes we term globalization, which provides terrorists access to resources and opportunities. Small non-state conspiracies gain capabilities that are disproportionate to their size and level of popular support. They can communicate, organize, and mobilize assets across borders with ease. The ideas and images that motivate terrorism are transmitted across the globe almost instantaneously. Second, specific historical events and circumstances have determined the trajectory of terrorism. Just as the civil war in Lebanon provided a hospitable environment for groups as disparate as the Japanese Red Army and the Armenian Secret Army for the Liberation of Armenia, as well as Palestinian and Lebnanese groups, so did the chaos that followed the Soviet withdrawal from Afghanistan which permited the establishment of al-Qaeda. Iraq may provide similar opportunities [Crenshaw: 2006].

Need to focus on liberal political Islam: After September 11, vitriolic rhetoric against the United States by prominent Muslims and the war against terrorism make one easily believe that Islam and Muslims are enemies of the West. Graham Fuller in his excellent analysis [Fuller: 2004] calls it a flawed thesis. His sweeping survey of trends in the Muslim world contends that the issue is not whether Islam plays a central role in politics, but what Muslims want. To focus on radicalism and extremism only and taking Osama bin-Laden and Taliban as the voice of all the Muslims blinds us from another trend: liberal political Islam. Proponents of liberal political Islam emphasize human rights and democracy, tolerance and cooperation. They face an uphill struggle as authoritarian regimes oppress opposition and use Islam to justify their undemocratic rule. As people are denied avenues to participate and criticize, as secular ideologies have failed, religion has come to play a central role in politics. The question today is whether Muslim majority will allow violence and anti-westernism to go ahead in their name and whether the ilk of Osama and Taliban will be allowed to shape the character of future relations between the Muslims and the West. And the answer lies partly in the hands of American policy-makers who are advised not to push their secularism and democracy agenda at the gun point in Islamic countries as the religion and politics are inextricably intertwined across a broad swathe of the globe throughout the Muslim world from northern Africa to Southeast Asia, and partly in the hands of Muslims themselves who are advised to look beyond the euphoria of invincibility of Islamist radicals who are obsessed with the triumph of their *jihad* against the powerful Soviets and dream of repeating these results elsewhere. The outcome of the struggle between extremists and liberals will determine the future of political Islam. To ease the current dilemmas and tensions experienced in the Muslim world today, the United States should look beyond the first phase of 'war on terror' which has more or less been won and should now address the deeper sources of political violence and terror there. Transforming the radical Islamist environment is not the task that can be fulfilled by rewriting history textbooks or demanding less anti-west media. It is rather possible by empowering the silent or 'silenced' Muslim majority that rejects radicalism and violence. The result could be political systems truly Islamist and truly democratic at the same time.

Need for an unprejudiced and unconstricted study of Islam: As Edward Said suggested, a hard and fast distinction has to be made between serious consideration of the Islamicate world and nearly everything that passes for Islam in the media and in all but a few places in the culture. One cannot look for help in promoting serious investigative discussion of Islam—even as a subject of academic inquiry—among traditional Orientalists or within the normally constructed programs of Middle-Eastern studies in today's Western universities. What is required is an unprejudiced and unconstricted study of Islam and of the problems of Islamic society and Islamic peoples by the segments of the population, Muslim and non-Muslim, who have a wider and more serious view of human problems in general: men and women who are committed not to Orient and Occident but to the cause of human rights, rather than lobbyists who act on behalf of human rights when they are paid to do so; students of comparative literature rather than Semitic philologists who know nothing about other literatures and who care little for the contemporary world; genuinely enterprising sociologists who know something about theory and care a great deal about issues confronting concrete societies, rather than specialists in the Islamic mind or in a monolithic thing called Islamic society. There can not be any substitute for a genuinely engaged and sympathetic—as opposed to a narrowly political or hostile—attitude to the Islamic world.

Edward Said would suggest that to dispel the myths and stereotypes of Orientalism, the world as a whole has to be given an opportunity to see Moslems and Orientals producing a different form of history, a new kind of sociology, a new cultural awareness. But, sadly, even with vast sums of money easily available, the Islamic world as a whole does not seem interested in promoting learning, building libraries, establishing research institutes whose main purpose would be modern scientific attention to Islamic realities and to seeing whether in fact there is something specifically Islamic about the Islamic world. In the great rush to industrialize, modernize and develop itself, the Islamicate world has become compliant about turning itself into a great consumers' market in West's image. The rhetorical attack upon neo-imperialism does not look convincing at a time when national governments and rulers openly espouse values that further the new style of imperialism without colonies. The distortion of the Arab-Islamic image in the Western media will go on unless any drastic attempt to correct distortions of Islam and the Arabs are taken seriously. As for Moslems who do not serve America's purpose, they will, as always, be portrayed as backward fanatics. [Said, Edward, "*Islam through Western Eyes*", *The Nation*, 26th April, 1980 issue].

War on Terrorism should not mean war on Islam: All care must be taken to see to it that war on terrorism does not turn, knowingly or unknowingly, into war on Islam or on Muslim countries as the Islamists are always in search of such an opportunity to convince their co-religionists that their religion is in danger at the hands of a culturally corrupt, secular West. In emotionally charged atmosphere such as the one that emerged after 9/11, even a single, minor mistake can be blown out of proportion to malign even the genuine efforts on the part of international community. Counter-terrorism efforts need to be balanced by promoting those willing partners in Muslim countries who have a moderate approach and want to challenge the radicals who have earned a bad name and negative image for the compassionate, liberal version of Islam. Even those Muslim groups and parties that have slightest respect for democracy, freedom and productive changes in their systems should be immediately roped in the process of peace building lest they would drift into the terrorists' trap with slightest provocation or laxity of approach on the part of the world leaders.

Both US and Islamic radicals need a soul-searching: Equally important is a little soul-searching by US in the realm of its policies and actions that are bound to have a greater impact upon the evolution of Islam and upon politics in the Muslim world over the next decades. Towards the end of 2006, the US faced what was basically a broad and disparate insurgency across huge parts of the Muslim world. The relentless application of US military force is not only failing to solve the problem but also exacerbating it. It will take at least a generation before entire cadres of young Muslims begin to moderate their bitterness and intense dislike of the US. Today, for Muslims, the Muslim identity overshadows all other identities: ethnic, sectarian, professional, class. This is an abnormal situation, even for Muslims for whom Islam normally forms only a part of their personal identity, along with other elements. Yet the salience of the Muslim identity is intensified by war, conflict and the image of Muslims under siege everywhere. What is urgently needed today is to help calm-down, depressurize and gradually normalize such super-charged atmosphere of the Muslim World to give room to moderates to start the political process once again, to allow alternate identities to re-enter the normal process of political and social life.

A sincere effort at creating conditions conducive for a peaceful evolution will diminish the threat of terrorism and permit the Muslim world to return to a state in which armed struggle, insurgency, civil war, large-scale political violence, super-charged Islamic rhetoric, and a dominating focus on the state of the ummah will recede from daily discourse and behavior.

The international bodies, along with US should help war-ravaged countries **build accountable, functioning governments** that can serve the needs of their populations, reduce violence within their borders, and eliminate the export of terror. As violence decreases and security improves, more responsibility can and should be turned over to local security forces, though the requirement of some US forces remaining for some time there in order to deter external threats cannot be ruled out. The consequences of failure can be predicted: Afghanistan would revert to being a safe haven for terrorists, and Iraq would become another one—larger, richer, and more strategically located. Parts of Iraq have the danger of falling under the sway of countries like Iran. The balance of power in the Middle-East would tip further toward terror, extremism, and repression. The terrorists and rogue states would be emboldened and would see further opportunities to weaken the international state system. The goal should be to see in Iraq and Afghanistan the emergence of stable governments and societies that can act as international allies against the terrorists and not as breeding grounds for expanded terrorist activities.

On Democracy: An analysis into the ground conditions that help terrorism to flourish is that democratic societies are both used and misused by terrorists in ways suitable to them. On the one hand, democracies are unsuitable to terrorists in that these societies have continuous and inbuilt corrective mechanisms to keep their house clean by addressing the general peoples' grievances and by accommodating their opinions through democratic channels, leaving no or little space for extremists to exploit the grievances and use these as justification for their violent activities. Yet on the other hand, it is also true that the openness and permissive socio-cultural and political milieu of democracies suits terrorists in that they exploit this democratic openness and permissiveness to 'swim like a fish in the water'—as goes the Chinese proverb—and exploit it to their advantage by mounting terrorists acts from the soils of these societies. Al-Qaeda's use of American pilot training schools and converting domestic planes into missiles that did hit World Trade Towers in New York in 2001 is a case of (mis)using democratic openness to execute terrorist plans. Al-Qaeda's attacks were part of bin-Laden's plans for a *jihad* aimed at reshaping the world in accordance with his extremist interpretation of Islam.

Tackling clash of US hegemony and Islamic fundamentalism: Rise of Islamist fundamentalism should not be taken as an emerging substitute for American hegemony; rather, both complement each other. Instead of anyone's hegemony, there should be intercivilizational equality and justice. Religion, instead of inciting conflict, should foster peace and harmony—its actual objective and domain.

Humanity must build bridges, not the faultlines between civilizations and cultures. Since peoples' sense of identity is determined more by their sense of belonging to their communities and ethnicity defined at local and regional level, particularly when they are exposed to another civilization where they consider themselves to be aliens up to many generations, it is superficial to impose loyalty without providing any substantial basis for their identification with their new nation-state [Tibi: 2000]. Politicization of religion is the greatest challenge to cultural modernity and gravest threat to peace and stability.

Need to recognize cultural and historical contexts of 'other' societies: In strategic studies, whereas counterinsurgency was the main thrust; now the focus is shifting to peace studies. Shultz and Dew argue that in many of the places where Western forces might find themselves confronting local insurgencies, they need to recognize the cultural and historical context, in which forms of warfare are interwoven with the organization of local society. They are likely to face warrior cultures led by local chieftains, who can compensate for their lack of firepower with local knowledge, popular authority, and an array of suitable tactics for which regular forces tend to be poorly prepared. All of this is a warning against the arrogance of Western countries that believe that at such an advanced stage of social and political development, they should be readily able to make others conform to their will. Shultz and Dew illustrate their point with detailed case studies of Somalia, Chechnya, Afghanistan, and Iraq, going back to colonial times to demonstrate the bloody consequences of a failure to understand local societies [Shultz and Dew: 2006].

Cassidy also takes up the theme of military culture and draws our attention to the persistent refusal by the US military to prepare for irregular warfare. The policy-makers in West should be dissuaded from getting the country involved in such wars in the first place. To demonstrate that Western countries need not be paralyzed by their military cultures, he finds evidence in past practice, including in the Indian wars and the Philippines that it is possible to deal with insurgents in a politically astute manner. The requirements are minimal but include credible force, close cooperation between civil and military agencies, indigenous forces employed where possible, and legitimate political processes, even when this means drawing in opposition elements.

Benefits of Democracy: Countries that lack functional law enforcement structures in part or in all of their territories provide lawless spaces in which terrorists can easily operate. Of course, terrorists have also succeeded in finding spaces to operate in democratic states. Democracies, however, have an advantage in their ability to rise to the occasion and implement needed reforms with popular consent. States with liberal institutions and democratic systems can thereby respond to terror with greater public support and, in the long-run, greater effectiveness than authoritarian states can. A post-9/11 attacks United States and a post-July 7, 2005 London bombings UK, both using the democratic process, could better handle the situation by promptly enacting new laws against incitement to terrorism and encouraging civic society to engage disaffected Muslim youth. Counterterrorism actions in democracies reflect the will of citizens, and citizens feel integrated into the overall actions of their government. In contrast, fighting terror with oppression eventually leads to more of both.

There are other global benefits accruing from democratization. The free flow of information within and among democracies builds stronger, more flexible, more dynamic societies that are better positioned to fight terrorism. In the global war on terror, global responses are required. Democracies working together enhance one another's responses far more effectively than do non-democratic states, where the flow of intelligence and trust is limited. New and emerging democracies not only provide viable, legitimate recourse for their own citizens' grievances, but also offer greater opportunities for counter-terrorism partnerships with other democracies. Interdependent, networked liberal institutions throughout the globe, reinforced by the structure of democratic governments, provide the best means to defeat the interdependent, networked terrorist cells of radical extremists who seek to destroy democracy and, in fact, the nation-state system itself. Not surprisingly, the terrorists oppose democracy, as it poses a threat to their plans. Al-Qaeda leaders have specifically rallied against the notion of democracy, seeking to label it as heretical, and terrorists have killed innocent Afghans simply for having voter registration cards. No doubt, American administration's counter-terrorism and democracy-promotion policies each have a unique agenda and emphasis, raising doubts in the minds of its critics, there is also synergy between them. The convergence of these policies is necessary, and the overlap is purposeful [Dobriansky and Henry Crumpton: 2006].

Multi-pronged strategies of counter-terrorism required: The 9/11 Commission report mentions about the two-pronged strategy: **disrupt the leadership of all the important and dangerous terrorist groups; and compete against those radical Islamist ideologies that inspire terrorism.** The long-term strategy is complex and aims at reducing the appeal of Islamism by strengthening national governments to provide their Muslim citizens with attractive alternatives so as to incapacitate the Islamists and religious ideologues of any dispensation to exploit any sense of alienation produced by any factor, be it globalization, corruption, breakdown of institutional machinery in a state or marginalization of the young generation. Any counter-terrorism strategy must be multilateral and should not be limited to law-enforcement or military methods alone, their utility in getting immediate results notwithstanding. Merely eliminating terrorist camps is not enough as they could be rebuilt or relocated with little effort and within a short time. In weighing the military options, all care should be taken in using foreign forces or on relying on the local ones, of carrying out operations covertly or overtly, and of notifying the local governments beforehand or conducting them without prior notification. Actions taken without local approval are bound to heighten suspicions and could well be regarded as acts of war. Therefore, counter-terror strategies using military options have to be carefully planned to avoid any further deterioration of relations of trust among the nations.

Building up the regional governments' institutional capacities for combating terrorist groups within their territories is very important. For this to achieve, a streamlining of their judicial system and law-enforcement agencies and building-up state-run schools is required so that the children have a better option than going to radical *madarassas* where extremist brands of Islam are propagated and by working with suffering countries to better manage their communal tensions and identifying religious flashpoints before they erupt. Regional counter-terrorism centers should be built, and frameworks for harmonious extradition agreements so as to deal with the criminals and terrorists more effectively and speedily and to prosecute them if found guilty. It will mutually help governments in investigations and data-sharing about potential terrorists, their groups and activities. The importance of multilateral intelligence sharing and extradition agreements is underscored by the apparent fact that many captured al-Qaeda and Jemmah Islamiyah terrorists have provided authorities with useful information that led to further arrests and the discovery of new plots that could be disrupted in time.

Disrupt terrorist finances: As the 9/11 Commission has also argued, the terrorists' finances must be tracked and disrupted to starve them of their oxygen. A number of countries, despite all international pressures, remain non-cooperative in fight against money laundering. Indonesia, Philippines and Burma in Southeast Asia and a large number of countries in the Middle-East and also in South Asia remain non-compliant in this regard. Informal financing mechanisms such as *hawala* transfers and cash donations are continuing and a number of front companies and businesses have links with terrorists. A regional clearinghouse to shut down such illegal and informal play of money is urgently required.

A Balanced Projection of Islam Required

Let Islam not be hijacked: In fact, Islam is not a monolithic faith as mullahs and other radical fundamentalists may like Muslims and non-Muslims to believe. On the contrary Islam can be an infinitely elastic religion, capable of endless adaptation to fit the temper and circumstances of its followers. Ironically, Muslims today often overlook or minimize these differences. Orthodox Islam seems frightened of its own richness and variety and falls prey to western stereotype: that Muslims should all march in lockstep, that experimentation must be punished. Any attempt to understand Quran and *hadith* should start with historical context. What was the practice in pre-Islamic Arabia, the so-called *Jahaliyya* or Age of Darkness? What was the specific context into which a revelation was given? What does the *Qur'an* actually say? What did Mohammad himself say or do? How did Muslims, then and later, understand his sayings or behavior? Is their behavior today consistent with the temper of individual *ayat* or *hadith*? If not, why not? Has the *Qur'an* been hijacked? Or simply misunderstood? Where does the *Shari'ah* fit into this scheme? One must therefore return to the original texts and seek answers in fourteen hundred years of Muslim history.

Search for a truly Islamic state: As for Islamists' obsession with establishing a purely Islamic state based on original texts, Rachid Ghannoushi (Tunisia) argues that where Muslims are secure and free to pursue their religion, such as the Muslim diaspora in the US and West and the largest minority of nearly fourteen crore Muslims living in India, there exists by definition *a state of Dar al-Islam* (the realm of Faith). Therefore, the full exercise of Islam does not require an Islamic state, organized along the lines of Medina. One can be a good Muslim in the fullest sense of the term in Britain or in the United States because the state gives Muslims the space and freedom to freely exercise their beliefs. For Ghannoushi an Islamic state is one which allows Muslims to live according to the principles of Islam. He then reverses the argument and says that in a state where Muslims form the majority, laws other than the *Shariah* must also be given equal weight. This is regrettably not the case in so-called Islamic states such as Saudi Arabia, Iran or Pakistan today. Of course, Islam in Medina and Mecca, was above all a practical religion, attempting to resolve everyday problems. While there are formidable orthodox objections to the idea that Islam is compatible with Popular Sovereignty, there are also valid intellectual arguments to suggest the two can be compatible. But this requires that much of the baggage of the past fourteen hundred years must be cleared out of the way. This holds even truer for the tangle of ignorance, prejudice and distortion surrounding Muslim ideas about women, and about gender equality [see: *Living Islam Project*: 2002].

The radical Islamists of today not only distort but also forget what Quran actually said about **gender equality** and gloss over the doubts and disagreements that existed between the Caliph Companions of Prophet after his death. Also they selectively choose from the four different Sunni schools of jurisprudence, which largely differed on gender issues, to downplay and discard women's rights regarding initiating divorce, to remarry or to force a husband to enter in a marriage contract in favor of monogamy. As Living Islam Project (2002) infers that **Quran only set certain basic principles and never intended them to be taken literally.** These could have been interpreted by successive generations to fit their specific circumstances and evolving cultures. But all this has been progressively forgotten by today's fundamentalists and terrorists in favor of a traditional, androcentric Islam based on self-interest and myth and on partial and biased reading of the Holy Scriptures.

Islam and Democracy: Contemporary Islamist thinkers, notably Hassan al-Turabi (Sudan) and Rachid Ghannoushi (Tunisia) have argued that **Islam and democracy can be reconciled**, but with some radical rethinking. Sovereignty can and must be vested in human beings, acting as God's representatives on Earth. An Islamic party has to accept the idea that it does not have a monopoly on virtue, that its Islamic program is essentially political, not divine, and can therefore be reversed, amended and rejected by the electorate. The model of an Islamic polity has to be de-linked from the Past. What worked in a small Arabian city will not necessarily work today; nor do the most pious and learned Muslims always make the best rulers. John L. Esposito says that Islam like all other religious traditions is an ideal. **It has taken many forms historically and is capable of multiple interpretations conditioned by reason and historical/social contexts.** For example, much of the debates over the relationship of Islam to women's rights must be seen not only in specific context of Islam but in context of other religions too which also suffer from the centuries of conditioning by the patriarchy which has always remained biased in favor of male. All religions of the world have always had male interpreters of religious texts and laws [see: *Esposito: Harvard International Review*]. It is due to multiple interpretations and reinterpretations that we see many acceptable differences existing between Shias and Sunnis and between Islam and Sunnis. After the controversial articulation of Clash of Civilizations thesis, it has become a habit of most of the analysts to see and interpret all political conflicts in religious terms, as 'Islamic-Christian' or 'Hindu-Muslim', or Protestant-Catholic or Shia-Sunni and so on. Although, the communities in these conflicts may be identified in religious terms, but their source of conflict has to do with politics than with religion. Most of the times, these religiously identified communities are asking for autonomy, independence, rights or are expressing socio-economic issues and grievances—addressed to political class, to policy-makers or to governments. It is another matter that the issues are hijacked by religious fundamentalists who also try to capitalize the situation politically in their favor and use peoples' religious sentiments to draw the faultlines. **There is a need to expose the political agendas of these self-appointed saviors of religion. Any religion for that matter; not only Islam.**

Secularists argue in favor of separation of religion and politics while the rejectionists maintain that Islam has its own form of governance and is incompatible with Western style of secular democracy. However, an accomodationist view can be least objectionable in that they suggest for using the traditional concepts and practices already existing in Islam, such as, *Shura* (consultation), *ijma* (consensus) and *jihad* (reinterpretation) to bring in the islamically acceptable forms of popular participation and democratization. In 1990s, in the countries like Kuwait, Yemen and Turkey, elections were successfully held and democratic Islamists emerged as mainstream political actors to the surprise and shock of many fundamentalists who charged them with 'hijacking democracy'. In fact, such charges represent the frustrations of those rulers who like to continue to rule without dissent.

A major question related with compatibility of Islam with democracy is whether those rulers who come to power through elections, in whatever limited sense it might be, would tolerate diversity and healthy opposition of their policies and actions. The past experiences raise doubts as opposition has been met with repression. But, as suggests Esposito, these questions and doubts are related more to the absence of a culture of political participation and of a civil society in Islamic countries than to the religion as such. Truly, opening up a so far closed system would release the forces of competing opposition parties. This process would challenge the monopoly of rulers. To win elections, they would have to compete for vote and if they come to power, they would have to strike a balance between diverse interests and would have to deliver on their promises. This would require them to adapt or broaden their ideology in response to domestic realities, diverse constituencies and interests and keeping in focus compulsions of international political environment. Hence, the process of democratization in Islamic countries is difficult but not incompatible. Projection of Islam as a threat to world peace and stability is part of Western propaganda machine that uses this as an excuse to suppress any Islamic movement, even the non-violent ones, and to keep access to Arab oil intact.

Proper projection of political Islam: Moreover, the tendency to identify political Islam, the mixing of religion and politics, as always leading to fundamentalism, extremism and terrorism might be partially true and is fraught with danger of being counter-productive. This point can be explained by giving the examples of other groups mixing religion and politics, such as Jews and Christians within Judaism and Christianity, but these are very often not termed as fundamentalists. Moreover, within Islam also, there are movements that are moderate and they participate within the system and seek change from below. The selective projection of only the radical, extremist movements as the mainstream voices in Islam will certainly not serve the interests of peace and security. It might give undue, unwarranted importance to radicals and extremists who need such publicity to sustain themselves. Esposito would term the tendency of reducing all the ills of society (government failures and under-development) on a single cause (radical fundamentalism) as 'reductionism'. The challenge is to move beyond the monolithic Islam as practiced in Saudi Arabia, Iran and Libya or pre-9/11 Afghanistan and to move towards more moderate brand of it as practiced by certain movements, such as, Muslim Brotherhood in Egypt and Jordan that eschewed violence and participated in electoral politics. The activities of a radical minority should not be allowed to serve as a convenient excuse for the governments' own failures to build strong, equitable, modern states where a family will not remain perpetually in power without question and holding the national assets as their private property.

In a global, interdependent world, both the conflicts and co-operation are the results of common national and strategic interests. The commonality of economic and strategic interests explains the cooperation between a socially primordial Saudi Arabia and a modern, secular America. Therefore, many would agree with Esposito that future global threats and wars would be less due to clash of 'civilizations' than to the clash of interests, economic or otherwise.

An unbiased media for breaking stereotypes: As mentioned in the Chapter on Islamic Fundamentalism, it is suggested that a biased media on both sides of the spectrum deepen the dividing line between the cultures and religions. Therefore, the regional media could work in the direction of building up public opinion in favor of human rights, social and political justice, scientific and technological development need to prescribe and consume a secular education and could help people come out of their mental ghetto and see the things not only religiously, but in the context of an independent analysis. Development of such media would not only provide an alternative source to the state-controlled dissemination of information that is bound to be one-sided, but would also foster greater self-examination, internal debate and exposure to the world of new ideas. Unfortunately, the new media in Islamic countries is, so far, reinforcing existing stereotypes and narratives of Arab victimization that play into radicals' agendas.

Is there a Clash of Civilizations?

The current clashes between the Muslim countries and United States is not due to civilizational or cultural clashes. Rather, these clashes and conflicts are due to negative portrayal of Islam in western media on the one hand and the ambitious foreign policy decisions and their unilateral execution on the other. Inherent in these two sides of the spectrum are the highly autocratic and repressive political systems in the Arab and Muslim world that try to justify their repression and autocratic ways under the cover of religion through their selective and out of context references from Quran and US obsession with it democracy project. Stiff position on both sides leads to conflictual situation. US Foreign policy towards the Muslim world clarifies some of these observations.

How to tackle Crisis and Dilemma facing Islam and the Muslim world today?

Drawing from previous chapters, a few facts pertaining to crises of Islam may be summarized here:

- (a) Islam is not only theology, it is sociology also;
- (b) There is no separation of church and state in Islam;
- (c) All Muslim states are not the same in their levels of efficiency in governance, in repression/or according rights to people, in their interpretation of *Sharia*; and
- (d) Its followers belong to different streams and ethos, such as, Shias and Sunnis and so on...

What we find today is that the **Islamic world is passing through a phase of struggle at two fronts**:

1. A struggle within itself to determine its identity, nature and values that have serious implications for its own future status and its relevance in the contemporary world without losing its original identity, and
2. A struggle with the outside world that feels threatened and affected as a result of Islam's struggle with itself.

It is time to workout a judicious and balanced approach that can drive it towards its evolution in a positive direction. For this, it is essential for Islam to get rid of the clutches of its hijackers—the radicals, extremist fundamentalists and terrorists—who have helped in projecting Islam, in the eyes of the rest of the world, as a religion sanctioning violence, blood-shed, hatred for non-believers, anti-women, anti-progress, primordial and backward. Rest of the task is a trick of a highly vocal and engineered western media and the specialized Centres of Islamic and Middle-Eastern studies that conduct state-funded research whose outcomes are, more often than not, biased and pro-western in their thrust. To clear Islam of the charges of being a religion of suicide-bombers, extremists and terrorists, it is essential to de-link Islam from these so-called self-appointed saviors of Islam and to address the deeper causes of extremism and radicalism that lie in deficiency in socio-economic and political development.

In fact, the religious bigotry of the religiously-driven leadership in Muslim countries is proving to be self-destructive. For example, the education imparted through *madrassahs* and other state-controlled schools teaching an overtly Islamic curriculum might at best prepare the future Taliban and religious fundamentalists suitable to be the *ulamas* in the mosques who in turn will produce the future zealots and will help keep the cycle of bigotry moving. Such an education will not prepare future scientists, doctors, engineers, architects, social scientists, geologists and researchers on whose intellect and skills a nation can be proud and developing with indigenous pool of rich resources. In other words, combine the rich oil assets and human-intellect deficit and you have a perfect recipe to lure those who are poor in oil resources but too rich in human-intellect and other scientific and technological resources. Lacking the skilled and laborious work force, oil-rich countries of the Gulf depend on nearly more than seventy-five percent expatriates, who one day will be nightmare for economists and statisticians studying the state of economy of these countries gathering mind-boggling figures of flight of capital and other resources from these countries to others.

Thus, **too much of dependence on religion and puritanical values not only restricts development and growth, it is anti-peace** also. A religion that can be a helpful, constructive force in character-building and personality-development and a source of inner strength and peace might prove to be dangerously destructive and prohibitive if allowed to enter all domains of a person's and nation's life, particularly the political life. Invasion of these countries, not only military but also economic, cultural, political and social will follow, which in turn will give reason to fundamentalists to resist it with what they have: religious appeal. Hence, the theses like Clash of Civilizations and of cultures will emerge. The current crisis and dilemma of Muslim countries is exactly this that they have chosen to remain underdeveloped and grounded in religion only despite immense human and material resources available to them. Hence, the volatility of the contemporary Islam engaged in an internal and external struggle over its values, identity and its place in the world should be rationally understood and addressed so as to avoid ambitious, outside powers to take advantage of their current volatile situation.

To make Islam compatible with democracy, the modernists and secularists in Islam, who are in a weaker position than traditionalists and fundamentalists in terms of their resources should be supported by giving them a powerful platform by which they can familiarize Muslims and non-Muslims about the positive, compassionate elements in Islam, by encouraging them to publish their interpretations of religion to the mass audience that help counter the fundamentalist interpretations, by assisting them financially to run their schools and institutions of higher learning with a secular, modern ethos as against the terrorized and fearsome images that have made not only the rest of the world but also a large majority of simple, God-fearing Muslims suspicious about their own religion. A consistent effort should be made to bring traditionalist close to modernists and secularists within Islam. Fundamentalists/terrorists should be targeted by exposing the inaccuracies and inconsistencies in their interpretations of Islam and showing to them and to the world their inhumane acts and approach that is in complete violation of the very script they pretend to restore in its original form.

Moreover, to isolate terrorists and to mainstream the ordinary simple Muslim believers of Islam, it is essential to **delink Islam from terrorists and terrorism,** as they and their acts do not have sanction either from religion or from the wider community. Despite the wide recognition of the fact, particularly in the non-Muslim world, that Islam is at the minimum compatible with tolerance, moderation, diversity and democratic way of life, calculated and concerted efforts need to be made to highlight the Quranic preachings regarding dignity of individual, freedom of conscience and love for all living creatures, People of the Book (Christians and Jews) and even those without a Book. As says Sachedina, the normative aspects of Muslim religious formulations should be uncovered and their application in diverse cultures of a pluralistic world order of 21st century should be revealed. Its aim should be to retrace, reinterpret and reconstruct it to make it relevant for today [see: Sachedina, Abdlaziz: 2001]. However, such an effort should not be allowed either to slip into an attempt to glorify the past or to allow its hijacking by the terrorists.

Celebrations in some circles over the attacks of 9/11 should be taken as an aberration and not the view of the entire global Muslim community. Certainly, this kind of 'religion-building', as the Bernard Cheryl's Rand Report would argue, is an arduous task, particularly with a religion as complex as Islam which has been complicated by many extraneous problems and issues getting entangled with the religion and its use by all actors on the political spectrum to 'Islamize' the issue.

Faith and reason are both divine gifts to further coexistence and therefore the moment one party assumes a moral high position it changes the dialogue to a monologue. Just as no self-righteous attitude among Muslim leaders can ever further dialogue with other communities, so is the case with self-righteous positions taken by all others. From this incident, what the Western and American think-tanks and scholars should learn is that they should avoid taking high moral positions and avoid any hurt to the religious sensitivities of Muslim community passing such a delicate phase in their history. Unnecessary controversies should be avoided in their initial stages themselves so as not to permit non-issues becoming important issues and inviting radicalists to take advantage. For example, the controversy over whether wearing head-scarf in western or American schools or in other official places might be endorsed or not can be taken as a minor matter of a community's preferences in dress-code, which in fundamentalists' scheme of things might acquire enormous importance if denied. But Benard of RAND would also caution at the same time of inherent implications of such a stand as it might be interpreted as aligning themselves with the fundamentalist/ radicalists end of the spectrum.

In the absence of a scientific and technological revolution in Islamic world, most of the countries with Muslim majority populations are mired in orthodoxy, backwardness, superstitions and lethargy facilitating the religion and religious leaders to usurp the central place what justifiably could have been the place of civil society institutions capable of handling its own social, economic and political matters and issues on a day-to-day basis. Therefore, it is suggested that instead of dwelling on the past and merely reflecting the erstwhile glory of *Islam,* it is most urgent that the Islamic societies concentrate on the present and on the concrete action programs for the resurgence of scientific thought and the spirit of discovery in Islamic Polity, with a view to bringing the Islamic countries at par with the West in the matter of science and technology, in as short a time as possible. Study history not just to glorify the past but also to understand the long period of decline.

Why Didn't the Scientific Revolution Happen in Islam? Reflecting upon this question, Pervez Hoodbhoy, in CHOWK, December, 23, 1997, says that though the roots of science are to be found in highly diverse cultural and temporal origins, including Chinese, Islamic and Hindu cultures, it is due to the cultural and political dominance of the West today that a selectively unilinear and inexorable march of Greco-Roman ideas into the European renaissance is projected as the universal truth. However, it is also true that a scientific revolution did not take place in Islamic civilization between 9th to 13th centuries. It was due to the dominance of the group lead by Ashrite dogmas that a belief was insisted upon the people that God is responsible for everything, denying any space for the scientific concept of the connection between cause and effect. The eventual preponderance of such fatalistic attitudes, the denial of independent judgment, and rejection of the rationalizing culture, made it harder for any important intellectual advance to occur, much less allow for a Scientific Revolution.

A second factor, which discouraged learning for learning's sake, was the increasingly **utilitarian character of post-Golden-Age Islamic society,** which reduced learning only to the materialistic ends. This resulted in inevitable denigration of theoretical knowledge, which permeated throughout Islamic society. This was coincident with growing rigidity of dogma and closing of the doors of theological inquiry. The lack of interest in "useless" theoretical knowledge among Muslims began around the 14th century, and is continuing till date. The lack of curiosity to learn permeated the Mediaeval Islamic thinking and continued later on. Even the gains made by Turkish Ottomans who, in the sixteenth century, had established an extensive and magnificent empire were not capitalized. The Saudis in the current phase on their part have made no secret of their liking for the comforts provided by the wonders of modern technology, but also of their dislike for theoretical, scientific knowledge. There is little doubt that they fear the liberating effect it has on the minds of men, and the dangers it holds in store for a rigidly hierarchical and dynastic society where the leaders derive their legitimacy by appeal to Divine sanction. Such attitudes and values in Muslim society do not augur well for the development of science.

In British India, when Lord Macaulay introduced European science and a system of modern management and accounting into the schools of the sub-continent in the beginning of the 19th century, the two major communities, the Hindus and the Muslims, reacted differently to this decision. While the Hindus welcomed it enthusiastically, and pressed the British to give more opportunities for secular education and establish more colleges and schools, the Muslims, on the other hand, looked upon the British decision with suspicion and resentment. European science was seen as a ruse of the enemy for subverting the Islamic religion and culture. A combination of hurt, pride, defiance, and conservatism led the Muslims to reject modern learning. Parents preferred to keep their children away from schools, preferring to either keep them at home or send them to *madrassahs.* Social pressure, including threats and derision, was applied against the small number of parents who defied the ban. Used to the bygone glories of the Mughal era, Muslims considered most intellectual work, including accountancy and book-keeping, as fit for low-caste Hindus only.

It should be clearly understood however that when a society evolves in complexity and desires to remain relevant and responsive to contemporary realities, it cannot rigidly adhere to the simplicity of past patterns and therefore, must search for solutions which satisfy the needs of progress while maintaining some level of historical and cultural continuity. The inability of the traditional system of education to respond adequately to a changing world may well have been the most critical factor, which denied to Muslims the chance of spear-heading the Scientific Revolution.

Nature of Islamic Law: The Scientific and Industrial Revolution of post-Renaissance Europe was a result of a very complex economic and social phenomenon. Advances in technology certainly gave rise to powerful new means of production, that created a class of what Marx would call 'bourgeois' but which ultimately signaled the end of the feudal society. Though Marx called the bourgeoisie as the exploiter and natural enemy of the working class, yet it was a class that was capable of coordinating the means of production and of bringing about fundamental structural transformations by making innovations and investments. But, on this front also, the nature and practice of Islamic Law was instrumental in discouraging the emergence of a bourgeoisie and nascent capitalism. According to the Weberian argument, the existence of a bourgeois class makes essential the existence of a legal system, which can resolve disputes on property rights, contractual obligations, banking and financial transactions, etc. **Legal judgments should be derivable from rational laws as opposed to arbitrary ones, and the scope of these laws should be broad enough to cover the wide range of problems and cases, which occur in a complex economic environment.** New laws are needed for new situations, and these must be consistent in spirit with the existing laws. Legal rationality is a pre-requisite for modern capitalism; without a systematized and comprehensive legal system, the economic system would not sustain. Such a system is incompatible with the nature of Islamic law, which is inseparable from ethics and religious belief, and as such is not rooted in clearly definable principles. It derives entirely from the revelations and traditions of the Prophet. The legal activity of a Qazi (judge) amounts to discovering a sacred legal tradition and holding that as applicable to the case in hand. The absence of a sharp distinction between Islamic ethics and law means that a systematized legal system cannot be brought to the service of the bourgeoisie in order to protect private property within a comprehensive, rational system. The four legal schools operative among the Sunnis have been taking care of all major problems of Islamic jurisprudence, but their content and interpretation has not changed since the time of Prophet, despite rapid changes taking place in society all through these centuries.

However, one can identify numerous instances where even religious authorities transparently violate the Shariat—in spirit if not in letter. For example, the Islamic prohibition on lending money at interest had never stopped the practice of usury on a large scale in Muslim society. The practical effect of the ban was to create ingenious methods of circumvention. These methods have a name in Arabic: *hiyal*, meaning ruses, or wiles. Rodinson's book *Islam and Capitalism* contains a fascinating account of the past and present practice of circumvention. It encompasses current legal and economic issues such as international trade, joint stock companies, loans from foreign donors, principles of taxation, etc. Nevertheless, all Islamic countries have definite rules governing such matters, which derive from secular, universalistic legal principles. It could be argued, for example, that Islamic law should bar Islamic countries from accepting loans with interest from non-Islamic or Islamic countries. But in practice the Shariat has not influenced the attitudes on this issue. In their internal policies, modern Islamic states pay only lip service to the Shari'at.

For example, the insistence of certain fundamentalists that all depictions of a human face be banned has not prevented the modern state from imposing the requirement that citizens possess identification cards with pictures, or led to a ban on television broadcasting. The need of the state to impose its control over the population is clearly the dominant force. In the border areas of Afghanistan and Pakistan, the self-appointed preservers of Sharia have blatantly violated it by refusing to condemn or put a ban on the trade in narcotic drugs, clearly showing how the material interests take precedence over moral, ethical or religious considerations. Thus, the refusal to adopt the rational principles of economic behaviour also contributed to the failure of Islamic societies to take advantage of the scientific and industrial revolution.

On economic front also, the Muslim societies, by and large, preferred to live in a state of frozen medievalism that prevented their transformation from feudal society to capitalist one whose advances in technology could place them in the league of advanced nations. Though there existed an urban class in Islamic society that could have substantially utilized the incentive for technological advances in production. **But two factors prevented them to take advantage of the capitalist innovations.** ***One,*** since it was based on a system of regular supply of revenue and food from the villages to urban cities, whatever the plight of the villagers, it substantially reduced the incentive for technological advances in production. In this pre-capitalist society, the aim of production was immediate consumption, albeit regulated by traditions and the prevalent hierarchical structure. A very small ruling class that possessed immense wealth had little inclination to invest in technologically advancing devices. The agrarian exploitation pursued successfully by the rulers in Islamic societies made their economy immune, by and large, to the temptations of imitating European technology until it was too late. ***Two,*** there was an absence of autonomous trade guilds in Islamic societies which were playing an important role in the development of European industrial society. Existence of autonomous institutions would have stimulated the growth of industry in Islamic lands and allowed it to maintain the lead which it possessed over the rest of the world until the fourteenth century. Whatever guilds existed, these were in fact created and controlled by the state which determined norms of work, organization, training, the type and quality of work, and the prices at which finished goods could be sold. Unfortunately, the industrial goods from Islamic lands could not compete with the rapidly industrializing West. So, the lack of desire to invest in technological advancement and absence of growth-stimulating economic factors had a negative impact on scientific/industrial revolution touching Islamic societies.

When the dawn of renaissance in Europe was weakening the authority of the church over the state and was facilitating the advancement in industrial and scientific innovations, Islamic societies chose to remain overly religious refusing to separate the church from the state which continues till date and is responsible not only for the social, economic and political backwardness of these societies, but also gives undue and unparalleled authority to those who are truly or spuriously religious.

Thus, the factors such as: the superiority of dogma over reason, tradition-based pattern of non-secular education, nature of Islamic law devoid of the requirements of a changing society, lack of desire to invest in industrial and scientific technology and religion's total grip over society and politics, have led to the scientific and industrial revolution bypassing Islamic societies, making them relatively incompatible with the modern state system. Hence, the often witnessed collision between the Muslim countries and the West. The current problem of Islamic radicalism can be tackled through the growth of a civil and democratic Islam that helps foster growth and development.

Robin Wright argues in *Islam and liberal Democracy* [Wright: 1996] that politicized Islam is not a monolith; its spectrum is broad. Only a few groups, such as the Wahhabi in Saudi Arabia, are in fact fundamentalist. Many of today's Islamic movements are trying to adapt the tenets of the faith to changing times and circumstances and attempt to use religious doctrines to transform temporal life in the modern world. Many streams amongst them are forward-looking, interpretive, and often as they seek to bring about a reconstruction of the social order. The common denominator of most Islamist movements, then, is a desire for change. The quest for something different is manifested in a range of activities, from committing acts of violence to running for political office. In Egypt, Islamists have provided health-care and educational facilities as alternatives to expensive private outlets and inadequate government institutions. In Turkey, they have helped to build housing for the poor and have generally strengthened civil society. In Lebanon, they have established farm cooperatives and provided systematically for the welfare of children, widows, and the poor. In Jordan, Yemen, Kuwait, and elsewhere, they have run for parliament. The specific motives vary from religiously grounded altruism to creating political power bases by winning hearts and minds. But in diverse ways, they are trying to create alternatives to ideas and systems that they believe no longer work.

The views of scholars like Abdul Karim Soroush and Rachid Ghannouchi, that the **sacred Texts do not change, but their interpretations can,** to suit the requirements of the age and place in which the believers live are quite progressive. Fundamentalists' and orthodox fundamentalists' enmity with secularism is unfounded, as it does not mean something western; it only means—think scientifically and act rationally. Popularizing such ideas would signify a positive shift for a major monotheistic religion that provides a highly specific set of rules by which to govern society as well as a set of spiritual beliefs. Separation of religion and politics would not harm Islam; rather, it would strengthen it.

As for bringing the outdated practices such as **polygamy,** which however are supported and endorsed by the fundamentalists such as Taliban in Afghanistan giving out of context references from Koran which supports it at various places in certain specific circumstances and on the condition of judicious and equitable treatment to all the wives, any modernist would condemn such a practice as outdated in the changed environment and conditions of today. As says Akbar Ahmad (2003) that Koran was ambivalent about it even then. The entirely different setting of the modern urban world no longer finds this or other practices such as treating wives shabbily or considering them as private property of husband relevant. If one concentrates on the essence of the Prophet's teachings and the examples set by him, one would surprisingly discover that his views and actions signified greater equality, justice and harmony as the guiding principles of social interaction and would conclude that he himself was a social reformer [Benard: 2003]. Thus, introducing reforms in society is in fact in accordance with the spirit of the Koran.

Though the political character of Islam can be accepted, any dogmatic interpretation of it as a solution to the crisis of the nation-state in Islamic civilization, be it purely secular *or* fundamentalist, has to be discouraged.

'East' is a creation of the 'West' and the vice-versa: To another controversial formulations expressed by Edward Said in his writings (which have been discussed in previous chapters) that East is the creation of West, it can be inferred that the desire to rebuild Islamic social life on the very principles of Islam has always been a salient feature of Muslim thinking. Preoccupation with the glorious Islamic past characterizes the views of both the secularists and the fundamentalists in the Middle-East. This preoccupation is a major obstacle to the unfolding of a worldview adjusted to the structure of the real, existing world of today. Muslim fundamentalists for the most part continue to be preoccupied with dichotomizing the world into an Islamic East and a Christian West, but the world can and should no longer be reduced to such a simplifying dualism. Although the war in Bosnia enforced this dualistic worldview, the realities of the modern world are characterized by far more complex structures, and the preoccupation with West/Islam dualities hampers the development of insights into these structures. If one agrees with Edward Said that the 'East' is a creation of the 'West' then one can hardly exclude modern Arab and Islamic writers from culpability for creating a "West" that is the enemy. The alleged holistic "West" does not exist; it is a creation of Islamic thought and it seems fruitless to continue such a debate.

Islamization of knowledge should be discouraged: The program of Islamization of knowledge is one of the discouraging aspects of current political Islam. There is a need to engage in dispassionate reasoning. Islamic fundamentalists insist that there is an Islamic knowledge that by its nature divorces them from the West, but surely political Islam can be understood and analyzed by thoughtful interpretation of its texts in the light of the concomitant historical context. The literary production of Muslim authors, like that of any group embedded in a given civilization, is generated from encounters with real issues and is based on human knowledge and experience. Studying and understanding these issues in their historical context is possible even for non-Muslims. To concede the distinctive character of a culture is not tantamount to contending that these specifics can be properly understood only by members of the culture itself.

Every religion has a scope for rethinking due to lessons learnt from history: Fundamentalist thought is clearly embedded in a new context, one that differs markedly from that of classical Islam. In short, the issue of whether Islam is basically a religious belief or is, rather, a system of government manifested as an expression of a divine order—whether domestic or global in purview—situates the faultlines between secularists, liberal Muslims, and fundamentalists in the Islamic world. The ordinary Muslims' expectations that the fundamentalist approach would deal with the real problems emanating from the crisis of the nation-state and they would find an appropriate political system for the people of Islamic civilization are belied by the behaviour of fundamentalists whose efforts seldom go beyond the self-congratulatory assertion that divine order in general is preferable to any secular one. In keeping with the classical tradition of Islamic thought, in which politics was inseparably linked with the *shari'a,* the political writings of Islamic fundamentalism are convictional in attitude and scriptural in method. One's first glimpse of this literature reveals that in addressing the current state of affairs, it neither offers a political analysis nor shows a way out. The Islamist fundamentalists view the notion of *al-nizam al-Islami* or Islamic order as a conviction lying beyond evidence and the need for firm grounds. Those who share it are viewed to be true Muslims; those who question it, whatever their sympathies are scorned as deviants of the Islamic *umma,* or even "infidels." Since the symbol of Islamic order so easily projects different patterns of meaning, it is not surprising that it is an issue in contention even among its own adherents. Islamic fundamentalists of various currents insist on the righteousness of their own views, and deny righteousness to their fellow Muslims, other fundamentalists included. Scriptural argumentations provide the logic of the debates, for only the scripture is conceded to uphold or to reject any given interpretation of the notion of an Islamic system of government. The lessons of history, the actuality of existing political structures, and the efficacy or inefficacy of institutions do not matter as a source or defense of argument, for if their reality does not accord with the precepts invoked, then they are merely refutable deviations from true Islam.

A liberal Muslim thinker, Mohammad Arkoun, recently denounced this sort of anti-intellectualism and made the point that every religion, Islam not excluded, is subject to rethinking, that is, to the lessons and forces of history. But sadly, the response by the representatives of political Islam has been mostly hostile. Within Islamic thought and the writings acclaimed as 'classics', there exists a difference of opinion on the unity or not of religion and state. For example, while Rashid Rida's *al-Khilafah wa al-imamah al-'uzma/The Caliphate and the Great Imamate* promotes the idea of the unity of religion and state, Ali 'Abd al-Raziq's *al-Islam wa usul al-hukm*/Islam and the Basis of Government says that although such a unity has been imputed to Islam, the notion lacks any justification in the primary sources, that is, in the Qur'an and the *hadith.*

Though all fundamentalists subscribe to concept of divine order or God's rule preferably used by Sunni Arab fundamentalists, there are writings such as by Shaltut and Al-Azhar, which formulate an Islamic alternative to fundamentalism. [al-Azhar: *Bayan li al-nas/ Declaration to Humanity*] and can be called significant effort to adjust the Islamic worldview to the changed conditions of modern age. Those Muslims who regard their religion more as an ethical than a political context, as a source of conduct and not a system of government, are considered by fundamentalists to be 'misguided Muslims', or even apostates. Thus, the fundamentalists draw the conclusion that they are justified in slaying these Muslims they arbitrarily declare apostates. In a widely read popular book by two exponents of political Islam, *Asalib al-ghazu al-fikri li al-'alam al-Islami/Methods of the Intellectual Invasion of the Muslim World*, some of the new fundamentalist attitudes are encountered.

A befitting reply to the exponents of political Islam, Muhammad 'Abduh (1849-1905), the foremost intellectual father of Islamic modernism, made an effort at a **synthesis between Islam and cultural modernity**. In the view of fundamentalists however, 'Abduh and all Islamic reformers are themselves products of the "intellectual invasion" by the West and are therefore indicted. For them, the only adequate understanding of "true Islam" is the one that first acknowledges that "the major task of the Qur'an is to govern and the unity of state and religion is the crucial part of this understanding, not such that religion is a partial dimension of the state, but on the contrary, such that religion is the major element of the state, a part of it, not a partner to it. Fundamentalists also assign the reason for the end of Caliphate to a colonial conspiracy, though in fact, the concept of Caliphate nowhere exists in Quran and the fact that the modern world is based on the concept of nation-state is described by fundamentalists like a return to *jahilliya,* that is, a state of pre-Islamic heedlessness. That is their logic to call for 'return to Islam', for Islamic awakening that would overcome setbacks suffered by Islam and the call for an Islamic system of government is a major feature of this awakening that would result in the return to Divine.

The fundamentalists, in their obsession with a government based on Sharia, forget that Sharia is a post-Qoranic construction. There is no justification in the Qur'an for such declarations as sanctioning the slaying of fellow Muslims. As says a liberal Muslim intellectual Bassam Tibi: The right to view religious belief as an ethics and to refuse to permit its politicization to be part and parcel of intellectual freedom ought not to be questioned. Freedom of belief is a basic human right and the Qur'an rigorously forbids the killing of Muslims. Algerian fundamentalists slew twelve leading Algerian intellectuals in 1993 alone and legitimized their killing of Muslims as *shari'a*-decreed executions. Islam as a religious belief is thereby distorted, and the fundamentalists defile Islam not only by legitimizing the totalitarian rule they seek but also by silencing reasoning liberal Muslims. An unbiased reading of the political writings of current fundamentalists leads one to conclude that we are dealing with a brand of political propaganda. The style, language, and method of argumentation of these works all reveal propagandistic tactics, not a theology or an intellectual discourse. Reasoning is for these zealots a heresy.

As regards the fundamentalists' obsession with ***Caliphate,*** the outspoken and liberal Muslim intellectuals like Al-Khartabuli and Said al-Ashmawi bluntly state that neither in the Qur'an, as the Islamic revelation, nor in the sayings of the Prophet *(hadith)* does there exist a religious justification for the political order of the caliphate. "The caliphate has no Islamic grounds ... it did disservice to Islam in confusing religious belief and politics". Ashmawi is among the Muslim intellectuals in Cairo hunted for his rational arguments by fundamentalists. al-Khartabuli refers to events of the tenth century, during which Muslims had simultaneously *three* caliphs, one in Baghdad, the second in Cairo, and the third in Cordoba (now Spain), *each one* claiming to be the true Islamic successor of the Prophet and describes this as bizarre state of affairs [see: al-Khartabuli, Ali Husni, Islam and the Caliphate, Beirut: Dar Beirut li al-Tiba'a wa al-Nashr, 1969 and Said al-Ashmawi, Muhammad, The Islamic Caliphate, Cairo: Dar Sina, 1990, as quoted in MEIC's report, *ibid.*].

Scholarly Islam needs to be promoted as a counterweight to the radical variety: There is an urgency to initiate a more careful readings of the Muslim faith from scholars who could use the weight of collective experience, accumulated over 14 centuries, to solve the dilemmas of life in the modern age and making the various legal schools of Islam respect each another and stop calling each other infidels. This exercise can be useful in stumping out extremism. Some may even suggest to promote, in non-Muslim countries such as in the West and in America, the state-sponsored clerics who will have official authority to issue *fatwas* and look into the mosque affairs as an effort to bring forward the 'good', 'moderate' and 'scholarly' Muslim clerics over the 'bad' ones. However, such efforts have their critics who cite the examples of similar political controls on religion in many Sunni countries such as Shia Iran where religion has top place: the supreme leader is always a cleric and is known for his orthodoxy. It also has the danger of dragging the state into the religious matters, a system that clearly runs counter to the western liberal and secular values. Though, even in states with no political freedom, no official imprimatur can drown the multiplicity of voices vying to influence Muslim mind and the age of electronics has broadened the Muslim market-place, yet there always will remain a possibility that those Muslims who are drawn to *jihadist* violence, or to strident forms of political Islam, are indifferent to, or ignorant of, the nuances of theology; that makes them susceptible to "amateur" *fatwas*, as they have a general aversion to the idea of elaborate theology.

What US-West can and should do?

One of the striking lessons of the years since the 9/11 attacks, and especially since the Madrid and London bombings, is that it is important for Western countries to ***know their minorities***—not just who they are but how they live and what they feel. Whereas in the Muslim world terrorists tend to be bred from the impoverished underclass, in the West they are typically disaffected immigrants. Countries that have serious problems with integration—high unemployment levels, discrimination and low levels of educational success—often, though not always, face a greater overall threat. Paul M. Barrett, who produced a series of page one stories for *The Wall Street Journal* in 2003, the articles that provided the basis for his *American Islam: The Struggle for the Soul of a Religion*, provides some thoughtful forays into Muslim America. He feels that Muslims in America are caught in the cross-currents. Though well–rooted in America, yet because they occupy leadership roles in a community with a large number of immigrants and others who feel removed from the nation's power structure, they live double lives. They simultaneously play the role of dexterous insider among their co–

religionists and outraged outsider with non–Muslims. Some of them try to press their fellow Muslims to adopt changes that are at odds with hundreds of years of tradition. Some among them are trying to champion a tolerant, ecumenical version of Islam against the forces of insularity and xenophobia. Happily, the second catastrophic attacks since 9/11 have not occurred despite multidimensional predictions. It is because American Muslims are not ghettoized. Muslims in America and west are, by and large, well educated and able to earn good salaries. Though they may be angry about US foreign policy, they are not alienated from American society or values [Barrett: 2007]. The Pew Survey with the title, *American Muslims: Middle Class and Mostly Mainstream*, indicates that generalizations about the socio–economic conditions of this group still stand. Marc Sageman, the author of *Understanding Terror Networks*, likes to say, "American Muslims have bought into the American dream". It is also because the basic foundations of American values are very much in compliance with true Islam—freedom of religion, freedom of speech, toleration. Due to this general American Muslim attitude towards America, the law enforcement officials feel that the American Muslim community will be the line of first defense against any act of terrorism. Trusting that despite the traumas of the post-9/11 period, most Muslims still believe it is possible to satisfactorily practice Islam in the US, they hope the community will dissuade members who are tempted by radicalism, failing which they will notify the authorities about individuals who may be inclined to violence [Sageman: 2004].

West should overcome its arrogance: Preoccupation with democracy is a characteristically Western way of looking at things, but in the ideology of Islamic fundamentalism—or, for that matter, in the minds of the Islamic peoples—democracy is not an important issue and no deeper understanding of Islamic fundamentalism is to be reached simply by pinpointing the anti-democratic direction of the movement. To understand the important currents in Islam, the study the ideology and major themes of Islamic fundamentalism is needed, as it is propounded by operatives within Islam. The pivotal concept of its ideology is order and their goal is an Islamization of the political order, which is tantamount to toppling existing regimes, with the implication of de-Westernization. In the West there have been great exaggerations of the impact of the Islamic Revolution in Iran on the rest of the World of Islam These exaggerations overlook the fact that Sunni Muslims have a concept of order different from that of the Shi'i Muslims in Iran. Mark Juergensmeyer's plea that the West ought to overcome its arrogance and be more tolerant toward non-Western cultures is quite appreciable. This must not amount, however, to consent to religious fundamentalism.

Need to change the provocative vocabulary: *The Economist* in its July 5, 2007 write-up raises the relevant issue of changing the vocabulary to confront terrorism as the cliché, the "global war on terror" (GWOT), America's response to the September 11th attacks, as GWOT implies, wrongly, that there exists a military solution to a problem. True, a few countries, such as Afghanistan, require a coordinated nation-building effort but for most of the situations, what is required is a patient police and intelligence work. **"War" should be the exception, not the focus of the effort against terrorists.** The language of war and its actual execution in Afghanistan and Iraq has alienated much of the Muslim world who believe, though incorrectly, that the West has been waging a "war on Islam" for decades, if not centuries. Such vocabulary reinforces the propaganda of al-Qaeda, which claims to be fighting a global *jihad* to defend Islam against "Crusaders and Jews". Language matters. As a positive sign of change in the usage of language, July 2007 car bombings in London and Glasgow, terrorists were termed as criminals, the mention of the words "Islam" and "Islamic" was avoided and the reports spoke about "communities" rather than "Muslims". Decoupling the word "Islam" from terrorism may make it easier to enlist the help of Muslims who might otherwise feel queasy about "betraying" co-religionists. But since the terrorists themselves claim to be acting in Islam's name, it will be difficult to purge the word "Islam" from the general public discourse on terrorism till *Jihadists*, who claim to be fighting a war against the West also change their vocabulary.

Need for stress on Development: A military approach alone will not defeat al-Qaeda or Taliban. In Afghanistan, as in the greater Middle-East, comprehensive reconstruction combined with an enhanced campaign against the regional sources of terrorist training and indoctrination will undermine al-Qaeda's ability to brainwash another generation of frustrated, impoverished young men. It is essential to contain al-Qaeda's resurgent leadership in Afghanistan's restive backyard.

Building effective state institutions in Afghanistan is the most cost-effective means of combating al-Qaeda and the Taliban. To this end, the Afghan National Army requires more trainers, better weapons, armored vehicles, and airlift capabilities to be a more effective and independent fighting force. More importantly however is required the development plans that can address the day-to-day needs of the ordinary Afghan or Iraqi as what they want is schools for their children, a well-placed health care delivery system, and healthy state institutions. International donors must work together to increase the pace and scope of current capacity-building efforts, empower the private sector as the country's key engine of growth, and get aid into the hands of languishing villagers as quickly as possible. Despite their many frustrations, what the Afghan people fear most is that the international community will once again abandon them. These fears need to be addressed.

Public-Private Intelligence: Beefed-up, better coordinated intelligence can be an effective counter-terror strategy. Writing on the eve of the 6th anniversary of 9/11 attacks in *Hindustan Times*, Pramit Pal Chaudhary tries to find out why America has not had a repeat of 9/11 catastrophe since September 2001. His argument is that though al-Qaeda was never as strong as it was exaggeratingly projected, the more important reason is better intelligence. "In past thirteen months, the US has broken at least three separate homegrown Islamic terrorist plots at the planning stage: a plot to bomb a Manhattan train tunnel was picked up by FBI monitoring of *jihadi* websites and encompassed police in six countries. A store clerk asked to copy a *jihadi* video on to a DVD tipped-off police about on a New Jersey military base. And two months ago, US security officials arrested a group wanting to blow up a fuel pipeline at JFK airport, thanks to a police informant" [see: Chaudhry, Pramit Pal, "How US foiled a Repeat of 9/11": *Hindustan Times*, September 7, 2007, p.16]. The intelligence expert Amy Zegart of the University of California, Los Angeles assigns the reason of al-Qaeda's long silence after 9/11 to the luxury of safe havens its leaders enjoy and can afford to a notoriously long planning cycle. According to him, the 1998 African Embassy bombings were at least five years in the making, as was 9/11. However, the experts like Rohan Gunaratana ascribe a few reasons for this long terror holiday to the US ability to avoid at least the low level terror attacks that trouble UK, Indonesia and India, such as: By investing in intelligence, the US, Canada, Europe and Australia have prevented over 100 terrorist attacks since 9/11. The

US, with 16 separate Intelligence agencies with a budget of at least $50 billion, maintains an easy global lead in electronic signal espionage. This has been combined with "AL Capone prosecutions" that disrupt the terror plots at the early stages of planning. Now, the terrorist planners are charged for the lesser crimes that are committed at the preparation stage of a terrorist attack such as credit card frauds, illegal weapon possession and immigration violations.

Common people should also be incorporated into the counter-terrorism network. Truck drivers, garbage collection men, sales clerks and building security staff are useful to report anything unusual to law-enforcement agencies. Even, some basic training should be provided on what to watch out for. Many breakthroughs in US homegrown terrorist cases have been reported by persons on the street. New York City pays for television ads thanking over 14,000 people who reported about the suspicious activities around to the city police in 2006. Such grass-root level intelligence network can avoid a significant number of big and small terrorist incidents, it does only a little damage to civil liberties of the people in liberal countries. James Carafano, a terrorism expert from Heritage Foundation says, however, that though American counter-terrorism has done a very good job of protecting the civil liberties of individuals.... It does a poor job of protecting itself against spurious allegations of abuse or incompetence. But, American public and courts do not sanction extra-legal actions like the one in Guantanamo Bay which they call counter-productive though Bush administration believed that the regular civilian judiciary could not handle terrorists and therefore some special legal set-up is required to deal with the terrorists. Post-2001 homeland security arrangements have successfully delivered the results [Carafano: 2005; 2008].

US Foreign Policy should take 21st century's realities in account: To address the changing realities of 21st century, the United States cannot return to state-centered realism, because non-state actors like al-Qaeda are more threatening than any single state's army or air force. Bill Richardson, writing in *Harvard International Review* mentions about the **six trends** that are transforming the world, these are: *fanatical Jihadism* bursting from an increasingly unstable and violent greater Middle-East. This trend had been growing for years, but the invasion and subsequent collapse of Iraq have fueled its growth. A **second trend** is *the growing power and sophistication of criminal networks* capable of disrupting the global economy and trafficking in weapons of mass destruction. Together, these two trends raise the terrible specter of nuclear terrorism, Pakistan's A.Q. Khan sold nuclear materials to rogue states. The existence of a black market for nuclear materials is well documented, and the proliferation of nuclear weapons to new countries has further increased the opportunities for *Jihadists* to obtain them.

A third trend transforming the world is the *extraordinarily rapid rise of Asian economic and military power,* particularly in China and India. The inclusion of these two countries, the most populous in the world, in international discussion has changed the nature of diplomacy, both through bilateral agreements and through international organizations. A **fourth trend** is the *re-emergence of Russia as an assertive global and regional player,* with a large nuclear arsenal and strong control over energy resources. The simultaneous rise of India, China, and Russia requires that these countries are co-opted in anti-terrorism efforts.

A **fifth trend** transforming the world is the *growth of both global economic interdependence and of global financial imbalances,* unaccompanied by the growth of institutional capacities to manage these realities. Globalization has made national economies more vulnerable to resource constraints and financial shocks originating beyond national borders, and growing global demand for energy has the potential to lead to geopolitical tensions or even a global energy crisis. The **sixth trend** is the *globalization of urgent health, environmental, and social problems.* Global warming and pandemics like AIDS do not respect national borders. Poverty, ethnic conflict, and overpopulation also spill over borders, feeding what Moises Naim has called the "five wars of globalization" (over drugs, arms trafficking, money laundering, intellectual property, and alien smuggling).

These six trends present the global community with problems, which are international in their origins and effects and will therefore require international solutions. They cry out for political leadership with futuristic vision, devoid of parochialism and narrowly defined nationalism. If the world succeeds in preventing nuclear terrorism, defeating *Jihadism,* integrating rising powers into a stable order, protecting global financial market stability, and fighting environmental and health threats, the world can be expected to be facing lesser terrorist violence and wars. If the world fails to meet these challenges, the United States will just as surely deserve much of the blame [Richardson, Bill, "A New Realism: Crafting a US

Foreign Policy for a New Century", *Harvard International Review, Vol. 29 (2)—Summer 2007*].

In current political milieu, while the importance of a strong military cannot be ruled out, equal importance of diplomacy and multilateral cooperation has to be understood and practiced. A new realist foreign policy will require that the United States repair its alliances with other nations to help them trust US leadership. US policy-makers need to restore respect and appreciation for US allies and for shared democratic values in order to coordinate international efforts for global problems. For this, commitment to international law and multilateral cooperation is a must, which means: promoting expansion of the UN Security Council's permanent membership to include Japan, India, Germany, and one country each from Africa and Latin America, promotion of some ethical reform at the United Nations so as to enable UN to help its many underdeveloped and destitute member states in meeting the challenges of the 21st century and also endorsement for expanding the G-8 to include new economic giants like India and China.

Beyond the United Nations, a commitment to international law means that the United States must be impeccable in its own human rights behavior. The US government must join the International Criminal Court and respect all international treaties, including the Geneva Conventions so as to attain moral authority to put global, multilateral pressure on such regimes such as Rwanda and Sudan whose leaders are responsible for committing crimes against humanity.

Efforts at nuclear non-proliferation required: Most urgently, the United States must focus on the real security threats, from which Iraq has so dangerously diverted its attention. This means doing the hard work to build strong coalitions to fight terrorists and to stop nuclear proliferation. There is a pronounced need for better human intelligence and better international intelligence and law enforcement coordination to prevent nuclear trafficking. US diplomatic leadership is needed to unite the world, including Russia and China, to sanction the nuclear ambitions of Iran and North Korea, and to provide these nations with positive incentives and face-saving ways to renounce nuclear weapons. If the United States wants other countries to take the NPT seriously, it must show that it takes it seriously itself. The United States should re-affirm its NPT commitment to the long-term goal of global nuclear disarmament, and it should invite the Russians to join in a moratorium on new weapons and further staged reductions in arsenals, beyond what has already been agreed, over the course of the next decade. It should also ratify the Comprehensive Test Ban Treaty, not only because it is good policy, but also to send a signal to the world that it has turned a corner and once again will be a global leader, not a unilateralist loner. The international community needs to present Arab and Muslim populations with a better vision than the apocalyptic fantasy of the *Jihadists*: a vision of peace, prosperity, tolerance, and respect for human dignity. There are a number of steps the United States can take to help accomplish this. If the United States wants Muslims to be open to it, it should start by closing Guantanamo.

Reform in education system of Muslim societies can help de-radicalization: A crucial effort in fighting terrorism must be support for public education in the Muslim world. Many Muslim students have no educational opportunities except for *madrassas*, some of which teach Jihad. It must be a major component of international aid policies to poorer Muslim countries, as well as of US diplomacy with all Muslim countries, to take education out of the hands of those who preach violence. Moderate Muslims must be given a louder, more systematic voice in US policy toward the Middle-East so that they can speak the truth about the West and be heard by their fellow Muslims. The Middle-East must be reengaged in peace process, as peace would deprive the *Jihadists* of their most effective propaganda tool. The sole superpower must use all its sticks and carrots to strengthen Palestinian moderates and to achieve a two-state solution which guarantees peace in West Asia. The challenges facing all nations today are the shared problems of an interdependent global society. Solutions necessarily also must be shared.

Empowered United Nations: The interplay of law and politics reveals that the United Nations has proved to be very weak in face of international challenges. It has been used to reinforce national sovereignty at the expense of international conflict resolution by nations. In this background, reform of the Security Council and other crucial UN agencies for strengthening the United Nation's credibility and authority is very essential. Hobbesian dilemma of making a choice between chaos and order in the 'State of Nature' and then resolving the dilemma by surrendering personal choice to a 'sovereign' can be a useful paradigm for resolving conflicts, at least in limited fields such as of the nature of international terrorism. It can apply to resolve conflicts beyond national borders, to the war on terror and to the hope for a more peaceful future.

Dalia Mogahed and Zsold argue that Muslim minorities in the West are often scrutinized through paradoxical prism of Muslims versus the West. A Pew poll found that given a choice of identifying as first Muslim or Christian or as first a citizen of their country, the majority of British, French, and German Muslims choose faith, while the majority of British, French, and German Christians choose country. Some have taken these results as witness to the danger of over-accommodating religious differences and even advocate that European Muslims be persuaded or forced to forsake their Islamic identity for a Western one. However, new findings from a Gallup study of Muslims in London, Paris, and Berlin, and the general public in each corresponding country, challenge the very legitimacy of such a trade-off and offer another way to reconcile citizenship and creed. In many such instances, the divide between European Muslims and the general public is inexistent or not as large as originally seemed. This fresh perspective indicates that national and religious identities are not mutually exclusive, but mutually enriching, and that integration is not defined by citizen conformity, but by citizen cooperation. One of the most pervasive assumptions in discourse on European Muslim integration is that Muslim religiosity threatens Europe. Those who believe in the irreconciliability of Western and Muslim identity generally argue that Muslim piety, expressed in religious symbols and moral conservatism, contrasts against the backdrop of an increasingly secular and sexually liberal Europe: a recipe for increasingly insular Muslim communities and profound alienation from European national identity. These isolated communities, the argument continues, represent illiberal islands corrupting Western society's liberal values; they are "cesspools" for radicalization. Integration, or conformity with majority culture, is therefore seen as vital defense against citizens with dual loyalties. However, Mogahed and Zyiri gallop poll results would show that though majority of European Muslims, like they are everywhere else, is highly religious, this piety does not lead to sympathy for terrorism, desires to isolate, or lack of national loyalty. Religion is an important part of their daily lives and they identify strongly with their faith. As compared to the public at large, Muslims overwhelmingly see homosexual acts, sex before marriage, and abortion as "morally wrong." However, religious and national identities are not mutually exclusive. Not only do urban Muslims identify strongly with their religion, but they are at least as likely as the general public to identify strongly with their countries of residence. As a British Muslim MP recently commented, "My nationality is first British and my religion is first Muslim." [Nyiri Mogahed and Zsold "Reinventing Integration: Muslims in the West", *Harvard International Review*, Summer, 2007].

While Muslims in Europe wish to hold on to their values, they also choose diversity over conformity. They define integration as mutual respect and cooperation between distinct cultures, not as the dilution of minority culture into a dominant mainstream, nor as the dilution of majority culture into a politically correct muck. The West should therefore not consider Muslims a threat, given their predilection for integration within their societies. Even in predominantly Muslim countries, a Pew study found, the majority of Muslims in Pakistan, Jordan, Egypt, and even Turkey consider themselves Muslims first, rather than citizens of their country. Significantly, the same poll found that Christian Americans—seldom accused of lacking national pride—were almost evenly split between those who said they were first Christian and those who said they were first US citizens. Clearly, expressing a religious primary identity does not necessarily mean rejecting one's country. The commonality among these communities is not a lack of patriotism, but a majority who considers religion important—and therefore at the heart of its identity.

In a Gallop poll in Indonesia, the largest Muslim majority country in the world, not a single respondent from the small minority who condoned the attacks of September 11 cited the West's perceived moral decay or the Qur'an for justification. Instead, this group's responses were markedly secular and worldly, mostly relating to US foreign policy. For example, one Indonesian respondent said, "The US government is too controlling toward other countries, seems like colonizing." Another said, "The US has helped the Zionist country, Israel, to attack Palestine." Similarly, one of the people responsible for Europe's own September 11, the July 7, 2005 London train bombings, cited political, not cultural or religious justifications for his horrific acts.

Moreover, while residents of predominantly Muslim countries are critical of the West's perceived breakdown of traditional values, this is neither the primary driver of extremist views nor the demanded change that Muslims cite for better relations with the West. To the question of what the West can do to improve relations with the Muslim world, the most frequent responses were neither for Western societies to be less democratic nor less liberal. Far from it—what Muslims said they admired most about Western societies were their democratic systems of government, and accompanying features like government transparency and freedom of speech. Instead, **to improve relations, Muslims called for Western societies to change their economic and political policies toward Muslim nations, but most of all to "stop thinking of Muslims as inferior and to respect Islam."** It is not that most Muslims believe there is nothing to respect about the West, but rather that they believe the West finds nothing to respect about them. Gallup results have shown this recurrent theme in the West-Islam relationship, with the divide a result of Western misperception of Muslims rather than Muslim antagonism of the West. In fact, instead of using religious doctrine to condone terrorism, many Muslims refer to their beliefs to condemn it "Killing one's life is as sinful as killing the whole world," paraphrasing verse 5:32 in the Qur'an. Far from reviving terrorism, the Qur'an is, for many Muslims, the inspiration for rejecting it, again challenging the perceived risk of Muslim religiosity in Europe.

A policy of strategic restraint is required of US: Instead of launching the extreme steps like 'war on terror, there is a need to develop a new paradigm that will focus less on cultural conformity, and more on the emergence of shared goals and a commitment to democracy to leverage untapped common ground. American internationalism should be reshaped in order to save it from drifting into extremes and for keeping its international legitimacy intact. A policy of strategic restraint is required to attain this. It should not only fulfil its international commitments, but should also try to revitalize the international institutions, law and behaviour by its own example. America has the wherewithal to achieve this kind of a peaceful world order only if it gives up the tendency of preemption just because it knows that it can get away with anything.

Though al-Qaeda was capable of wreaking havoc, particularly if it could acquire nuclear or biological weapons, which happily did not happen, United State's 'war on terror' was an over reaction of sorts as al-Qaeda is neither Nazi Germany nor Soviet Russia. Raising *jihadis* to the status of a global 'totalitarian' is not only counter-productive but also raised unwarranted fear psychosis in Americans' mind who out of fear supported a costly war in Iraq and faced a stronger state at home undermining democratic values. Perhaps most destructively, the war on terrorism worsens some of the factors that contribute to Muslim wariness of the West. Israeli hardliners were extolled as a model for dealing with terrorism, and the American refusal to recognize Hamas's electoral victory in Palestine belied Washington's talk about democratization. The war in Iraq has been carried out callously and confrontation with Iran appears to be a case of nuclear "orientalism." Continued US backing for repressive Arab regimes remains a sore point with Arab democrats, and repressive regimes are being bolstered in Central and South Asia. American government should stop referring to the anti-terror effort as another epic episode of America's triumphal battle against totalitarianism. The analogy is weak, and it is leading the country to support poor and even catastrophic policies in the anti-terror effort.

Thoughtful aid and technical assistance to Afghanistan: In Afghanistan, international assistance remains ineffective as far too much aid is supply-driven and focused on wasteful technical assistance programs. When Afghans see poorly built schools and clinics collapsing in their provinces (or a lack of such facilities), they lose faith in the Afghan government and the international community. Too much aid is delivered outside the government's budgetary framework and guided by the priorities of donors rather than the national government. Currently, only five percent of international assistance funds are given to the Afghan government, and 83 percent are disbursed without the government's knowledge or approval. The consequence of this imbalance is a weak sense of ownership of the development process among Afghans themselves. In the absence of proper coordination and oversight, sovereign government will be ill-equipped to address local grievances; meanwhile, the forces like al-Qaeda and the Taliban would have reasons to capitalizing on the public's frustration. International donors must work together to increase the pace and scope of capacity-building efforts, empower the private sector as the country's key engine of growth, and get aid into the hands of languishing villagers as quickly as possible. Despite their many frustrations, what the Afghan people fear most is that they will once again be abandoned by the international community.

In the long-run, the United States and its allies are far more likely to win the war on terror than al-Qaeda, not only because liberty is ultimately more appealing than a narrow and extremist interpretation of religion but also because they learn from mistakes, while al-Qaeda's increasingly desperate efforts will alienate even its potential supporters. But victory in the war on terror will not mean the end of terrorism, the end of tyranny, or the end of evil, utopian goals that have all been articulated at one time or another. Terrorism, after all (to say nothing of tyranny and evil), has been around for a long time and will never go away entirely. From the Zealots in the first century AD to the Red Brigades, the Palestine Liberation Organization, the Irish Republican Army, the Tamil Tigers, and others in more recent times, terrorism has been a tactic used by the weak in an effort to produce political change. Like violent crime, deadly disease, and other scourges, it can be reduced and contained. But it cannot be totally eliminated.

US must learn some bitter lessons: Only an enhanced budget of national security and curtailing liberties cannot solve problem of terrorism. Nor the over-reaction as is asserted by David Frum and Richard Perle [2003] that there is "no middle ground" and that "Americans are not fighting this evil to minimize it or to manage it"; the choice, they say, comes down to "victory or holocaust" can sole it. Thinking in these terms is likely to lead the United States into a series of wars, abuses, and over-reactions more likely to perpetuate the war on terror than to bring it to a successful end. The war on terror will end with the collapse of the violent ideology that caused it The ideology will not have been destroyed by US military power, but when its adherents will have decided that the path they chose could never lead them where they wanted to go. The extremist Islamism in the future will have a few adherents here and there.

The terrorism needs to be reduced to such a level that its global networks and sanctuaries are destroyed, its financial sources blocked, its communications interrupted, and, most important, its supporters persuaded to find other ways to pursue their goals. Terrorism will not be over, but its central sponsor and most dangerous executor should be. The terrorists need to be made to see that the result of any attack they carry out is not the over-reaction they sought to provoke but rather the stoic denial of their ability to elicit a counterproductive response and their violent acts be seen as the heartless criminals they are rather than as the valiant soldiers they seek to be. In this way, over time, the risk of terrorist attacks can diminish even further as the terrorists will no longer be serving their intended purpose. Unlike the current state of affairs after 9/11 when curbing terrorism remains the main driver of US foreign policy, it needs to be reduced to an ordinary policy matter.

Fundamentalist Islamism also has poor long-term prospects as a broader political ideology. Indeed, far from representing a political system likely to attract increasing number of adherents, fundamentalist Islamism has failed everywhere it has been tried. In Afghanistan under the Taliban, in Iran under the mullahs, in Sudan under the National Islamic Front, different strains of Islamist rule have produced economic failure and public discontent. Indeed, the Taliban and the Iranian clerics are probably responsible for creating two of the most pro-US populations in the greater Middle-East. Opinion polls show that there is even less support for the kind of fundamentalist Islamic government proposed by bin-Laden. "Many people would like bin-Laden ... to hurt America," says the political scientist and pollster Shibley Telhami, "but they do not want bin-Laden to rule their children." Fundamentalist Islamism has not yet run its course and cannot be expected to in less than a generation [Gordon: 2007].

Finally, there are good reasons to believe that the forces of globalization and communication that have been unleashed by changing technology will eventually produce positive change in the Middle-East. This will especially be true if there is successful promotion of economic development in the region, which would produce the middle classes that in other parts of the world have been the drivers of democratization. Even in the absence of rapid economic change, the increasingly open media environment created by the Internet and other communications technologies will prove to be powerful agents of change. Although only around ten percent of households in the Arab world have access to the Internet, that percentage is growing rapidly, having already risen five-fold since 2000. Even in Saudi Arabia, one of the most closed and conservative societies in the world, there are over 2,000 bloggers. Cable news stations such as the independent Qatar-based al Jazeera and the Dubai-based al Arabiya reach tens of millions of households throughout the Arab world, often with information or perspectives the repressive governments in the region would rather not be heard. According to the Arab media expert Marc Lynch, "The conventional wisdom that the Arab media simply parrot the official line of the day no longer holds true. Al Jazeera has infuriated virtually every Arab government at one point or another, and its programming allows for criticism, and even mockery. Commentators regularly dismiss the existing Arab regimes as useless, self-interested, weak, compromised, corrupt, and worse." Lynch points out that one al Jazeera talk show addressed the issue "Have the existing Arab regimes become worse than colonialism?" The host, one of the guests, and 76 percent of callers said yes—"marking a degree of frustration and inwardly directed anger that presents an opening for progressive change" [see: Gordon: 2007].

The victory in the war on terror may take quite a long time. But falling prey to the illusion that this is World War III—and that it can be won like a traditional war—runs the risk of perpetuating the conflict. The US experience in Iraq suggests the perils of trying to win the war on terror through the application of brute military force. The victory in the war on terror can come but only when the terrorists' ideology loses support and when potential adherents of that ideology see viable alternatives to it. To achieve this, the countries around the world, including US, would not overreact to threats but instead would demonstrate confidence in its values and its society—and the determination to preserve both. The international community needs to expand its efforts to promote education and political and economic change in the Middle-East, which in the long run will help that region overcome the despair and humiliation that fuel the terrorist threat.

Dependency on Arab oil needs to be reduced, thus freeing the constrains on the foreign policies of different countries and in turn making oil-dependent Arab autocracies to diversify their economies, more evenly distribute their wealth, and create jobs for their citizens. The fact that the conflict between Israel and its neighbors has a direct effect on the outbreak of terrorism has to be accepted and a diplomatic offensive designed to bring an end to this conflict has to be launched which stands as a key source of the resentment that motivates many terrorists. The common threat of terrorism can be combated by acting in close co-operation with each other keeping the mutual legitimate interests of all the allies in view.

With time and experience, Muslims themselves will turn against the extremists in their midst and will seek to put their civilization on a path toward restoring the glory of its greatest era—when the Muslim world was a multicultural zone of tolerance and intellectual, artistic, and scientific achievement. The agents of change might come from within the Muslim societies themselves.

In today's increasingly interconnected world, weak and failed states pose an acute risk to global security. Indeed, they present one of the most important foreign policy challenges of the contemporary era. States are most vulnerable to collapse in the time immediately before, during, and after conflict. When chaos prevails, terrorism, narcotics trade, weapons proliferation, and other forms of organized crime can flourish. Left in dire straits, subject to depredation, and denied access to basic services, people become susceptible to the exhortations of demagogues and hatemongers. It was in such circumstances that in 2001 one of the poorest countries in the world, Afghanistan, became the base for the deadliest terrorist attack ever on the US, graphically and tragically illustrating that the problems of other countries often do not affect them alone. The international community is not, however, adequately organized to deal with governance failures. There is a need to develop the tools to both prevent conflict and manage its aftermath when it does occur. Such efforts will entail not just peace-keeping measures, but also influencing the choices that troubled countries make about their economies, their political systems, the rule of law, and their internal security. Weak countries are unable to take advantage of the global economy not just because of a lack of resources, but also because they lack strong, capable institutions. To promote sustainable peace, the international community of states must thus commit to making long-term investments of money, energy, and expertise.

There is a ***crisis of governance*** in a large number of weak, impoverished states, and this crisis poses a serious threat to international peace and security. The foreign policy architecture of the United States and of most of the countries was created for the threats of the twentieth century—enemies whose danger lay in their strength. Today, however, the gravest danger to the nation lies in the weakness of other countries—the kind of weakness that has allowed opium production to skyrocket in Afghanistan, the small arms trade to flourish throughout Central Asia, and al-Qaeda to exploit Somalia and Pakistan as staging grounds for attacks. Terrorism, conflict, and regional instability are on the rise throughout the developing world, and the repercussions will not just be felt locally. Weak and failed states and the chaos they nurture will inevitably harm the global security economy that provides the basis for international stability. The world needs a new, comprehensive strategy to reverse this trend and turn back the tide of violence, humanitarian crises, and social upheaval that is sweeping across developing countries from Afghanistan to Zimbabwe—and that could engulf the rest of the world. An effective strategy will embrace a four-pronged approach focused on crisis prevention, rapid response, centralized decision-making, and international cooperation. A plan of such scope must first recognize that the roots of the weak-state crisis, and any hope for a long-term solution, lie in development: fostering stable, accountable institutions in struggling nations—institutions that meet the needs of the people, empowering them to improve their lives through lawful, not desperate, means. The state building is not an act of simple charity but a smart investment in the states' own safety and stability.

Need to develop an integrated defensive system: Michael Levi, in his article "Stopping Nuclear Terrorism: The Dangerous Allure of a Perfect Defense" [*Foreign Affairs*, January/February 2008] feels that Nuclear terrorism poses a grave threat to global security, but seeking silver bullets to counter it does not make sense. Instead of pursuing a perfect defense, US policy-makers should create an integrated defensive system that takes advantage of the terrorists' weaknesses and disrupts their plots at every stage, thereby chipping away at their overall chances of success. Initiatives to secure nuclear weapons and materials are vital, but they will always fall short, too. Rather than search for a perfect defense, which will never exist, counter-terrorism strategists must use the many imperfect tools at their disposal to confront the many imperfect terrorist groups that they face. To pull off a nuclear attack, a group would need to acquire nuclear materials or a weapon, build a bomb or unlock an existing one, move that weapon to its target, and detonate it. Securing nuclear weapons and materials, although critical, confronts only one part of a plot and cannot eliminate the threat entirely. Strategists must build on this one defense to develop an integrated defensive system that also draws on border security, law enforcement, intelligence operations, military and diplomatic initiatives, and emergency response efforts. To do so properly, they must develop a more realistic picture of nuclear terrorism that draws on a careful understanding of how terrorist groups work and how their plots can fail.

In counter-terrorism operations, effective protective measures are the *sine qua non* of success. Make it impossible for the terrorist to hijack or blow up a plane or to kidnap an individual. If despite all security measures he succeeds, stand firm and refuse to concede his demand. Deny him the aura of martyrdom, new recruits, funds and weapons, the theatre he needs for publicizing his actions. If all this is done in an effective and sustained manner, his organization will start withering away. The greater the anger in the community from which the terrorists have arisen, the greater the flow of new recruits and the stronger the motivation. It is, therefore, important to ensure that counter-terrorism operations do not add to the already existing anger. bin Ladens do not go round looking for recruits. Enraged elements in the Islamic world go to Pakistan and Afghanistan looking for bin-Ladens and their ilk to help them in giving vent to their anger appropriately.

Conclusion

To deal with fundamentalist absolutisms, we *need* a cross-cultural consensus on values, the promotion of cultural pluralism and the rejection of cultural relativism. The revival of norms and values believed to be authentically primeval leads to according them an absolute validity. Whereas cultural modernity means agreeing that every thought is revisable, neo-absolutism means permitting no revision of sanctioned thought. Moreover, fundamentalism is *not* simply the voice of another culture. For those who are concerned with human rights as universal rights, the idea of cultural pluralism (honoring legitimate differences) must not become tantamount to a self-defeating cultural relativism, which permits itself to excuse even abhorrent differences. Thus, the acceptance or even sanctioning of practices constituting violations of human rights (for instance torture, as the suppression of freedom of expression, or the genital mutilation of women) as being simply expressions of a different culture cannot be tolerated by an enlightened world. The pluralism of cultures presupposes a variety of means to an end, but not a variety of the ends themselves. In particular, there can be no compromise so far as human rights are concerned. In this sense we may speak of the cross-cultural validity of a program of human rights, a program that is coexistent and consonant with cultural varieties of how to create these rights on local but cross-cultural foundations. Further, we may argue that cultural pluralism in the realm of human rights cannot be allowed to mean more than a cultural indigenization of basic individual human rights in local cultures.

Despite the fact that due to the contribution of the west in science, technology, economics and culture, the imprints of the west will remain. But at the same time, the resurgence of the much larger non-western world would inevitably result in the inevitable decline of the western hegemony. The end of cold war has already initiated a change in the outlook. The much talked about 'new world order' will have to reflect the whole world, not just the Western/American world. In his book *From Globalism to Regionalism*, Patrick M. Cronin has expressed his belief that we are moving into a multipolar world in which we can hope for a more coherent order based upon consensus rather than on western domination which will allow a balance of power between the regions. Many of world's growing ethnic, religious, and national problems will have to be settled at regional level, without giving space to the outsiders to intervene. The recent US interventions, on whatever ground, have revealed that it is unrealistic in political, economic and military terms for any single nation to guarantee world security. Rather than presuming to know what is good for the whole world, it is essential to respect others' objectives and values. What is required is a shift from dominance to partnership [Cronin: 1993].

REFERENCES

Ahmed, Akbar S., *Islam Under Siege*, New Delhi: Vistaar, 2003.

Alexander, Yonah (eds.), *Countering Terrorism: Strategies of Ten Countries*, Michigan: University of Michigan Press, 2002.

Barrett, Paul M., *American Islam: The Struggle for the Soul of a Religion*, FSG Books, 2007.

Benjamin, Daniel and Steven Simon, *The Age of Sacred Terror*, NY: Random House, 2002.

Benjamin, Daniel, *The Next Attack*, New York, Henry Holt/Times Books, 2005.

Benard, Cheryl, *Civil Democratic Islam: Partners, Resources and Strategies*, Santa Monica,: RAND, 2003.

Bull, Hedley, *Anarchical Society*, 1977.

Bruce, Lawrance B., *Shattering the Myth: Islam Beyond Violence*, Princeton University Press, 2000.

Carafano, James, *Private Sector, Public Wars: Contractors in Combat—Afghanistan, Iraq, and Future Conflicts*, (2008).

Carafano, James, *Winning the Long War: Lessons from the Cold War for Defeating Terrorism and Preserving Freedom* (2005).

Cassidy, Robert M., *Counterinsurgency and the Global War on Terror: Military Culture and Irregular War*, Praeger Security International, 2006.

Crenshaw, Martha, *Consequences of Counter-terrorism*, Russell Sage, 2010.

Cronin, A.K., "Behind the Curve: Globalization and International Terrorism", *International Security*, 27 (3): 30-58, 2002.

Cronin, Patrick M., *From Globalism to Regionalism: New Perspectives on US Foreign and Defense Policies*, Diane Publishing, 1993.

Dobriansky, Paula J. and Henry A. Crumpton, "Tyranny and Terror", *Foreign Affairs*, Jan./Feb. 2006.

Duffy, Helen, *The War on Terror and the Framework of International Law*, Cambridge University Press, 2005.

Esposito, John L., "Claiming the Center: Political Islam in Transition", *Harvard International Review.*

Frum, David and Richard Perle, *An End to Evil: How to Win the War on Terror*, Random House, 2003.

Fukuyama, Francis, *America at Crossroads: Democracy, Power and the Neoconservative Legacy*, Yale University Press, 2006.

Fuller, Graham E., *Future of Political Islam*, Palgrave Macmillan, 2004.

_____, *A World Without Islam*, Little, Brown & Company. 2010.

Gordon, Phillip H., "Can the War on Terror be Won", *Foreign Affairs*, November/December, 2007.

Gunaratna, Rohan, *Inside al-Qaeda: Global Network of Terror*, Berkley, (2003).

Hashmi, Sohail H., (ed.), *Islamic Political Ethics: Civil Society, Pluralism and Conflicts*, Princeton Oxford UK: Princeton University Press, 2002.

Herbert, Bob, "Masters of Deception", *New York Times*, Oct. 5, 2004.

Holsti, Kalevi Jaakko, *Taming the Sovereigns: Institutional Change in International Politics*, Cambridge University Press, 2004.

Juergensmeyer, Mark, *Global Rebellion: Religious Challenges to the Secular State, from Christian Militia to al-Qaeda*, Berkeley: University of California Press, 2008.

Juergensmeyer, Mark, *Terror in the Mind of God: The Global Rise of Religious Violence*, Oxford University Press, 2001.

Kagan, Robert and William Kristol, "The Right War for the Right Reasons", *Weekly Standard*, 9(23), Feb. 23, 2004.

Layne, Christopher, "The Unipolar Illusion Revisited: The Coming End of United States' Unipolar Moment", *International Security*, Vol. 1, No. 2, Fall 2006.

Leiken, Robert S. and Steven Brooke, "The Moderate Muslim Brotherhood", *Foreign Affairs*, March/April, 2007.

Living Islam Project, 2002.

Nafali, Timothy, *Blind Spot: The Secret History of American Counter-Terrorism*, Basic Books, 2005.

Nye Joseph S. Jr., *Bound to Lead: The Changing Nature of American Power*, New York: Basic Books, 1990.

Nye Joseph S. Jr., *The Paradox of American Power: Why the World's Only Superpower Can't Go It Alone*, New York: Oxford University Press, 2002.

Nye Joseph S. Jr., "Soft Power: The Means to Success in World Politics," New York: *Public Affairs*, 2004.

Parwez, G.A., "Holy Quran According to Our Traditions", G.A. Parwez, *The Status of Hadith*, translation by Aboo B. Rana, 2002.

Pew Survey, *American Muslims: Middle Class and Mostly Mainstream*, 2007.

Rabasa, Angel *et al*, *Ungoverned Territories: Understanding and Reducing Terrorism Risks*, RAND Project Air Force, 2007.

Ramdan, Tariq, *In the Footsteps of Prophet: Lessons from the Life of Muhammad*, Oxford, 2007.

Reuter's Report, "Saudi Terror Plot Foiled", *Hindustan Times*, 28 April, 2007, p. 12.

Risen James and Judith Miller, "No Illicit Arms Found in Iraq, US Inspectors Tell Congress", *New York Times,* Oct. 3, 2003.

Rodinson, Maxime, *Islam and Capitalism*, London: Saqi Books, 2007.

Sachedina, Abdlaziz: *Islamic Roots of Democratic Pluralism*, New York: Oxford University Press, 2001.

Sageman, Marc, *Understanding Terror Networks*, University of Pennsylvania Press, 2004

Schmid, Alex P. (ed.), *Countering Terrorism Through International Co-operation*, Rome: International Scientific and Professional Advisory Council of UN, 2001.

Shultrz, Richard H. Jr. and Andrea J. Dew, *Insurgents, Terrorists, and Militias*, Columbia University Press, 2006.

Takeyh, Ray, "Time for détente with Iran", *Foreign Affairs*, March-April, 2007.

Tibi, Bassam, "Post-Bipolar Order in Crisis: The Challenge of Politicized Islam", *Millennium Journal of International Studies*, 2000.

Tibi, Bassam, The *Challenge of Fundamentalism: Political Islam and the New World Disorder*, University of California, Berkeley, 1998.

Walt, Stephen M., *Taming American Power: The Global Response to US Primacy*, New York: W.W. Norton, 2005.

Waltz, Kenneth, "The Emerging Structure of International Politics," *International Security*, Vol. 18, No. 2 (Fall 1994), pp. 44-7.

Wright, Robin, "Islam and Liberal Democracy: Two Visions of Reformation," *Journal of Democracy* 7.2 (1996) 64-75.

News Papers/Journals

Encyclopedia of Afghan Jihad (10 volumes).

Chicago Tribune.

Country Reports on Terrorism, US Department of State, April 30, 2007.

Final Report of the National Commission on Terrorist Attacks Upon the United States, July 22, 2004.

(popularly known as The 9/11 Commission Report).

Foreign Affairs (Council on Foreign Relation).

Foreign Policy.

Harvard International Review, Jan.15, 2007 (Book Peview of: Fuller, Graham E., Future of Political Islam: 2004).

Jane's Terrorism Watch Report.

Journal for the Scientific Study of Religion.

Journal of Cold-War Studies.

New York Times.

RAND Report on Civil and Democratic Islam.

Stanford Journal of International Relations (SJIR).

Time (a magazine on partnership with CNN).

United States Institute of Peace, Special Report No.113, January 2004.

Washington Post.

Website of: *Terrorism Knowledge Base* (TKB) developed by the *Memorial Institute for the Prevention of Terrorism* (MIPT).

Bibliography

Abuza, Zachary, "Funding Terrorism in Southeast Asia: The Financial Network of al-Qaeda and Jemmah Islamiyah", in *National Bureau of Asian Research (NBR) Analysis,* Vol. 14, No. 5, December, 2003.

——, *Militant Islam in Southeast Asia: Crucible of Terror*, Lynne Rienner Publishers, 2003.

Ahmed, Akbar S., *Islam Under Siege*, New Delhi: Vistaar, 2003.

Ahmed, Imtiaz (ed.), *Understanding Terrorism in South Asia: Beyond Statist Discourses*, New Delhi: Manohar, 2006.

Alexander, Yonah (eds.) *Countering Terrorism: Strategies of Ten Countries*, Michigan: University of Michigan Press, 2002.

Almond, Gabriel A., R. Scott Appleby (eds.), *Strong Religion: The Rise of Fundamentalism Around the World*, Chicago and London: University of Chicago Press, 2003.

Al-Rasheed, Madawi, *Contesting the Saudi State: Islamic Voices From a New Generation*, Cambridge University Press, 2006, p. 332.

Ansari, Ali M., *Confronting Iran: The Failure of American Foreign Policy and the Next Great Crisis in the Middle-East*, 2006.

Antoun, Richard T., *Understanding Fundamentalism: Christian, Islamic and Jewish Movements*, New York, Oxford: Alta Mira Publishers, 2001.

Appleby, R. Scott, Gabriel Abraham Almond, and Emmanuel Sivan *Strong Religion*, Chicago: University of Chicago Press, 2003.

Bahgat, Gawdat, "Nuclear Proliferation: The Islamic Republic of Iran", in *Iranian Studies*, Vol. 39, No. 3, September 2006.

Bajpai, K. Shankar, "Understanding India and Pakistan" in *Foreign Affairs,* May/June 2003.

Baker, Anni P., *American Soldiers Overseas: The Global Military Presence*, Praeger, 2004.

Basevich, Andrew J., *American Empire: The Realities and Consequences of US Diplomacy*, Cambridge, Harvard University Press, 2003.

——, *The New American Militarism: How Americans Are Seduced By War*, New York & London, Oxford University Press, 2005.

Bayat, Asef, *Making Islam Democratic: Social Movements and the Post-Islamist Turn*, Stanford University Press, 2007.

Benard, Cheryl, *Civil Democratic Islam: Partners, Resources and Strategies*, Santa Monica, RAND, 2003.

Berger, Peter, *The Sacred Canopy*, New York: Doubleday, 1967.

Berke, Jason, *Al-Qaeda: Casting a Shadow of Terror*, NY: I B Tauris, 2003.

Bernard Lewis (1916->) Jewish English, prolific author, lately a modern partisan insider, *Arabs in History* (1950); *Muslim Discovery of Europe* (1982, 2001); *What went Wrong? The Clash between Islam and Modernity in the Middle-East* (2002).

Biddle, Stephan D., *American Grand Strategy after 9/11: An Assessment, A Monograph*, Strategic Studies Institute, 2005: website: www.carlisle.army.mil/ssi

Bjorgo, Tore, *Root Causes of Terrorism*, London and NY: Routledge, 2005.

Blum, William, *Killing Hope: US Military and CIA Interventions Since World War II.*

Boot, Max, *The Savage Wars of Peace: Small Wars and the Rise of American Power*, Basic Books, 2002.

Booth, Ken and Tim Dunne (eds.), *Worlds in Collision: Terror and Future of Global Order*, NY: Palgrave Macmillan, 2002.

Borum, R., *Psychology of Terrorism*, University of South Florida, 2004.

Bostom, Andrew (ed.), *Legacy of Jihad: Islamic Holy War and Fate of Non-Muslims*, Prometheus Boos, 2005.

Brands, H.W., *What America Owes to the World?: The Struggle for the Soul of Foreign Policy*, Cambridge University Press, 1998.

Brill, E.J., *Encyclopaedia of Islam* (*EI*), Brill Publishers, 1913-36; Reprinted, 1987 (it is the standard encyclopaedia of the academic discipline of Islamic Studies and not a Muslim or an Islamic encyclopedia).

Briody, Dan, *The Halliburton Agenda: The Politics of Oil and Money*, May 2004.

Brown, L. Carl, *Religion and State: The Muslim Approach to Politics*, Columbia University Press, 2000.

Bruce, Lawrance, *Shattering the Myth: Islam Beyond Violence*, Princeton: Princeton University Press, 1998, 2000.

Bruce, Steve, *God is Dead: Secularism in the West*, Blackwell: Oxford, 2002.

Brzezinski, Zbigniew, *The Grand Chessboard: American Primacy and Its Geostrategic Imperatives*, New York: Basic Books, 1997.

_____, *Second Chance: Three Presidents and the Crisis of American Superpower*, New York, Basic Books.

Bull, Hedley, *The Anarchical Society: A Study of Order in World Politics*, New York: Columbia University Press, 1977 [reviewed in 1999, 2nd edition, forward by Stanley Hoffman].

Bunt, Gary R., *Islam in the Digital Age: E-Jihad, On-Line Fatwas and Cyber Islamic Environments*, London: Pluto, 2003.

Burleigh, Michael, *Sacred Causes: The Clash of Religion and Politics, From the Great War to the War on Terror*, HarperCollins, 2007, p. 576.

Byman, Daniel, *Deadly Connections: States That Sponsor Terrorism*, New York: Cambridge University Press, 2005.

_____, "Iran, Terrorism and Weapons of Mass Destruction", remarks before the Sub-Committee on Nuclear and Biological Attacks on Homeland, September 8, 2005.

Byrne, Malcolm and Jeffrey Richelson eds., "Terrorism and U.S. Policy, 1968-2002: From the Dawn of Modern Terrorism to the Hunt for bin-Laden" *Chadwyck-Healey/Proquest*, 2002.

Cameron, Fraser, *US Foreign Policy After the Cold-War: An Introduction*, UK: Routledge, 2002.

Carbaugh, John E. Jr., "Pakistan-North-Korea Connections Creates Dilemma for US", Pakistan-Facts.Com, updated April 2004.

Cassidy, Robert M., *Counterinsurgency and the Global War on Terror: Military Culture and Irregular War*, Praeger Security International, 2006.

Chaliand, Gerard, Arnaud Blin, *The History of Terrorism: From Antiquity to al-Qaeda*, University of California Press, 2007.

Chellany, Brahma, "Fighting Terrorism in Southern Asia", *International Security*, Vol. 26, No. 3 (Winter 2001/02), pp. 94-116.

Chomsky, Noam, *Rogue States: The Role of Force in World Affairs*, New Delhi: Indian Research Press, 2000.

Chossudovsky, Michel, War and Globalisation: The Truth behind September 11, *Global Outlook and Center for Globalisation (CRG)*, 2002.

Chung, Chien-Peng, "China's War on Terror: September 11 and Uighur Separatism" in *Foreign Affairs*, July/August, 2002.

Clarke, William R., *Petrodollar Warfare: Oil, Iraq and the Future of the Dollar*, New Society Publishers, 2005.

Coakley, John (eds.), *The Territorial Management of Ethnic Conflicts*, London: Frank & Cass, 2003.

Cogan, Charles, "Partners in Time: The CIA and Afghanistan since 1979," *World Policy Journal*, Summer 1993.

Cohen, Marjorie, *The Deadly Pipeline War: US Afghan Policy Driven by Oil Interests.*

Cohen, Stephen Phillip, "Pakistan's South Asia Strategy", in *India Abroad*, May 11, 2007.

Cohen, Stephen Phillip, *The Idea of Pakistan*, Brooking Institute Press, 2004.

Cole, Juan, *Sacred Space and Holy War: The Politics, Culture and History of Shiite Islam*, London and New York: I.B. Tauris, 2002.

Cole, N. Scott, "Hugo Chavez and President Bush's Credibility Gap: The Struggle Against US Democracy Promotion" in *International Political Science Review*, 28:4, September, 2007.

Conry, Barabara, "Loose Cannon: The National Endowment for Democracy", in *Cato Foreign Policy Briefing No. 27,* November 8, 1993.

Cooley, John K., *Unholy Wars: Afghanistan, America and International Terrorism*, ND: Penguin, 2001.

Cooper, John, *et al.*, (eds.), *Islam and Modernity: Muslim Intellectuals Respond*, London: I.B. Tauris, 2000.

Country Profile: Saudi Arabia, Federal Research Division, Library of Congress, September 2006.

Crenshaw, Martha, *Terrorism in Context*, Pennsylvania, PSU Press, 1995.

____, "Counter-Terrorism in Retrospect" in *Foreign Affairs*, July/August, 2005.

CRF paper, "Strengthening the US-Egyptian Relations", A *CRF paper* (Council on Foreign Relations), May, 2002.

Cronin, A.K., "Behind the Curve: Globalization and International Terrorism", *International Security*, 27 (3): 30-58, 2002.

Cronin, Patrick M., *From Globalism to Regionalism: New Perspectives on US Foreign and Defense Policies*, Diane Publishers, 1994.

Crotty, William (ed.), *Democratic Development and Political Terrorism: The Global Perspective*, Boston: NE University, 2005.

Daalders, Ivo H., James M. Lindsay, *America Unbound: The Bush Revolution in Foreign Policy*, Brookings Institution Press, 2003.

Daniel, Benjamin, *The Age of Sacred Terror*, NY: Random House, 2002.

Dershowitz, Alan M., *Why Terrorism Works: Understanding the Threat, Responding to Challenge?*, New Haven: Yale University Press, 2005.

Dixon, A.C., *The Fundamentals: A Testimony To The Truth,* Bible Institute of Los Angeles, 1910-15.

Dobriansky, Paula J. and Henry A. Crumpton, "Tyranny and Terror" in *Foreign Affairs*, Jan./Feb. 2006.

Dobson, Alan P. and Marsh, Steve, *US Foreign Policy Since 1945*. 1. Auflage, Routledge: 2000, 2006.

Dolan, Chris J. and Betty Glad (eds.), *Striking First: The Preventive War Doctrine and the Reshaping of US Foreign Policy*, New York & London, Palgrave Macmillan, 2004.

Dolan, Chris J., *In War We Trust: The Bush Doctrine and the Pursuit of Just War*, Burlington, VA, Ashgate, 2005.

Domke, David, *God Willing? Political Fundamentalism in the White House, the "War on Terror," and the Echoing Press*, Pluto Press, 2004.

Duffy, Helen, *The War on Terror and the Framework of International Law*, Cambridge University Press, 2005.

Ehrehfeld, Rachel, *Funding Evil: How Terrorism is Financed and How to Stop It*, Chicago: Bonus Books, 2003.

____, "Funding Terrorism: Sources and Methods", NY*: Centre for the Study of Corruption*, 2002.

____, *Funding Evil: How Terrorism is Financed and How to Stop It*, Chicago: Bonus Books, 2003.

Elan, Evan, *The Empire Has No Clothes: US Foreign Policy Exposed*, The Independent Institute, 2004.

Engineer, Asghar Ali, "Islam, Globalization and Fundamentalisms", in *Islam and Modern Age*, August, 2002.

Esposito, John L., "Claiming the Center: Political Islam in Transition, in *Harvard International Review* (criticizing Esposito as an apologist, Martin Kramer writes that Esposito "would have remained obscure even by the standards of Middle-Eastern studies" if it were not for Edward Said. It was Said's successful 1978 book *Orientalism* that spurred a large demand "for sympathetic texts on Islam" that were "uncontaminated by anti-Americanisms" and "preferably even written by an American." Esposito's rise, Kramer writes, can be solely attributed to his ability to meet this demand. In 2002, Campus Watch, an American think tank, described Esposito "as one of the foremost apologists for Islamism in recent years".)

Falk, Richard, *The Great Terror War*, Northampton, Mass.: Olive Branch Press, 2002.

Feldman, Noah, *After Jihad: America and the Struggle for Islamic Democracy*, NY: Strauss, 2003.

Freedman, Lawrence, *A Choice of Enemies: America Confronts the Middle-East, Public Affairs*, 2008.

Fukuyama, Francis, *America at the Cross Roads: Democracy, Power and Neo-Conservative Legacy*, 2006.

Fukuyama, Francis, *End of History and the Last Man*, New York; Free Press, 1992.

Fukuyama, Francis, *End of History and the Last Man*, NY; Simon & Schuster, 1996.

____, *State Building*, London: Profile Books, 2004.

Fuller, Chris, *Renewal of Priesthood*, Princeton: Princeton University Press, 2003.

Fuller, Graham E., *The Future of Political Islam*, NY, Palgrave Macmillan, 2006.

Gaddis, John Lewis, *The US and the Origins of the Cold War, 1941-47*, 1972.

——, *Surprise, Security and the American Experience*, 2004.

Ganguly, Rajat and Mac duff Ian, *Ethnic Conflict and Secessionism in South and Southeast Asia*, New Delhi: Sage, 2003.

——, "Israel, Conflict and Peace", *Israel Ministry of Foreign Affairs*, 5 November 2003.

Ganguly, Sumit, *Conflict Unending: India-Pakistan Tensions since 1947*, New York: Columbia University Press, 2001.

Gendier, Irena, "Who Rules the Middle-East Agenda?" in Crotty, Williams (ed.), *op. cit.*, 2005.

Gill, K.P.S. and Ajai Sahni, *The Global Threat of Terror: Ideological, Material and Political Linkages,* Delhi: The Institute for Conflict Management, 2002.

Gordon, Philip H., *Winning the Right War: The Path to Security for America and the World*, Brookings Institution Press, 2007.

Gordon, Michael and General Bernard E. Trainer, *The Generals' War: the Inside Story of the Conflict in the Gulf*, Little, Brown, 1995.

——, *Cobra II: The Inside Story of the Invasion and Occupation of Iraq*, New York: Knopf, 2006.

Greenspan, Alan, *The Age of Turbulence: Adventures in a New World*, Penguin Press, 2007.

Griest, P. and S. Mahan, *Terrorism in Perspective*, London: Sage, 2003.

Griffin, David Ray, *The New Pearl Harbor: Disturbing Questions About Bush Administration and 9/11*, Massachusetts: OliveBranch Press, 2004.

Griset, Pamela L. and Sue Mahan (eds.), *Terrorism in Perspective*, Thousand Oaks, C.A.: Sage, 2003.

Gunaratana, Rohan, *Inside al-Qaeda: Global Network of Terror*, New Delhi: Roli Books, 2002.

Haass, Richard N., *The Opportunity: America's Moment to Alter History's Course,* 2006.

Hahn, Peter L., *Crisis and Crossfire: The United States and the Middle-East Since 1945*, 2005.

Halliday, Fred, *Two Hours That Shook the World: September 11, 2002: Causes and Consequences*, London: Saqi Books, 2002.

Hanson, Eric O., *Religion and Politics in the International System Today*, Cambridge University Press, 2006.

Harmon, Christopher C., *Terrorism Today*, London: Frank & Cass, 2000.

Harrison, Selig S., "Will Pakistan Break Up?" Remarks by Selig S.,

Director, Asia Program, *Center for International Policy, Carnegie Endowment for International Peace,* June 09, 2009.

____, "A New Hub for Terrorism", in *Washington Post,* August 2, 2006.

Hashmi, Sohail H., (ed.), *Islamic Political Ethics: Civil Society, Pluralism and Conflicts,* Princeton and Oxford UK: Princeton University Press, 2002.

Herbert, Bob, "Masters of Deception", *New York Times,* October 5, 2004.

Herring, George, *America's Longest War* (4th edition), New York, 2001.

Hilladey, Fred, *Islam and the Myth of Confrontation,* NY: I.B. Tauris, 2003.

____, *100 Myths About the Middle-East,* University of California Press, 2005.

Hinde, Robert A., *Why Gods Persist: A Scientific Approach to Religion,* Routledge, 1999.

Hiro, Dilip, *War Without End: The Rise of Islamic Terrorism and Global Response,* Delhi: Roli Books, 2002.

Hodgson, Godfrey, *More Equal than Others: America from Nixon to the new Century,* Princeton University Press, 2004.

Hoffman, Bruce, *Inside Terrorism,* Columbia University Press, 1999.

Huntington, Samuel P., *The Clash of Civilizations and the Remaking of the World Order,* Simon and Schuster, 1996.

Huntington, Samuel P., *The Clash of Civilizations and the Remaking of the World Order,* Simon and Schuster, 1996.

Husain, Mir Zohair, *Global Islamic Politic,* New York: Harper Collins College Publishers, 1995.

Hussain, Zahid, *Frontline Pakistan: The Struggle with Militant Islamism,* New York: Columbia University Press, 2007.

Hutchison, John and Anthony D. Smith, *Ethnicity,* New York: Oxford Readings, 1996.

Ikenberry, John (ed.), *America Unrivaled: The Future Balance of Power,* NY: Cornell University Press, 2002.

Jaco, Charles D. *et al., The Complete Idiot's Guide to the Politics of Oil,* Alpha Books, 2003.

Jarisha 'Ali M. and Muhammad Sh. Zaibaq, *Methods of Intellectual Invasion of the Islamic World,* second printing (Cairo: Dar al-I'tisam), 1978.

Jervis, R., "Why The Bush Doctrine Cannot be Sustained" in *Political Science Quarterly,* 120 (3), 2005.

Jha, Prem Shankar, *The Twilight of the Nation State: Globalization, Chaos and War,* London: Pluto Press, 2006.

Joffe, Josef, *Uberpower: The Imperial Temptation of America,* New York and London: Norton, 2006.

John L. Esposito (1940-) U.S.A., *Islam. The Straight Path* (Oxford, 1988); editor-in-chief, *Oxford Encyclopedia of the Modern Islamic World* (4 volumes, 1995); *Islam and Civil Society* (European Univ. Inst., 2000).

Johnson, Charmer, *Blowback: The Costs and Consequences of American Empire*, New York: Metropolitan Books, 2002, 2005.

____, *Nemesis: The Last Days of American Republic* (reprint): Metropolitan, 2006, 2007.

____, *The Sorrows of Empire: Militarism, Secrecy and the End of the Republic*, Metropolitan Books, 2004.

Juergensmeyer, Marc, *Terror in the Mind of God: The Global Rise of Religious Violence*, University of California Press.

____, *The New Cold-War?: Religious Nationalism Confronts Secular State*, University of California Press, 1994.

Kabbani, Muhhammad Hisham, "Islamic Extremism: A Viable Threat to US National Security", at Open Forum at the US Department of State, January 7, 1999.

Kagan, Robert, *The Return of History and the End of Dreams*, Knopf, 2008.

Kagan, Robert and William Kristol, "The Right War for the Right Reasons", *Weekly Standard*, 9(23), February 23, 2004.

Kaplan, Esther, *With God On Their Side: How Christian Fundamentalists Trampled Science, Policy and Democracy in George Bush's White House*, NY: Free Press, 2004.

Kaplan, Lawrance F., William Kristol, *The War Over Iraq*, San Francisco: Encounter Books, 2003.

Kaplan, Robert D., *Warrior Politics: Why Leadership Demands a Pagan Ethos*, New York: Random House, 2001.

Kepel, Gilles, *Jihad: The Trail of Political Islam*, London: I.B. Tauris, 2000.

____, *Revenge of God: Resurgence of Islam, Christianity and Judaism in the Modern World*, Cambridge: Polity: 1994.

Khatri, Sridhar K. and Kueck G., *Terrorism in South Asia: Impact on Development and Democratic Process*, Delhi: Shipra, 2003.

Kissinger, Henry A., *Does America Need a Foreign Policy? : Toward A Diplomacy for 21st Century*, Simon & Schuster, 2002.

Klare, Michael T., *Blood and Oil: The Dangers and Consequences of America's Growing Dependency on Imported Petroleum*, Henry Holt, 2004.

Kohan, Hans, *The Idea of Nationalism: A Study in its Origin and Background*, New Brunswick: Transaction Publishers, 2005 (New edition with an introduction by Craig Calhoun to the original 1944 edition).

Kohut, Andrew et al, *The Diminishing Divide: Religion's Changing Role in American Politics*, Washington D.C.: Brookings Institution Press, 2000.

Krauthammer, Charles, "Democratic Realism: An American Foreign Policy for a Unipolar World", 2004 Irving Kristol Lecture, *American Enterprise Institute (AEI)*, Feb. 10, 2004.

Kronstadt, K. Alan, "International Terrorism in South Asia", *CRS Report for Congress*, November 3, 2003.

Krueger, Alan B. & Jitka Maleckova, "The Economics and the Education of Suicide Bombers: Does Poverty Cause Terrorism?", *The New Republic Online*, 24.06.2002.

Kupchan, Charles and Joanne J. Myers, *The End of the American Era: U.S. Foreign Policy and the Geo-politics of 21st Century*, Carnegie Council, 2003.

Kux, Dennis, *The United States and Pakistan, 1947-2000: Disenchanted Allies*, Washington: Woodrow Wilson Center Press, 2001.

Kux, Dennis, *The United States and Pakistan: Disenchanted Allies*, Woodrow Wilson Press, Washington D.C., 2000.

Lapidus, Ira M., *A History of Islamic Societies*, Cambridge University Press, 2002.

____, *Islam, Politics and Social Movements*, 1988.

Laquer, Walter, *No End to War: Terrorism in 21st Century*, 2003.

____, *A History of Zionism*, NY: Schocken Books, 2003.

Larsson, J.P., *Understanding Religious Violence: A New Framework for Conflict*, Ashgate Publishers, 2004.

Lawrance, Bruce B., *Shattering the Myth: Islam Beyond Violence*, Princeton University Press, 2000.

Leiken, Robert S. and Steven Brooke, "The Moderate Muslim Brotherhood", in *Foreign Affairs*, March/April, 2007.

Levering, Ralph B., *The Cold War: A Post-Cold War History*, 2nd edition, 2005.

Lewis Bernard, *What Went Wrong?: Western Impact and Middle-Eastern Response*, 2002.

____, *The Crisis of Islam: The Holy War and Unholy Terror*, NY; Modern Library, 2003.

Libschutz, Ronnie D., *After Authority: War Peace and Global Politics in 21st Century*, New York: Suny Press.

Lincoln, Bruce, *Holy Terrors: Thinking About Religion After September 11*, Chicago: Chicago University Press, 2003.

Lining Islam, a Project of Independent Broadcasting Associates Inc. (IBA), USA, 2002 (Funded by Ford Foundation).

Lippman, Walter, *The Cold War: A Study in US Foreign Policy* (1947).

Mackay, Neil, "Bush Planned Iraq 'Regime Change' Before Becoming President", *Sunday Herald* (Scotland), September 15, 2002.

Mage, John, "The Nepali Revolution and International Relations" in *Economic and Political Weekly,* May 19, 2007.

Maley, William ed., *Fundamentalism Reborn?: Afghanistan and the Taliban*, New York University Press: New York, 1998.

Maley, William, *The Afghanistan Wars*, London and NY: Palgrave, 2002.

Mamdani, Mahmood, *Good Muslim, Bad Muslim: America, The Cold War and the Roots of Terror*, Pantheon Books, 2005.

Mamdani, Mahmood, *Good Muslim, Bad Muslim: America, The Cold-War and the Roots of Terror*, New York: Pantheon, 2004.

Manbiot, George, "War on the Third World", *The Guardian*, March 5, 2002.

Mani, Rama, "In pursuit of an Antidote: The Response to September 11 and the Rule of Law", an Opinion in *Conflict, Security and Development,* 3:1, April, 2003.

Manyin, Mark, "Terrorism in Southeast Asia", in *CRS Report for Congress*, 27th Sept, 2005 (updated).

Maria, C., *Terrorism and Game Theory : Coalitions, Negotiations and Audience Costs*, Ireland: University of Limerick, 2003.

Markey, Daniel, "A False Choice in Pakistan", in *Foreign Affairs*, July/ August, 2007.

Martin, C., The Nature of the Beast: Defining Terrorism, California State University, Sage: 2003.

Marty, Martin E. and R. Scott Appleyby (eds.), *The Fundamentalism Project* (Five Volumes), University of Chicago, 1991-95.

Mastanduno, Michael, "Preserving the Unipolar Moment: Realist Theories and U.S. Grand Strategy after the Cold War", *International Security*, Vol. 21, No. 4 (Spring, 1997), pp. 49-88.

McCoy, Alfred W., *The Politics of Heroin in Southeast Asia. CIA Complicity in the Global Drug Trade*, 1972.

Mead, Walter Russell, *Mortal Splendor: The American Empire in Transition*, Houghton Miffin, 1988.

Mearsheimer John J., Stephan M. Walt, *The Israel Lobby and US Foreign Policy,* Farrar, Straus, & Giroux, 2007.

Mearsheimer, John J., *The Tragedy of Great Power Politics*, NY: W.W. Norton & Co., 2001.

Mehta, Bhanu Pratap, *The Burden of Democracy*, New Delhi: Penguin India, 2003.

Menotti, Victor, The Other Oil War: Halliburton's Agenda at the WTO; A Policy Brief on the Energy Services Negotiations in the World Trade Organization (WTO), June 2006.

Miller, Greg, "Influx of al-Qaeda, money into Pakistan is seen", in *Los Angeles Times*, May 20, 2007.

Mondal, Anshuman A., "Liberal Islam?", *Prospect Magazine*, January 2000.

Morgenthau, Hans J., *Politics Among Nations: The Struggle for Power and Peace*, Fifth Edition, Revised, New York: Alfred A. Knopf, 1978.

Muni, S.D. (ed.), *Responding to Terrorism in South Asia*, Delhi, Manohar, 2006.

Munthe, P., "Terrorism: Not Who But Why?", *Royal United Service Institute Journal*, August, 2005: 8-12.

Musharraf, Pervez, *In the Line of Fire: A Memoir*, Free Press, 2006.

Nafali, Timothy, *Blind Spot: The Secret History of American Counter-Terrorism*, Basic Books, 2005.

Nassar, Jamal, *Globalization and Terrorism*, New York: Rowman & Littlefield, 2005.

Neack, Laura, *The New Foreign Policy: US and Comparative Foreign Policy in 21st Century*, NY: Rowman & Littlefield, 2002.

Norris, Pippa and Ronald Inglehart, *Sacred and Secular: Religion and Politics Worldwide*, Cambridge: Cambridge University Press, 2004.

Nye, Joseph Jr., *The Paradox of American Power: Why the World's Superpower Can't Go It Alone.*

____, "Soft Power: The Means to Success in World Politics," *Public Affairs*, 2004.

O'Donnell, Tom, *Global Political Economy of Oil: US Policy in the Persian Gulf.*

Parwez, G.A., "Holy Quran According to Our Traditions" in *The Status of Hadith*, translation by Aboo B. Rana, 2002.

Peters, John E., *et. al*, *War and Escalation in South Asia*, Santa Monica: Rand, 2006.

Phadnis, Urmila, *Ethnicity and Nation Building in South Asia*, New Delhi: Sage, 2001.

Phares, Walid, *Future Jihad: Terrorist Strategies Against the West*, 2006.

Phillips, Kevin, *American Theocracy*, NY: Viking, 2006.

Phillips, Kevin, *American Theocracy: The Peril and Politics of Radical Religion, Oil and Borrowed Money in 21st Century*, New York: Viking, 2006.

Pipes, Daniel, *In the Path of God: Islam and Political Power*, 1983.

____, *The Long Shadow: Culture and Politics in the Middle-East*, 1990.

____, *Militant Islam Reaches America*, Norton, 2002.

Post, Jerrold M., "Addressing the Causes of Terrorism", *Madrid Agenda, Club de Madrid Series on Democracy and Terrorism*, Vol. 1, March 2005.

Prados, John, "Notes on the CIA's Secret War in Afghanistan" in *The Journal of American History*, Vol. 89, No. 2, Sept. 2002.

Prince of Wales, *Islam and the West: a lecture given in the Sheldonian Theatre, Oxford on 27 October 1993* (Oxford: Oxford Centre for Islamic Studies, 1993).

Pryce-Jones, David, *The Closed Circle: An Interpretation of the Arabs*, 1989.

Rabasa Angel M. *et. al*, *Muslim World After 9/11*, Rand Project Air Force, 2004.

Ramadan, Tariq, *In the Footsteps of the Prophet: Lessons From the Life of Muhammad*, Oxford University Press, 2007, p. 242.

Raman, B., "Pakistan and Terrorism", a Paper by *South Asia Analysis Group*, January 2, 2000.

Rashid, Ahmed, *Jihad: The Rise of Militant Islam in Central Asia*, Yale University Press, 2002.

_____, *Taliban: Militant Islam, Oil and Fundamentalism in Central Asia*, Yale University Press, 2000.

_____, *Al-Qaeda in 2007: Striving to Regain the Initiative*, 2006.

_____, *Taliban: The Story of Afghan Warlords*, London: Pan Books, 2001.

_____, *Taliban: Islam, Oil and the New Great Game in Central Asia*, I.B.Tauris, 2002.

Ray, Ashwini Kumar, *Western Realism and International Relations: A Non-western View*, Foundation Books, 2004.

Reuter's Report, "Saudi Terror Plot Foiled" in *Hindustan Times*, 28 April, 2007, p.12.

Richard E. Rubenstein, U.S.A., professor of conflict resolution, *Alchemists of Revolution. Terrorists in the Modern World* (New York, 1987); *Aristotle's Children. How Christians, Muslims, & Jews rediscovered ancient wisdom & illuminated the Dark Ages* (Orlando, 2003).

Riedel, Bruce and Karl Inderfurth, "NATO Must Do More in Afghanistan", in *International Herald Tribune*, February 5, 2007.

Riedel, Bruce, "Al-Qaeda Strikes Back" in *Foreign Affairs*, May-June 2007.

Risen, James, *State of War: The Secret History of the CIA and the Bush Administration*, New York: Simon & Schuster (Free Press imprint), 2006.

Risen James and Judith Miller, "No Illicit Arms Found in Iraq, US Inspectors Tell Congress", in *New York Times*, October 3, 2003.

Rosenberg, J., *Globalization Theory: A Post Mortem*, UK: Pagrave Macmillan, 2005.

Roy, Olivier, *The Failure of Political Islam*, Cambridge, Harvard University Press, 1994, 2001.

Roy, Olivier, *Globalized Islam: The Search for a New Ummah*, CERI Series in Comparative Politics and International Studies, 2004.

Rubin, Berry and Judith Rubin, *Anti-American Terrorism and the Middle-East*, NY: Oxford, 2002.

Sachedina, Abdulaziz, *The Islamic Roots of Democratic Pluralism*, Oxford University Press, 2001.

Sageman, Marc, *Understanding Terror Networks*. Philadelphia: University of Pennsylvania Press, 2004.

Saha, Santosh C. (ed.). *Religious Fundamentalism in the Contemporary World: Critical Social and Political Issues*, New York: Lexington Books, 2004.

Sahni, Ajai, "The Dynamics of Islamist Terror in South Asia", *The Journal of International Security Affairs*, Fall, 2005, No. 5.

Said Aburish, *A Brutal Friendship: The West and the Arab Elite*, London: Indigo, 1998.

Said, Edward E., *Orientalism*, New York: Random House, 1978.

_____, "Islam Through Western Eyes" in *The Nation*, 28th April, 1980 issue or visit: http://www.thenation.com/19800426/19800426said

_____, *Covering Islam: How the Media and the Experts Determine How We See the Rest of the World*, New York: Pantheon; London: Routledge & Kegan Paul, 1981.

_____, *End of Peace Process*, NY: Pantheon Books, 2000.

Schacht, Joseph, *An Introduction to Islamic Law*, London, OUP, 1964.

Schmid, Alex P. (ed.), *Countering Terrorism Through International Co-operation*, Rome: International Scientific and Professional Advisory Council of UN, 2001.

Scott, Peter Dale, *Drugs, Oil, and War: The United States in Afghanistan, Columbia and Indochina*, Rowman & Littlefield, 2003.

Sen, Amartya, *Identity and Violence: The Illusion of Destiny*, W.W. Norton & Company, Inc, 2006.

Shadid, Anthony, "Reconstruction: For the Sufis, Taliban's Fall Means a Revival", in *Boston Globe*, January 23, 2002.

Shaffer, Brenda (ed.), *Limits of Culture: Islam and Foreign Policy*, MIT Press, 2006.

Shapiro, Ian, *Containment: Rebuilding a Strategy Against Global Terror*, Princeton: Princeton University Press, 2007.

Shultrz, Richard H. Jr. and Andrea J. Dew, *Insurgents, Terrorists, and Militias*, Columbia University Press, 2006.

Skaine, Rosemarie, *The Women of Afghanistan Under Taliban*, London: McFarland & Co., 2002.

Smith, Grant F., *Deadly Dogma*, Washington, DC, Institute for Research: Middle-Eastern Policy, 2006.

Sood, Vikram, "Uneasy lies the Head" in *Hindustan Times*, May 28, 2007.

Sperry, Paul E., *Crude Politics: How Bush's Oil Cronies Hijacked the War on Terrorism*, Thomas Nelson: 2003.

Staten, Clifford, *US Foreign Policy since World War II: An Essay on Reality's Corrective Qualities.*

Stern, Jessca, "Pakistan's Jihad Culture" in *Foreign Affairs*, Nov./Dec. 2000.

Stern, Jessica, *Terror in the Name of God*, Harper Collins, 2003.

Stern, Jessica, *The Ultimate Terrorists*, Cambridge, Massachusetts: Harvard University Press, 2001.

Stiglitz, Joseph E. and Linda J. Bilmes, *The Three Trillion Dollars War: The True Cost of the Iraq Conflict*, W.W. Norton, 2008.

Suskind, Ron, *The Price of Loyalty*, NY: Simon & Schuster, 2004.

Sutton, Fillip W., Stephan Vertigaus, *Resurgent Islam: A Sociological Approach*, Blackwell, 2005.

Takeyh, Ray, "Time for détente with Iran", in *Foreign Affairs*, March-April, 2007.

Takeyh, Ray, Hidden Iran: Paradox and Power in the Islamic Republic.

Tanzer, Michael, *The Political Economy of International Oil and the Underdeveloped Countries*, Boston: Beacon, 1969.

Telhami, Shibley, *The Stakes: America and the Middle-East*, Cambridge: Westview Press, 2002.

Tellis, Ashley J., *India as a New Global Power: An Action Agenda for the United States*, Washington, D.C., Carnegie Endowment for International Peace, 2005.

Tellis, Ashley J., *South Asian Seesaw: A New US Policy on the Subcontinent*, Washington, D.C., Carnegie Endowment for International Peace, May 2005.

Thundyil, Matt, "Pakistan is Not an Ally Against Terrorism", Pakistan-Facts.Com, updated July 2004.

Tibi, Bassam, *The Challenge of Fundamentalism: Political Islam and the New World Disorder*, Berkeley, University of California Press, 1998, updated edition, 2002.

——, *Islam Between Cultures and Politics*, Palgrave, 2005.

Upadhyay, R., "Islamic Institutions in India: Protracte Movement for Separate Muslim Identity", *South Asia Analysis Group,* Paper No. 599, February 2003.

Valentino, B.A., *Final Solution: Mass Killing and Genocide in 20th Century*, Ithaca: Cornell, 2004.

Varshney, Ashutosh, *Ethnic Conflict and Civic Life: Hindus and Muslims in India*, New Haven, Yale University Press, 2003.

Vicziany, Marika, "State Responses to Islamic Terrorism in Western China: Impact on South Asia" in *Contemporary South Asia*, Vol. 12, issue 2, June 2003, p. 243.

Walt, S., *Taming American Power: The Global Response to US Primacy*, New York: Norton, 2005.

Watson, Peter, *Modern Mind: An Intellectual History of the 20th Century*, 2001.

Weinbaum, Marvin, *Pakistan and Afghanistan: Resistance and Reconstruction*, Westview Press, Boulder, Colorado, 1994.

Wessels, Antonie, *Muslims and the West: Can They Be Integrated?* : Peeters Publishers, 2006 (Translated by John Bowden).

West, Cornel, *Democracy Matters: Winning the Fight Against Imperialism*, New York: Penguin, 2004.

Wilkinson, Paul, *Terrorism vs. Democracy: The Liberal State Response*, London: Frank and Cass, 2001.

Woodward, Bob, *Plan of Attack: The Definitive Account of the Decision to Invade Iraq*, Simon and Schuster, 2004.

Wordsmith Compilation: *The Jihad Fixation: Agenda, Strategy, Portents*, Delhi: Wordsmith Compilation, 2001.

Wright, Steven, *The United States and Persian Gulf Security: The Foundations of the War on Terror*, Ithaca Press, 2007.

Yousaf, Mohammed, *Afghanistan the Bear Trap: The Defeat of a Superpower*, Casemate, 2001.

Zakaria, Farid, "Vengeance of Victors", in *Newsweek*, January 8, 2007.

_____, *The Post-American World*, W.W. Norton and Company, Inc., 2008.

News Papers/Journals

Chicago Tribune.

Country Reports on Terrorism, US Department of State, April 30, 2007.

Encyclopedia of Afghan Jihad, (10 volumes).

Final Report of the National Commission on Terrorist Attacks Upon the United States, July 22, 2004. (popularly known as The 9/11 Commission Report).

Foreign Affairs (Council on Foreign Relation).

Foreign Policy.

Harvard International Review.

Jane's Terrorism Watch Report.

Journal for the Scientific Study of Religion.

Journal of Cold-War Studies.

Muslim World (Blackwell Synergy).

National Bureau of Asian Research (NBR) Analysis (published 5 times annually on current issues), Seattle, Washington, USA.
New York Times.
Stanford Journal of International Relations (SJIR).
Time (a magazine in partnership with CNN).
United States Institute of Peace, Special Report No.113, January 2004.
Washington Post.
Website of: Terrorism Knowledge Base (TKB) developed by the Memorial Institute for the Prevention of Terrorism (MIPT).
Weekly Standard (considered to be a journal of neoconservatives in US).
www.crisisgroup.org

Indian News Sources

Asian Age (New Delhi).
Economic Times (Mumbai).
Frontline.
Hindu.
Hindustan Times (New Delhi).
Indian Express.
Times of India (New Delhi).
Pioneer (New Delhi).

Indian Journals

Economic and Political Weekly (Mumbai).
Faultlines (Institute of Conflict Management, ICM, New Delhi).
Foreign Affairs Online.
International Relations Resource Center.
MIPT Terrorism Knowledge Base.
South Asia Terrorism Portal (SATP), ICM, New Delhi.
South Asia Intelligence Review, ICM, New Delhi.
Strategic Analysis (Institute for Defense Studies and Analysis, IDSA, New Delhi).
Strategic Digest (IDSA, New Delhi).
Terrorism Research Center.

Additional Notes

Failed States: The *Foreign Policy and Fund for Peace* issued a The Failed States Index in July/August 2005, according to whose findings, about 2 billion people live in countries that are in danger of collapse. Failed states have made a remarkable odyssey from the periphery to the very center of global politics. During the Cold War days, state failure was seen through the prism of superpower conflict and was rarely addressed as a danger in its own right. In the 1990s, "failed states" fell largely into the province of humanitarians and human rights activists and became central issues of concern for the analysis of geopolitics. The dangerous exports of failed states—whether international terrorists, drug barons, or weapons arsenals—are the subject of endless discussion and concern.

The factors such as: a government's loss of control of its territory or of the monopoly on the legitimate use of force, lack of authority to make collective decisions or the capacity to deliver public services, existence of a flourishing black market, failure to pay taxes, civil disobedience, outside intervention the presence of foreign military forces on its soil, or other military constraints—can also be the determinants of state failure. How many states are at serious risk of state failure? The World Bank has identified about 30 "low-income countries under stress," whereas Britain's Department for International Development has named 46 "fragile" states of concern. A report commissioned by the CIA has put the number of failing states at about 20.

About 2 billion people live in insecure states, with varying degrees of vulnerability to widespread civil conflict. While Somalia and Democratic Republic of Congo count for loss of governmental control over their territory, Philippines and Russia suffer from conflicts for autonomy and secession. Instability takes the form of episodic fighting, drug mafias, or warlords dominating large swaths of territory as in Afghanistan, Colombia, and Somalia. State collapse sometimes happens suddenly, but often the demise of the state is a slow and steady deterioration of social and political institutions such as in Zimbabwe and Guinea. Some countries emerging from conflict may be on the mend but in danger of backsliding like Sierra Leone and Angola. The World Bank found that, within five years, half of all countries emerging from civil unrest fall back into conflict in a cycle of collapse like Haiti and Liberia.

What was believed in 1970s—calling the Middle-East as a "Muslim Crescent" extending from Afghanistan to the "Stans" in the southern part of the former Soviet Union—is no more valid as the 'instability' factor extends from Moscow to Mexico City, far wider than the Muslim Crescent, not limited to the Muslim world only.

Israel Lobby: The Israel-Lobby is the convenient short-hand term for the loose coalition of individuals and organizations who actively work to shape U.S. foreign policy in a pro-Israel direction. The Lobby's membership includes neoconservative gentiles such as John Bolton, the late Wall Street Journal editor Robert Bartley, former Secretary of Education William Bennett, former U.N. Ambassador Jeane Kirkpatrick, and columnist George Will. Over the past 25 years, pro-Israel forces have established a commanding presence at the American Enterprise Institute, the Brookings Institution, the Center for Security Policy, the Foreign Policy Research Institute, the Heritage Foundation, the Hudson Institute, the Institute for Foreign Policy Analysis, and the Jewish Institute for National Security Affairs (JINSA). Jewish-Americans have formed an impressive array of organizations to influence American foreign policy, of which the American Israel Public Affairs Committee (AIPAC) is the most powerful and well-known and also a controversial American special interest group that lobbies the US Congress and the White House in favor of maintaining a close US-Israel relationship. Describing itself as America's Pro-Israel Lobby, it is a mass-membership organization including Democrats, Republicans and Independents and also the prominent Christian evangelicals. AIPAC was formed during the Eisenhower administration and since then has helped secure American aid and support to Israel. Its hard-liners generally support the Likud Party's expansionist policies, including its hostility to the Oslo Peace Process. The Israel Lobby's most powerful weapon is the charge of anti-Semitism: anyone who criticizes Israeli actions or says that pro-Israel groups have significant influence over U.S. Middle-East policy—an influence that AIPAC celebrates—stands a good chance of getting labeled an anti-Semite. The [Iraq] war was due in large part to the Lobby's influence, especially the neoconservatives within it.

Jihad **: The Concept of Holy War in Islam:** Regarded as one of the fundamentals of Islam, the Qoranic term *Jihad* means the holy fighting in the cause of Allah or any other kind of effort to make Allah's word superior, though the term has come to mean different things to different people. Muslims refer to a 'Greater' *Jihad* meaning the personal struggle against the sin and all that is against God and the teachings of the Quran. It is the personal struggle each Muslim wages to be a true believer and follower. The Quran urges one to stay on the straight path and to strive in Allah's cause. (22:78; 49:15).

And there is a 'lesser' *jihad* meaning holy fight against the enemies of Allah and the Islam. Thus, *jihad* is both a personal and community commitment to defend and spread the religion of Islam. In this way, while many Muslims speak out against terrorist acts committed in the name of Islam, others approve of such acts under certain conditions. Muslim scholars, however, differ on whether Islam teaches it to be unholy to start war although some wars are inevitable and justifiable and present jihad as coercive and violent. These scholars understand it to be an effort or struggle to bring righteousness and peace on the earth. Yet others believe that the Qoran urges to fight for the cause of Allah and kill pagans wherever they are found. The Quranic injunction to the Islamic nation to take as a duty "to enjoin good and forbid evil." (3:104) is a duty which is not exclusive to Muslims but applies to the human race who are, according to the Quran, God's vicegerent on earth. Muslims, however, cannot shirk it even if others do. The means to fulfil it are varied, and in our modern world encompass all legal, diplomatic, arbitrary, economic, and political instruments.

Monotheism: In theology, monotheism is the belief in the existence of one deity or God, or in the oneness of God. In a Western context, the concept of "monotheism" tends to be dominated by the concept of the God of the Abrahamic religions. It can be defined in contrast with polytheistic religions. Ostensibly monotheistic religions may still include concepts of a plurality of the divine, for example, the Christian Trinity, or the veneration of Saints, as well as the belief in "lesser spirits" such as angels or demons which Islam considers as a distortion of Jesus's original message of oneness of God. Islam means "submission to God". In Islam monotheism is unambiguously absolute, not relative or pluralistic in any sense of the word, as emphasized for example in surah Al-Ikhlas. Islam accepts as its fundamental tenet the oneness and uniqueness of God, the Arabic word for monotheism is Tawhid which means 'being one', i.e. alone, only one in number. Monotheism is usually contrasted with polytheism, which is the belief in many gods, and atheism, which is the absence of any belief in gods. The known monotheistic religious systems are Judaism, Christianity, Islam, and Sikhism. As a consequence of their belief in exclusivity, monotheistic religions have historically displayed less religious tolerance than polytheistic religions. The latter have been able to incorporate the gods and beliefs of other faiths with relative ease; the former can only do so without admitting it and while denying any reality or validity to others' beliefs [Wikipedia Search on Monotheism].

Islam teaches that God (called Allah in Arabic) is the source of all creation and that human beings are the best of His creation. He communicates by inspiring them towards goodness and by sending Prophets who deliver God's message. Muslims believe that the first Prophet was Adam followed by a long chain of Prophets to guide humanity. The Holy Qur'an, according to Muslim belief, is the word of God revealed to Prophet Muhammad. It mentions many other

Prophets like Adam, Noah, Abraham, Isaac, Ishmael, Moses, Jacob, Joseph and Jesus.. All the Prophets had brought the same message, i.e., belief in one God, upright human conduct and belief in the accountability of human acts at the end of time. Islam is the final monotheistic religion revealed to human beings through the last Prophet who was called Muhammad. He was born in 570 AD in present Saudi Arabia. God's message to humanity was delivered in the Holy Qur'an which was revealed to Prophet Muhammad. The Holy Qur'an, which is the holy book for Muslims, contains 114 chapters (called Suras). Muslims believe that it is the pure word of God, unadulterated over 14 centuries. It deals with issues which affect human beings in their earthly lives; issues like piety, upright human conduct, worship, the creation of a just and virtuous society and the practice of ethics (islamicoccasions.com).

Muslim Brotherhood (MB) is a world-wide Sunni Islamist movement founded by the Sufi schoolteacher Hassan al-Banna in1928. It has spawned several offshoot organizations in the Middle-East, dedicated to the credo: "God is our objective, the Quran is our Constitution, the Prophet is our leader, struggle is our way. The MB is the world's largest, most influential Islamist group and the largest political opposition organization in many Arab nations, especially Egypt. It ultimately seeks to reestablish the Caliphate. The Brotherhood strongly opposes Western colonialism, and helped overthrow the pro-western monarchies in Egypt and other Muslim nations during the early 20th century.

It expresses its interpretation of Islam through a strict religious approach to social issues such as the role of women where its founder called for "a campaign against ostentation in dress and loose behavior." "segregation of male and female students," a separate curricula for girls, and "the prohibition of dancing and other such pastimes". The brotherhood does not restrict women from voting, working, or taking an active role in politics and public life. It preaches that Islam enjoins man to strive for social justice, the eradication of poverty and corruption, and political freedoms provided they do not violate the laws of Islam. [Wikipedia search]

Muslim, Islamist and secular: According to Stephen Cohen's views, in his book *Idea of Pakistan*, 'Muslim state' refers to a state whose citizens are entirely or predominantly Muslim; 'Islamic' refers to the belief that Muslim state can be made to follow Islamic guidelines, however defined. "Islamist" refers to the group that advocates an Islamic state, and 'Islamism' as an Islamic flavored version of totalitarianism, seeking to impose a sustained program of various Islamic practices on a society. 'Secular' is used to describe the belief that even Muslim states can borrow from other cultures and societies, especially the west, and reduce Islam to the private sphere [Cohen: 2004: p. 162]

Neoconservatives, in post-9/11, are the kind of right-wingers. The original neocons were a band of liberal intellectuals who rebelled against the Democratic Party's leftward drift on defense issues in the 1970s. The neocons, in the famous formulation of one of their leaders, Irving Kristol, were "liberals mugged by reality." Noeconservatism was a movement founded on foreign policy issues and it is still here that neoconservatism carries the greatest meaning, even if its original *raison d'être*—opposition to communism—has disappeared. Pretty much all conservatives today agree on the need for a strong, vigorous foreign policy. There is no constituency for isolationism on the right.

One group of conservatives believes that we should use armed force only to defend our vital national interests, narrowly defined. They believe that we should remove, or at least disarm, Saddam Hussein, but not occupy Iraq for any substantial period afterward. The idea of bringing democracy to the Middle-East they denounce as a mad, hubristic dream likely to backfire with tragic consequences. This view, which goes under the somewhat self-congratulatory moniker of "realism," is championed by foreign-policy mandarins like Henry Kissinger, Brent Scowcroft and James Baker III.

Many conservatives think, however, that "realism" presents far too crabbed a view of American power and responsibility. They suggest that we need to promote our values, for the simple reason that liberal democracies rarely fight one another, sponsor terrorism, or use weapons of mass destruction. If we are to avoid another 9/11, they argue, we need to liberalize the Middle-East—a massive undertaking, to be sure, but better than the unspeakable alternative. And if this requires occupying Iraq for an extended period, so be it; we did it with Germany, Japan and Italy, and we can do it again.

The most prominent champions of this view inside the Bush administration were Vice-President Dick Cheney and Deputy Defense Secretary Paul Wolfowitz. Their agenda is known as "neoconservatism," though a more accurate term might be "hard Wilsonianism" who want to use American might to promote American ideals. Neocons are closer to the mainstream of the Republican Party today than any competing faction, particularly after 9/11. The National Security Strategy that George W. Bush released in September 2002 calls for "encouraging free and open societies on every continent".

Palestinian Authority: The Palestinian National Authority (PNA or PA) is an interim administrative organization that nominally governs parts of the West Bank and the Gaza Strip. The Palestinian National Authority was established in 1994, pursuant to the OSLO Accords between the Palestinian Liberation Organization (PLO) and the government of Israel, as a 5-year transitional body during which final status negotiations between the two parties were to take place. According to the Accords, the Palestinian Authority was designated to have control over both security-related and civilian issues in Palestinian urban areas (referred to as "Area A"), and only civilian control over Palestinian rural areas ("Area B"). The remainder of the territories, including Israeli settlements, the Jordan Valley region, and bypass roads between Palestinian communities were to remain under exclusive Israeli control ("Area C"). East Jerusalem was excluded from the Accords.

Index